Early Exploration of the Moon

Ranger to Apollo, Luna to Lunniy Korabl

More information about this series at http://www.springer.com/series/4097

Tom Lund

Early Exploration of the Moon

Ranger to Apollo, Luna to Lunniy Korabl

 Springer

Published in association with
Praxis Publishing
Chichester, UK

Tom Lund
San Diego, CA, USA

SPRINGER-PRAXIS BOOKS IN SPACE EXPLORATION

Springer Praxis Books
ISBN 978-3-030-02070-5   ISBN 978-3-030-02071-2   (eBook)
https://doi.org/10.1007/978-3-030-02071-2

Library of Congress Control Number: 2018959133

Cover design: Jim Wilkie. Image credit: NASA
Project Editor: David M. Harland

This Springer imprint is published by the registered company Springer Nature Switzerland AG
The registered company address is: Gewerbestrasse 11, 6330 Cham, Switzerland

Early Exploration of the Moon

Surveyor 3 and Lunar Module 6

Tom Lund

This book is dedicated
to my wife Barbara
and to my children
Ann, Tom, and Colin.

Contents

Introduction

The machines known as spacecraft that enabled early exploration of the moon were ingenious and reflected the best efforts of talented people working with the technology of the day. Those moon-bound spacecraft, designed in the 1960s, were remarkable for their performance, efficiency, and ruggedness. It is instructive to examine these machines and see just how capable they were and how best performance was wrung out of the technology available.

This book covers early lunar exploration efforts by the United States and by the Soviet Union. Russia was the major entity of the Soviet Union at the time, and the development of spacecraft was a Russian endeavor. Early exploration of the moon by the United States involved the Ranger, Lunar Orbiter, Surveyor, and Apollo spacecraft. The exploration advanced from taking photographs as the Ranger hurdled to impact the moon to the impressive manned lunar landing and exploration missions of Apollo. Russian spacecraft that explored the moon included lunar impactors, lunar flyby spacecraft, lunar landers, lunar orbiters, lunar sample return spacecraft, and the capable Lunokhod lunar rovers. The first five of those spacecraft were simply given the name Luna followed by a number. Russian hardware for a manned lunar landing did not rise to the task.

The author had significant responsibility for landing radars for both the Surveyor and Apollo programs. As a result, he had keen interest in all of the space programs during those pioneering years. Writing this book provided opportunity to relive some of those heady times and a chance to use material from his files.

The early space programs took place at a time when there were impressive aeronautical programs in the United States. The SR-71 Blackbird was cruising at Mach 3.2 at 80,000 feet with its crew of two for over 3,000 miles without breaking a sweat, and the X-15 was rocketing along at Mach 6.7 and reaching altitudes of 354,000 feet. The challenges of the difficult Apollo program did not seem insurmountable in that era.

The pinnacle of lunar exploration was the mission of Apollo 17 that saw the exploration of the moon by a trained geologist, Harrison Schmitt. Schmitt and Gene Cernan, commander of the mission, traveled 21 miles around the surface of the moon in a dune buggy-type vehicle, stopping frequently to explore. A striking photograph of Dr. Schmitt examining a large boulder on the surface of the moon is shown below (Fig. 1).

Fig. 1 Geologist Harrison Schmitt examining a boulder on the lunar surface (NASA photograph)

Early US hardware built for lunar exploration was dimensioned in English units. The author stuck with that treatment of units for US hardware in this book. An exception is the US Lunar Orbiter that used metric units, and that convention was retained. Russian hardware was dimensioned in metric units, and that convention was retained as well.

1

The Nature of the Moon

The moon is the most imposing feature of the night sky. Much is known about the moon today, thanks in part to exploration by the spacecraft discussed in this book. A few interesting facts about the moon are presented below to set the stage for the discussion of lunar exploration spacecraft that follows.

The moon is a satellite of the earth with an orbital period around the earth of 27.3 days with respect to the stars. The orbit is elliptical with an apogee of 405,504 km (252,022 miles) center-to-center from earth and a perigee of 363,396 km (225,852 miles). Perigee is the distance of closest approach, and apogee is the farthest distance from earth in the orbit. The plane of the moon's orbit around the earth is displaced 5.15 degrees from the ecliptic, or plane of the earth's orbit around the sun. The axis of rotation of the moon is displaced 6.68 degrees from perpendicular to the plane of the lunar orbit.

Interestingly, the rotation of the moon about its axis is locked to the earth such that the same face of the moon is always presented to the earth. This comes about because of a tidal bulge in the surface of the moon due to the gravity of the earth. This circumstance allows continuous communications from earth to spacecraft on the nearside of the moon. Deep space communications facilities at various locations around the earth allow this continuous communication.

The equatorial diameter of the moon is 3,476.2 km (2,160.5 miles) as compared to an equatorial diameter of 12,756.2 km (7,928.0 miles) for earth. The mean density of the moon is about 60% of that of earth, and the gravity at the surface of the moon is 1.62 m/s^2 (5.3 ft/s^2) compared with 9.80 m/s^2 (32.1 ft/s^2) at the surface of the earth. The factor of six reduction of gravity on the moon allowed the Apollo astronauts to easily move about on the lunar surface while carrying 139 pounds of life-support equipment on their back.

© Springer Nature Switzerland AG 2018
T. Lund, *Early Exploration of the Moon*, Springer Praxis Books,
https://doi.org/10.1007/978-3-030-02071-2_1

The orbital period of the moon around the earth of 27.3 days is known as the sidereal period. Since the earth, carrying the moon's orbit, is revolving around the sun, the moon must rotate more than 360 degrees for the sun to appear at the same elevation in the sky. Thus, it takes 29.5 earth days (708 hours) from sunrise-to-sunrise on the moon. The lunar day consists of 354 hours of light followed by 354 hours of darkness. The transition from light to darkness is abrupt since the moon has essentially no atmosphere.

The temperature of the lunar surface was measured in detail by an infrared radiometer on the US Lunar Reconnaissance Orbiter spacecraft. The surface temperature near the equator was measured at about 117°C (242° F) at lunar noon and about −179° C (−289° F) at the coldest time during the lunar night.

Composite photographs of the nearside and the far side of the moon are shown on the next two pages (Figs. 1.1 and 1.2). The composites were assembled from photographs taken by NASA's Lunar Reconnaissance Orbiter.

The nearside lunar surface contains low-lying maria as well as more rugged highlands. The maria, so-called because ancient astronomers equated the dark, smooth surface to seas (maria is plural for mare, the Latin name for sea), are relatively smooth and thought to be a result of lava flows. They are generally flat to gently rolling with numerous small craters. The highland regions are heavily cratered with many craters exceeding 20 km (12.5 miles) in diameter. The far side lunar surface is heavily cratered with only a few small patches of maria.

Like earth, the moon is composed of a crust, mantle, and core. The crust is the outer layer, and NASA data indicate that the thickness ranges from about 70 km (43 miles) to 150 km (93 miles). The core is made up of a solid, iron-rich center core about 240 km (149 miles) in radius, a molten layer 90 km (56 miles) thick, and a partially molten layer 150 km (93 miles) thick. The mantle extends from the top of the partially molten core layer to the bottom of the crust. It is made up of an upper rigid layer and a lower molten layer.

An important difference between the earth and moon is that the iron core accounts for only one to three percent of the mass of the moon, whereas the iron core of earth accounts for 32.5% of earth's mass.

The origin of the moon has been conjectured for centuries. Today, the most accepted theory is that a fledgling planet about the mass of mars impacted the earth with a glancing blow about 4.5 billion years ago. The moon formed from the remnants of the impacting body and debris from earth. This violent creation model is referred to as the impact theory. Other theories have been advanced, but most have been discarded for not agreeing with one or more known facts. One stubborn fact is that samples of rock returned from the moon by the Apollo missions have nearly identical chemical makeup as rocks on earth, even to identical ratios of the three isotopes of oxygen.

Fig. 1.1 Nearside of moon as imaged by Lunar Reconnaissance Orbiter (NASA photograph)

A formation theory that is presently receiving attention asserts that the collision between the earth and impactor was so violent that it vaporized the impactor and the upper mantle and crust of the earth. The vaporized material from the impactor and earth became homogeneous throughout its extent. The vapor condensed around the remnant of the earth's core to form a new earth, and the moon condensed around moonlets in the outer region.

The fascination about the origin of the moon continues with ongoing discussion and study.

So, what is the moon really made of? This age-old question was the objective of the capable and clever series of early spacecraft launched to the moon by the United States and the Soviet Union in the 1960s and early 1970s. At the end of this early exploration period, the world had a good idea of the composition of the moon.

Fig. 1.2 Far side of moon as imaged by Lunar Reconnaissance Orbiter (NASA photograph)

In total, the six Apollo missions that landed on the moon brought back 842 pounds of lunar rocks and soil for analysis on earth. The Soviet Union conducted three successful unmanned landing and sample return missions that returned a total of about 0.8 pounds of material from three lunar sites.

Chemical analysis of the returned material from Apollo missions varies from site to site, but the average compositions of the lunar surface of the two most common elements by weight were oxygen at 43% and silicon at 20%. For comparison, on average the two most common elements in the earth's crust by weight are oxygen at 46.6% and silicon at 27.7%.

Hamish Lindsay, writing about the Apollo Lunar Surface Experiments Package, states that the early moon was covered by a deep magma ocean. The present lunar highlands formed from low-density rocks that floated to the surface. The main composition of those rocks is feldspar, a mineral rich in calcium and aluminum.

The cooling magma ocean was bombarded by massive asteroids whose impact left huge basins that filled with lava. These basins, now referred to as maria, are made up of basalt rocks that are rich in magnesium and iron.

The rock on the surface of the moon has been ground up by impacts of comets and asteroids over millions of years. Much of the debris, or regolith, on the surface is a very fine powder. There was a concern that a landing spacecraft might sink a considerable distance into the powder, but this proved not to be the case.

Bibliography

Boyle, Rebeca, *What Made the Moon? New Ideas Try to Rescue a Troubled Theory*, https://www.quantamagazine.org
Earth's Moon, NASA Science website
Lindsay, Hamish, *ALSEP Apollo Lunar Surface Experiments Package*, Apollo Lunar Surface Journal, 2008
Taylor, S. R., *The Origin of the Moon: Geochemical Considerations*, Proceedings of the Origin of the Moon Conference, October 1984
Williams, David R., *Moon Fact Sheet*, NASA Science website

2

The Ranger Lunar Photography Mission

The Ranger project was designed to send a spacecraft to the moon and take a series of photographs of the lunar surface as the spacecraft descended toward impact. The project was fraught with failures, but it was finally successful, and high-quality pictures were transmitted to earth.

A photograph of an engineering model of the successful Block III series of Ranger spacecraft is shown next page (Fig. 2.1).

EARLY US VENTURES INTO SPACE

A preceding program to gather information about the moon, Pioneer, had a troubled history as well, and it will be mentioned briefly here to give the reader a feel for some of the angst that accompanied early space exploration.

The Pioneer program, which was begun in March 1958, was managed by the newly formed Advanced Research Projects Agency (ARPA). ARPA was and still is an agency of the Department of Defense. Pioneer was intended to gather information about the moon, including pictures of the far side, during a flyby. It was a rather disjointed program consisting of three US Air Force launches of a spacecraft and two US Army launches of a different spacecraft. The Air Force used a Thor ballistic missile first stage and Vanguard liquid fueled second stage. The US Army used a Jupiter-C ballistic missile first stage and a solid fuel upper stage developed by the Jet Propulsion Laboratory (JPL). The spacecraft for the Air Force was developed by Space Technology Laboratories, and the spacecraft for the Army was developed by JPL.

The first launch of the Air Force Pioneer took place in August 1958. Pioneer 0, as it would later be called, ended shortly after launch in a spectacular explosion.

T. Lund, *Early Exploration of the Moon*, Springer Praxis Books,
https://doi.org/10.1007/978-3-030-02071-2_2

Fig. 2.1 Model of Block III Ranger spacecraft (NASA photograph)

The next two Air Force launches of Pioneer 1 and Pioneer 2 failed due to problems in the upper stage rocket.

The first launch of the Army system, Pioneer 3, failed due to premature cutoff of the first rocket stage. The second Army launch in March 1959 was a success, and Pioneer 4 became the first US spacecraft to escape earth's gravity. After four failures in a row, it was gratifying to have a success although the spacecraft passed about 37,000 miles from the lunar surface, about a factor of two more distant than intended. Pioneer 4 did not carry a camera because priority had been given to obtaining additional data on the radiation belts around the earth and radiation in the vicinity of the moon. Radiation measurements were successfully made.

The Ranger program that followed Pioneer had a turbulent beginning with controversy over the scope of the program and experiments to be carried by the spacecraft. It was the first lunar program to be conducted by the newly formed National Aeronautics and Space Administration (NASA).

NASA grew up with the early spacecraft covered in this book, and it is interesting to look at its early history.

EARLY HISTORY OF NASA

Several organizations were conducting space-related programs in the United States in the 1950s. The Naval Research Laboratory was launching the Vanguard satellite, the US Army and the US Air Force were developing intercontinental ballistic missiles (ICBMs), and ARPA was supporting a heavy-lift rocket program as well as developing a Lunar Orbiter. The National Advisory Committee for Aeronautics (NACA) was supporting research in all phases of aerodynamics and conducting flights of the X-15 manned rocket-powered manned aircraft that flew to the fringes of space. The X-15 reached altitudes of 354,000 feet (67 miles) and speeds of Mach 6.72 (4,517 miles per hour).

President Eisenhower favored consolidating all of the space programs into one civilian-controlled organization. This would include the Army and Air Force space activities that were not directly tied to military applications. Congress took up the challenge to bring about the National Aeronautics and Space Administration (NASA). Lyndon Johnson, Senate Majority Leader, and John McCormack, House Majority Leader, shepherded the bill establishing NASA through congress. Johnson and McCormack were democrats, but they embodied bipartisan support for national programs that is rare today. Republican President Eisenhower signed the bill establishing NASA into law on 29 July 1958.

Formulating NASA was also looked on as a way to respond to the burgeoning space program of the Soviet Union. Soviet space programs became big news upon their orbiting of the Sputnik satellite in October 1957. The Soviet program extended to the moon with the Luna 1, Luna 2, and Luna 3 spacecraft that were all launched in 1959.

The new NASA organization incorporated the venerable National Advisory Committee for Aeronautics (NACA), Langley Aeronautical Laboratories, Ames Aeronautical Laboratory, Lewis Flight Propulsion Laboratory, the Army Ballistic Missile Agency in Huntsville, and the Jet Propulsion Laboratory. The personnel and programs of NACA became the nucleus of the new NASA organization.

The newly incorporated organizations were directed from NASA Headquarters in Washington, DC. The first top executives for NASA Headquarters were Dr. Keith Glennan, NASA Administrator, and Dr. Hugh Dryden, Deputy Administrator.

The first location of NASA Headquarters was in the Dolley Madison House on Lafayette Square in Washington, DC. Headquarters occupied that historic house from 1958 to 1961. The ballroom of the house was used to introduce the first astronauts to the press in April 1959. NASA Headquarters later moved to larger quarters in Federal Office Buildings FOB-6 and FOB-10B. It is now housed in a new large building on E Street, just south of the National Mall. The building is shown below (Fig. 2.2).

Fig. 2.2 NASA Headquarters in Washington, DC, (NASA photograph)

Directors of the early field centers were:

Center	First director
Marshall Space Flight Center	Wernher von Braun
Langley Research Center	Henry Reed
Ames Research Center	Smith DeFrance
Goddard Space Flight Center	Harry Goett
Flight Research Center	Paul Bikle
Lewis Research Center	Edward Sharp
Jet Propulsion Laboratory	William Pickering
Launch Operations Center	Kurt Debus

OVERVIEW OF THE RANGER PROJECT

The original goal of the Ranger project was to gather information about the moon by several scientific instruments on the spacecraft and to obtain close-up pictures of the lunar surface. A series of pictures would be taken and transmitted to earth as the spacecraft descended toward impact on the moon. After failure of the first five Ranger missions, the direction of the program was changed, and effort was concentrated on obtaining high-quality close-up pictures of the lunar surface to support the upcoming Apollo program. The scientific instruments, some of which were quite complex, were removed from the spacecraft.

The Ranger program had its beginning in December 1959 when NASA Headquarters assigned the Jet Propulsion Laboratory (JPL) seven flights to reconnoiter the moon. The flights were planned to occur during 1961 and 1962. JPL would develop and build the various Ranger spacecraft as well as manage the project. Design concepts for the Ranger spacecraft were released by JPL in February 1960.

Top management personnel on the initial Ranger program at NASA Headquarters were Abe Silverstein, Space Flight Programs Director, and Ed Cortright, Lunar and Planetary Programs Chief. Oran Nicks, Chief of Lunar Flight Systems, was the hands-on manager for Ranger for the Space Flight Program Office.

Key management personnel initially for the Ranger program at JPL were Clifford Cummings, Lunar Program Director; James Burke, Ranger Project Manager; and Gordon Kautz, Ranger Project Assistant Manager. From accounts related in NASA Report SP-4210, all were capable and energetic persons.

Management relationships between NASA Headquarters and JPL were contentious as Headquarters tried to impose their will on the independent-minded Jet Propulsion Laboratory. JPL particularly resented technical direction. By the time of the successful Block III Ranger phase, three years into the program, management personnel had changed at JPL and at NASA Headquarters, and management issues were less contentious. Author Cargill Hall in *Lunar Impact: A History of Project Ranger* discusses the management discourse at some length.

The spacecraft in the project were organized into blocks that corresponded to phases of the project. Block I was a proving phase for the spacecraft and launch vehicles and integration of the two. Operation was to be confined to earth orbit.

Two flights were planned for Block I. The two spacecraft, Ranger 1 and Ranger 2, carried ten scientific instruments along with solar panels and vehicle stabilization equipment. The spacecraft would conduct scientific measurements as the spacecraft traveled in a highly elliptical orbit around the earth with perigee of 37,500 miles and apogee of 685,000 miles. The orbit would take the spacecraft behind the moon and return to the vicinity of earth. Each mission was expected to last about 5 months.

Block II would consist of three spacecraft, and they would travel to the moon and take photographs as the spacecraft descended toward impact. The Block II spacecraft would carry fewer scientific instruments than Block I, but it would carry a camera and a small lander. The lander would be detached from the main spacecraft and employ a retrorocket to slow it for a survivable landing. The scientific instrument of the lander was mounted inside of a crushable balsa wood sphere to enhance survival. The only instrument inside the lander would be a seismometer to measure lunar quakes. It would also contain a small transmitter to send seismic measurements back to earth.

Block III would consist of four spacecraft, Ranger 6, 7, 8, and 9. The spacecraft configuration would depend on the results of flights of Block I and Block II spacecraft.

The first of the two Block I spacecraft, Ranger 1, was launched in August 1961. The planned parking orbit around the earth was achieved, but the Agena-B upper stage only fired briefly leaving the spacecraft in a low earth orbit. The spacecraft itself performed all of its functions, but the orbit soon decayed, and the spacecraft burned up reentering the atmosphere. Ranger 2 was launched in November 1961, but again the Agena-B upper stage failed, and the spacecraft soon burned up reentering the earth's atmosphere.

The first of the Block II spacecraft, Ranger 3, was launched in January 1962. It failed to achieve the desired trajectory and missed the moon by 22,860 miles (36,785 km). Ranger 4 was launched in April 1962, and its trajectory was good toward the moon, but an electronic failure left the spacecraft unresponsive to commands from earth. Ranger 5 was launched in October 1962, but a short circuit in the solar panel circuits left only battery power and that soon ran out leaving the spacecraft dead.

Five failures in a row could not go unanswered. Investigation boards were set up, all aspects of the project were probed, and there was call for an overhaul of the Ranger program with greater emphasis on reliability and quality control. Additional redundancy of critical functions was also incorporated. In keeping with common practice of firing the coach of a losing sports team, the JPL Lunar Program Director, Clifford Cummings, and Ranger Project Manager, James Burke, were replaced. Robert Parks became JPL Lunar and Planetary Program Director, and Harris Schurmcier became Ranger Project Manager.

The direction of the Ranger project was changed for the Block III missions to give priority to photographing the lunar surface in support of the upcoming Apollo program. The scientific instruments and lander that were present on the Block II spacecraft were eliminated, and a set of cameras were installed instead. Block III included four spacecraft: Rangers 6, 7, 8, and 9. Those spacecraft carried a very capable set of six cameras to image the lunar surface as the spacecraft descended toward impact.

A photograph of a model of the Block III Ranger spacecraft was shown on the second page of this chapter. A viewing port for the cameras was located in a cutout located part way up on the conical vertical structure shown in the photograph. The top edge of a large steerable parabolic antenna can just be seen in the photograph behind the body of the spacecraft.

The six cameras carried by Block III spacecraft had different fields of view and resolution. The last picture would be taken less than a second before impact at altitudes of a few thousand feet. The pictures were transmitted to earth in near real time.

Ranger 6 was launched in January 1964. The launch vehicle was a US Air Force Atlas missile with an Agena-B upper stage. Initially, all aspect of the flight looked good, and the spacecraft impacted the moon within 19 miles of the aim point. However, when the cameras were turned on, only noise was received. The problem was traced to the cameras being inadvertently turned on for about a minute shortly after launch while still in the earth's upper atmosphere, and their high-voltage power supplies arced and burned out.

Ranger 7 was launched in July 1964 and the mission was a complete success. A total of 4,308 good-quality photographs were transmitted to earth as the spacecraft descended over the Mare Nubium (Sea of Clouds) region. The first photograph of the lunar surface was taken when the spacecraft was 1,311 miles above the surface. Photographs continued to be taken down to an altitude of about 1,440 feet. The last picture taken had a resolution of about 1.6 feet.

Ranger 8, launched in February 1965, was also a complete success. It returned 7,137 good-quality photographs of the Sea of Tranquility region.

Ranger 9, also a success, was launched in March 1965. It returned 5,814 good-quality photographs of the Alphonsus crater region.

Promising looking landing sites for the Apollo spacecraft that were apparent in the photographs taken by Ranger 8 influenced selecting the Sea of Tranquility landing site for the Surveyor 5 and Apollo 11 spacecraft. Indeed, the Apollo 11 astronauts established Tranquility Base just 44 miles from the impact site of Ranger 8.

The total cost of the Ranger project, including development, launch, and support, was $170 million.

LAUNCH OF RANGER SPACECRAFT AND FLIGHT TO THE MOON

The Ranger spacecraft was launched toward the moon from Cape Kennedy, Florida, by Atlas LV-3/Agena-B launch vehicles. The spacecraft, with solar panels folded up and parabolic antenna folded under the spacecraft, was fit into the nose cone attached to the Agena-B upper stage. A photograph of the launch of Ranger 8 in February 1965 is shown on the next page (Fig. 2.3).

The Agena-B stage is the upper portion of the vehicle in the photograph. It extended down to the nose fairing of the Atlas first stage. The Agena-B engine nozzle extended down past the black cylindrical area in the photograph to inside the Atlas nose fairing. The overall launch vehicle was about 100 feet tall.

Fig. 2.3 Launch of Ranger 8 in February 1965 (NASA photograph)

The Atlas was 10 feet in diameter, and the Agena-B was 5 feet in diameter. The weight of the overall vehicle including the Block III Ranger spacecraft was about 276,800 pounds at liftoff.

The Atlas LV-3 first stage was built by General Dynamics. Five engines powered the Atlas at launch: two booster engines, one sustainer engine, and two vernier engines. The engines all burned rocket propellant-1 (RP-1), which was highly refined kerosene. The oxidizer was liquid oxygen.

The two booster engines generated 150,000 pounds of thrust each, the sustainer engine generated 57,000 pounds of thrust, and the two vernier engines provided 1,000 pounds of thrust each. All of these thrust levels were at sea level. The thrust was higher in a vacuum. The total thrust at liftoff was about 359,000 pounds, well above the 276,800 pound weight of the overall vehicle.

The booster engines were mounted on gimbals that allowed each engine to pivot 5 degrees in pitch and 5 degrees in yaw with respect to the centerline of the Atlas. The pivoted booster engines were used to steer the vehicle to a prepro- grammed trajectory after launch. The trajectory followed an arc that gradually tilted from vertical at launch toward the horizontal as the vehicle gained altitude and speed.

The powerful booster engines were shut off and jettisoned along with their associated fuel pumps about 145 seconds after liftoff. The vehicle was about 36.4 miles above the earth at that time. The sustainer and vernier engines of the Atlas continued to burn until they were cutoff about 92 miles above the earth and near orbital velocity.

The sustainer engine was also gimballed, and it could be pivoted 3 degrees in pitch and 3 degrees in yaw about the centerline. It thrusted along the centerline while the boosters were firing. Its pivoting ability was used for steering after the booster engines burned out and were jettisoned. The vernier engines of Atlas could be positioned within a 140 degree arc in pitch and 50 degree arc in yaw. This posi- tioning capability allowed the launch vehicle to be rolled to the desired orientation and to be controlled in pitch and yaw.

The total amount of propellant (fuel and oxidizer) carried by the Atlas was about 114.8 tons. Of this, about 74.6 tons was used during the booster firing, and the remainder, 40.2 tons, was available for use by the sustainer and vernier engines.

The Agena-B upper stage was built by the Lockheed Missiles and Space Company. It contained one engine that burned unsymmetrical dimethyl hydrazine as fuel and fuming nitric acid as oxidizer. This combination was hypergolic, ignit- ing upon contact with one another. The engine was developed and built by Bell Aerosystems as their Bell model 8091. The engine generated 16,000 pounds of thrust in a vacuum, and it could be shut down and restarted twice in orbit. Agena-B carried about 6.1 tons of fuel and oxidizer, and that gave a total burn time of 240 seconds. The Agena was 5 feet in diameter in the propellant and equipment areas and 23.7 feet long.

The launch and subsequent thrusting and maneuvering to guide Ranger 8 toward impact at a targeted spot on the moon are described below.

Ranger 8 was launched from Cape Canaveral, Florida, on 20 February 1965. All three engines of the Atlas burned normally. The booster engine cutoff (BECO) command was given at the proper time, and the booster engines were jettisoned. Steering commands for the sustainer engine were generated by the large digital computer at the Cape and sent by radio link to the Atlas. The computer also deter- mined the proper time to cut off the sustainer engine, and this was transmitted to Atlas at the appropriate time.

The Agena-B upper stage with Ranger attached was established in a parking orbit around the earth at an altitude of 115 miles and 7 minutes after launch. The spacecraft velocity at this time was about 17,500 miles per hour. At 21 minutes after launch, the Agena-B was ignited for a 90 second burn that put Agena/Ranger into an injection trajectory to impact the moon. The spacecraft velocity was about 24,475 miles per hour after the burn.

Ranger was then separated from Agena. The solar panels were deployed, and the spacecraft was oriented in space with the longitudinal axis pointed at the sun. The orientation was performed by gas jets controlled by the stabilization system from inputs from the sun sensors. The high-gain antenna was then deployed, and the spacecraft was rolled about the longitudinal axis until the earth sensor sensed the earth and provided signals to stop the roll and lock to the direction of earth. The antenna was then rotated on its hinge to align with the earth.

A midcourse correction was made when the spacecraft was about 99,440 miles from earth to bring the impact site on the moon close to the target. The spacecraft was first oriented to an altitude where the burn of Ranger's rocket engine would produce the desired correction. A 59 second burn of the rocket was then made to perform the correction. The rocket motor generated a thrust of 50 pounds.

Parameters of the midcourse correction had been calculated very well, and the correction resulted in the spacecraft impacting the moon within 14 miles of the initial aim point in the Sea of Tranquility. The accuracy was commendable given that the moon was over 224,000 miles away from earth at the time of launch.

Deep Space Tracking Network

Key to guiding the spacecraft close to the targeted impact point on the moon was trajectory measurements from the Deep Space Network. Operators at mission control were able to determine the trajectory of the spacecraft very accurately by using inputs from that network and send up instructions for midcourse corrections to refine the trajectory.

The Deep Space Network, at the time of flight of the Ranger spacecraft, was made up of stations at Goldstone Dry Lake in California, Island Lagoon (a dry lake bed) in Australia, and in a valley near Johannesburg, South Africa. The three stations allowed continuous tracking and communicating with the spacecraft from the rotating earth. Each station had a large parabolic antenna 85 feet (26 meters) in diameter that could be steered very accurately in azimuth and elevation to track the spacecraft. Once the spacecraft was in space, radio signals received from the spacecraft were used to automatically position the antenna to track the spacecraft.

Uplink commands were sent to the spacecraft by a 10 kilowatt, 890 MHz transmitter that fed the big antenna. The Ranger spacecraft had a transponder that phase locked to the uplink signal, modulated it with telemetry data, and translated it to 960 MHz for transmission back to earth. The phase-lock process allowed the two-way Doppler shift of the communication link to be measured very accurately, and this in turn gave very accurate measurement of relative velocity between the spacecraft and the earth-based antenna. The combination of very accurate velocity measurement and antenna angle measurement allowed accurate determination of the spacecraft trajectory.

CONFIGURATION OF RANGER SPACECRAFT

The spacecraft built for Block I, Block II, and Block III phases of the Ranger program were substantially different. The Block III spacecraft were successful in sending back high-quality photographs of the lunar surface. The Block I and Block II spacecraft all failed in their missions for various reasons. Details of those spacecraft will not be covered in this book.

Block III Ranger

Two views of the Block III Ranger spacecraft are shown in the drawings below and on the next page. The Block III spacecraft was less complicated than either the Block I or Block II spacecraft. The only scientific payload carried was a very capable set of six cameras.

The basic frame of the spacecraft was a hexagonal structure 5 feet across. A truncated conical structure clad with polished aluminum was attached to the top of the frame of the spacecraft. The conical tower was 27 inches at its base and 16 inches at the top. A cylindrical omnidirectional antenna was mounted at the top of the tower. The cameras were mounted within the tower, and a cutout in the side of the tower provided a viewing port (Figs. 2.4 and 2.5).

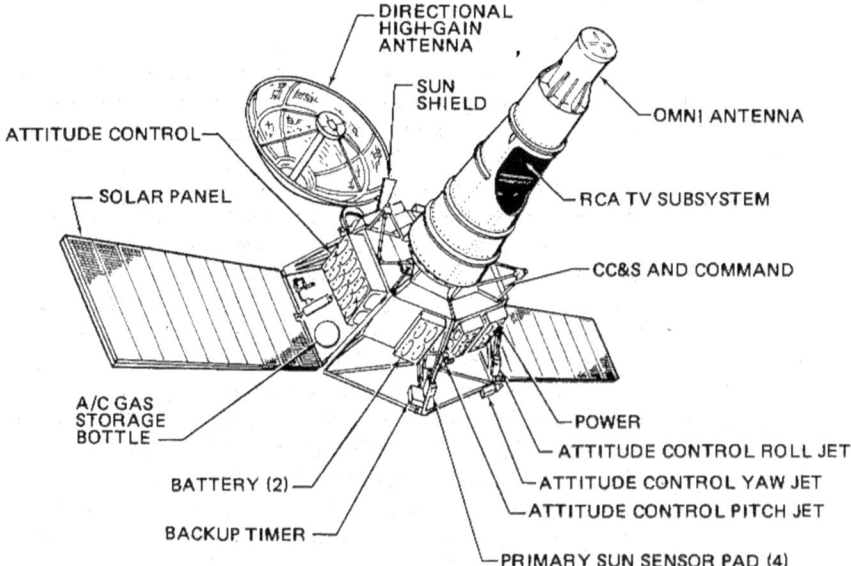

Fig. 2.4 View of the Ranger III spacecraft from the top (NASA graphic)

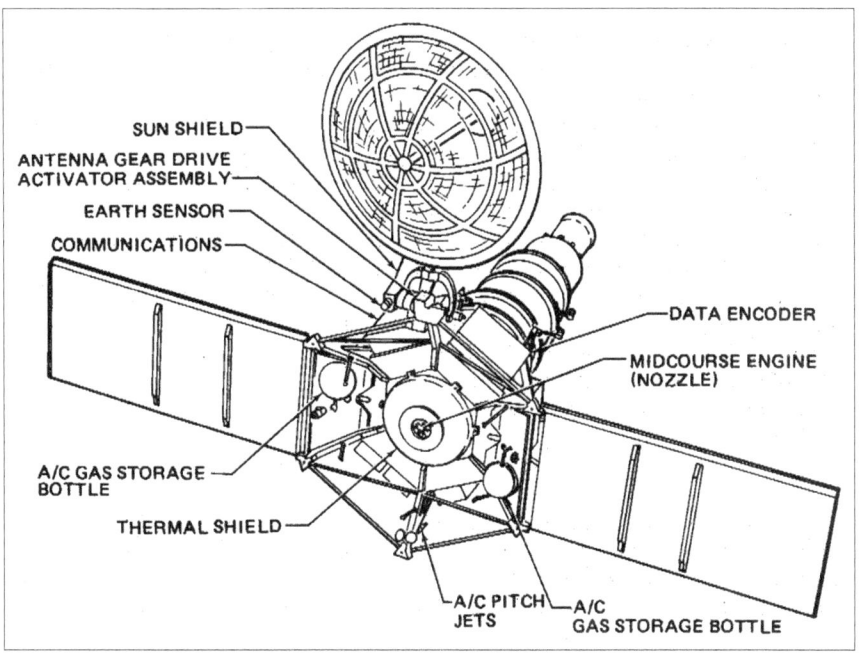

SUN SHIELD
ANTENNA GEAR DRIVE
ACTIVATOR ASSEMBLY
EARTH SENSOR
COMMUNICATIONS
DATA ENCODER
MIDCOURSE ENGINE
(NOZZLE)
A/C GAS STORAGE
BOTTLE
THERMAL SHIELD
A/C PITCH
JETS
A/C
GAS STORAGE BOTTLE

Fig. 2.5 View of the Ranger III spacecraft from the bottom (NASA graphic)

The cameras are labeled as RCA TV SUBSYSTEM in the top view of the spacecraft. The spacecraft contained two independent camera channels, the F-channel and the P-channel. Each channel had a separate battery, power supply, camera control electronics, and transmitter. The two batteries, two power supplies, and two sets of camera electronics were mounted inside the tower.

The hinged solar arrays were rectangular and 28.9 inches wide and 60.5 inches long. The total span across the two arrays was 15 feet. The total height of the spacecraft was 11.8 feet. The solar arrays generated about 200 watts of power for the spacecraft. Two batteries, each with a 1,000 watt-hour capacity, were available for backup power before the solar arrays were deployed. Each of the camera channels had a 1,200 watt-hour battery that was capable of powering the cameras, camera control electronics, and the 60 watt transmitter for 9 hours.

Cameras for Block III Rangers

The main purpose of the Block III Ranger spacecraft was to take photographs of the lunar surface to ascertain if it would be feasible to land the manned Apollo spacecraft on that terrain. To that end, a capable set of six cameras were mounted in the spacecraft. A photograph of the lenses of six cameras is shown on the next page (Fig. 2.6).

Fig. 2.6 View of lenses of the six cameras in Ranger III spacecraft (NASA photograph)

The six cameras had different fields of view, different look angles, and different scan times. The vidicon camera tube used was the same for all cameras. It was 1 inch in diameter (25.4 mm), and that allowed a square raster, 11 mm on a side, to be used for the image with ample margins.

The image from the camera lens was focused on the face of the vidicon tube. The inner face of the vidicon was coated with a photoconductive material of antimony sulfide oxysulfide (ASOS). The photoconductive material was initially charged by scanning an electron beam in a raster pattern. A slit-type shutter located just forward of the face of the tube was opened for a few milliseconds to present the image to the photoconductive surface. The surface was scanned again with the electron beam, and a charge current proportional to the brightness of the image at the location of the beam flowed through a load resistor. The varying voltage developed across the load resistor as the beam moved along was amplified and became the video signal.

The cameras were grouped in two channels known as the full scan or "F"-channel and the partial scan or "P"-channel. The electron beam scanned an area 11 mm square on the face of the tube for the full-scan cameras and an area of 2.8 mm square for the partial-scan cameras.

There were two full-scan cameras labeled A and B in the F-channel. The image on the 11 mm square area on the face was scanned by 1150 lines in 2.5 seconds. During the 2.5 second scan time of one camera, the image on the other camera was erased, and the photoconductive surface prepared for another image. The old image was erased by exposing the face to a high-intensity flash from a flash lamp. The face was then scanned by an out-of-focus electron beam to flood the face with charge. The shutter was then opened for 5 milliseconds to present a new image to the photosensitive surface. The total frame time was 2.56 seconds so that there was a space of about 5.12 seconds between successive frames from the same camera.

The A camera was fitted with a 25 mm focal length f/0.95 lens that resulted in a camera field of view of 25 degrees. The B camera was fitted with a 76 mm focal length f/2.0 lens that resulted in a camera field of view of 8.4 degrees.

The P-channel included four partial-scan cameras labeled P_1, P_2, P_3, and P_4. The scanned raster on the face of P-channel vidicon tubes was 2.8 mm square. The raster, which consisted of 300 lines, was scanned in 0.2 seconds. The shutter opened for 2 milliseconds. The shorter scan time allowed photographs to be taken closer to the lunar surface just before impact. The total time between frames of the same camera was 0.84 seconds.

Cameras P_1 and P_2 were equipped with the same 76 mm f/2.0 lens as the B camera, and cameras P_3 and P_4 were equipped with the same 25 mm f/0.95 lens as the A camera. Because of the smaller raster area, the field of view of the P_1 and P_2 cameras was 2.1 degrees, and the field of view of the P_3 and P_4 cameras was 6.3 degrees.

The fields of view of all the cameras were centered within 0.5 degrees of the X-Z plane of the spacecraft. The reference axis of the set of cameras was displaced 38 degrees from the +Z axis.

FLIGHT OF THE RANGERS

The two flights of Block I Rangers were unsuccessful as were the three flights of the Block II Rangers. After five Ranger failures in a row, new project objectives and new management were put in place at JPL. The Block III spacecraft that followed was less complex, and it had a higher degree of redundancy. The scientific payload consisted of only an assembly of six cameras.

The flights of the Ranger Block III spacecraft 7, 8, and 9 were successful and produced thousands of high-quality photographs of the lunar surface. Photographs were taken from distances of about 1,500 miles to just before impact as the spacecraft hurdled to the moon. Pictures taken at altitudes below 2,000 feet had resolution of about 1 foot. The photographs gave good assurance that it would be possible to land the Apollo Lunar Module in the vicinity of some of the sites photographed.

A summary of flights of the Block III Rangers is given below.

Flight of Ranger 6

Ranger 6 was launched from Cape Canaveral on 30 January 1964, 14 months after the last launch of a Ranger. The launch and insertion of Ranger into a translunar trajectory went well. Ranger 6 separated from Agena extended the solar panels and high-gain antenna and was stabilized in space after locking on to the sun and earth. From all indications, a healthy spacecraft was speeding toward the moon. The midcourse correction brought the predicted impact point to within a few miles of the target location.

The cameras were switched to a warmup period when the spacecraft neared the moon. The cameras were to switch on automatically a few minutes later and begin transmitting pictures from the two 60 watt transmitters. No pictures were received before the spacecraft impacted the moon.

Analysis of the failure disclosed that the cameras had been inadvertently switched on for 67 seconds about the time of separation of the booster stages from Atlas. The nominal time of separation of the booster was 145 seconds after liftoff, and the nominal altitude was 190,000 feet. Separation of the booster rockets from the Atlas on Ranger 6 occurred at 140 seconds after liftoff which likely corresponded to an altitude slightly less than 190,000 feet. It turns out that this is about the altitude where an arc can most easily occur in a high-voltage circuit.

For example, at sea level the voltage required for an arc between two spherical conductors separated 1 centimeter is about 33,000 volts. The atmospheric pressure at sea level is 760 millimeters of mercury (mm Hg). The minimum voltage required for arcing for a 1 centimeter spacing decreases to only 327 volts at an atmospheric pressure of 0.6 mm Hg. This pressure occurs at an altitude of about 170,000 feet for a standard atmosphere. The voltage required for breakdown increases very rapidly as the atmospheric pressure further decreases by going higher in altitude. The breakdown voltage for 1 centimeter spacing is above practical voltage limits at altitudes above 250,000 feet.

Engineers at JPL concluded that the high-voltage power supplies likely arced and failed when turned on at the critical altitude near 170,000 feet. Typical vidicon cameras of the day used about 300 volts for the accelerating voltage for the electron beam. The maximum voltage within the power supplies would be higher.

The open question was: what turned on the power supplies at the time of jettisoning the booster engines? The answer took months to uncover. It was determined that a quantity of fuel and oxidizer was released at the time of jettisoning of the booster. Photographs showed that this release often ignited in an explosive burn. The resulting plasma traveled up the launch vehicle and past an umbilical plug on Agena where a series of male pins were used to operate the camera from a patch cable during testing. One pin carried a voltage of 20 volts, and it was located about 0.25 inches from a sensitive pin that required only 3 volts to command turn-on of the cameras. The umbilical connector had a metal cover, but it was not airtight, and it was concluded that the plasma streamed into the receptacle, bridged across the gap between the pins, and turned on the cameras.

Early space programs had many failures and often from unusual causes. Ranger 6 experienced one of many unforeseen and far-out occurrences that would bring down a mission.

Flight of Ranger 7

Ranger 7 was launched from Cape Canaveral on 28 July 1964. The launch, earth orbit coast, and injection into the lunar transfer trajectory all went well. The midcourse correction was also successful, and Ranger 7 impacted only 8 miles from the intended target in Mare Nubium.

Originally, it was planned to perform a terminal maneuver to align the camera reference axis with the velocity vector before turning on the cameras. This would allow nesting of the images as the spacecraft descended, and it would avoid smearing of photographs taken in low altitudes. However, it was decided that the parameters of the trajectory were acceptable for picture taking, and to minimize risk to the mission, the terminal maneuver was not attempted.

A total of 4,308 photographs were taken by Ranger 7 as it descended toward the moon. The first picture was taken 16 minutes and 56 seconds before impact while the spacecraft was at an altitude of 1,311 miles. The slant range along the camera reference axis was 1,520 miles. That first photograph is shown on the next page (Fig. 2.7). The picture spans about 224 miles top to bottom.

The last full-scan picture from Camera A was taken 2.5 seconds before impact with the spacecraft at an altitude of 3.6 miles above the moon. That photograph is shown on the following page (Fig. 2.8). The distance on the lunar surface between the center reticle and the reticle directly above it is 1,970 feet. The transfer of the picture to earth was interrupted by impact with the moon, and that resulted in receiver noise on the right side of the photograph.

Fig. 2.7 The first photographs of lunar surface taken by Ranger 7

The last photograph taken by the partial frame camera P_4 was 0.6 seconds before impact at an altitude of about 4,800 feet. The last picture taken had a resolution of about 1.6 feet.

Flight of Ranger 8

Ranger 8 was launched from Cape Canaveral on 17 February 1965. The launch, earth orbit coast, and injection into a trajectory to impact the moon all went well. The midcourse correction was successful, and Ranger 8 impacted the moon only 14.3 miles from the intended target in Mare Tranquillitatis.

There had been a loss of data from the spacecraft for a short time early in the flight. As a result, it was decided not to risk performing the terminal maneuver to align the camera reference axis with the velocity vector to minimize smearing of the images.

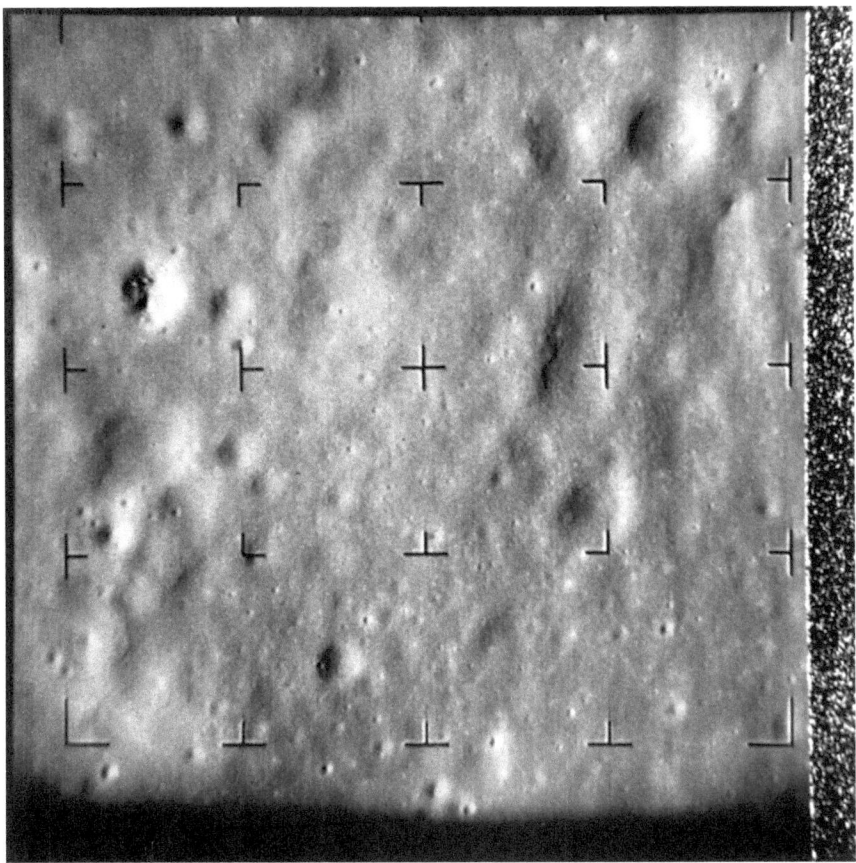

Fig. 2.8 Last picture taken by Camera A before impact of Ranger 7

A total of 7,137 photographs were taken by Ranger 8 and transmitted back to earth. The first photograph was taken 23 seconds before impact at an altitude of 1,560 miles above the moon. The spacecraft descended over the Sea of Tranquility just north of the landing site of Apollo 11.

The last full-scan picture taken by Camera B about 4.5 seconds before impact at an altitude of 5.1 miles is shown on the next page (Fig. 2.9). The distance on the surface between the central reticle and the next reticle above it is 1,312 feet (Fig. 2.9).

Flight of Ranger 9

Ranger 9 was launched from Cape Canaveral on 21 March 1965. Both Ranger 7 and Ranger 8 had photographed mare areas of the moon, and to expand the database, Ranger 9 was targeted for Alphonsus crater in the highland area of the moon. Ranger 9 impacted the moon just 2.7 miles from its target location in the Alphonsus crater.

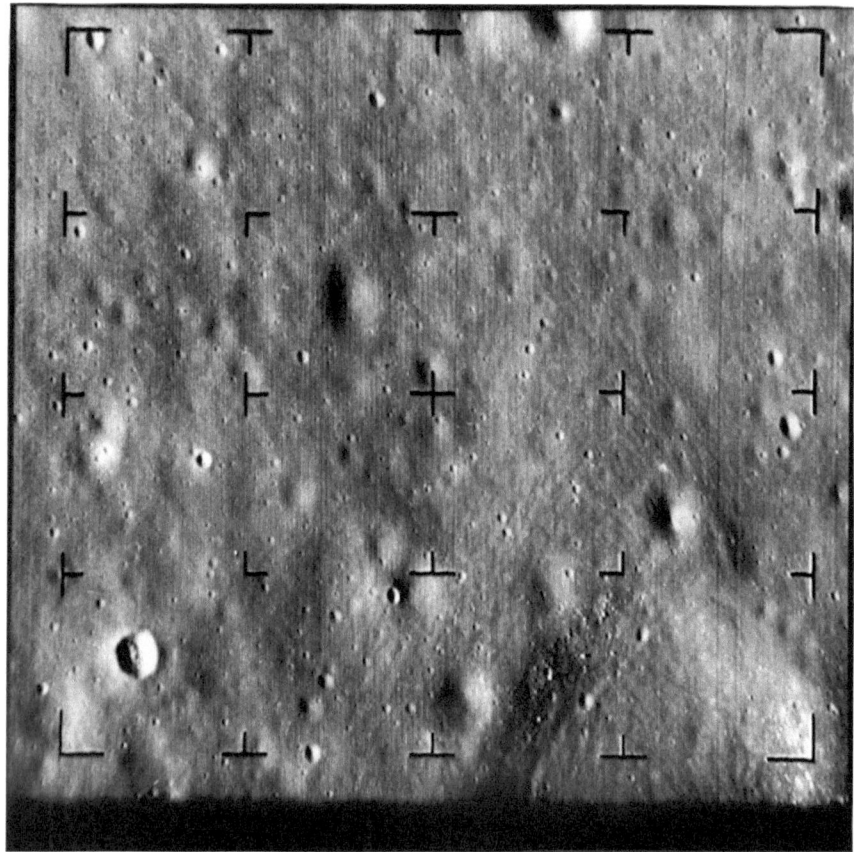

Fig. 2.9 Last full-scan photograph of Camera B before impact of Ranger 8 with the moon (NASA photograph)

The terminal maneuver to align the reference axis of the cameras with the velocity vector was begun at an altitude of about 4,475 miles above the moon. The maneuver was successful, and smearing of the photographs at low altitude was reduced.

A photograph taken by the A camera at an altitude of 251.4 miles above the Alphonsus crater is given on the next page (Fig 2.10). The center peak of the crater is very visible. NASA indicated that a resolution of 0.3 meters (1 foot) was provided by the last photograph from Ranger 9.

Although the Ranger program had a turbulent beginning, three spacecraft at the end were successful in returning good-quality, high-resolution photographs of the lunar surface. Those photographs were an aid in selecting landing sites for Apollo.

Fig. 2.10 Photograph by camera A of Ranger 9 from 251 miles above Alphonsus crater (NASA photograph)

Bibliography

Dick, Steven J., *50 Years of NASA History,* http://www.gov/50th_magazine/historyLetter.html

Hall, R. Cargill, *Lunar Impact: The NASA History of Project Ranger*, Dover Publishing, 2010

Heacock, Raymond L, *Lunar Photography: Techniques and Results*, Kluwer Academic Publishers, articles.adsabs.harvard.edu

Jet Propulsion Laboratory report *Ranger IX Photographs of the Moon,* December 1965

Jet Propulsion Laboratory report *Ranger VIII Photographs of the Moon*, August 1964

Liepack, O, International Astronautical Congress, IAA-98-IAA.2.3.08, *The Ranger Project, September 1998*

NASA – NSSDA/COSPAR ID:1962-001A, Ranger 3

NASA – NSSDA/COSPAR ID:1964-041A, Ranger 7

NASA-NSSDCA ID: ABLE 1 Pioneer 0, PION2 Pioneer 2

NSSDCA/COSPAR ID: 1958-008A Pioneer 3, 1959-013A Pioneer 4

Ulivi, Paolo and Harland, David M., *Lunar Exploration: Human Pioneers and Robotic Surveyors*, Springer-Praxis, 2004

3

The Lunar Orbiter Photography Mission

The purpose of the Lunar Orbiter program was to photograph potential Apollo landing sites in good detail from lunar orbit. The program launched five spacecraft into lunar orbit to photograph lunar terrain between August 1966 and August 1967. All five spacecraft were successful. The photography operation of each mission was also largely successful. Resolution of some of the photographs was as fine as 1 meter (3.28 feet).

Two of the spacecraft were directed into polar orbit of the moon, and they photographed nearly all of the front side and backside of the moon as the moon rotated under the plane of the orbit.

A photograph of an engineering model of the Lunar Orbiter is shown below (Fig. 3.1).

The first three missions of the Lunar Orbiter photographed potential Apollo landing sites with orbits near the lunar equator and with perilune (lowest altitude) of about 40 km (24.9 miles). The last two missions were flown in high-altitude polar orbits and resulted in photographs of nearly the entire surface on both the nearside and far side of the moon.

The many excellent quality photographs of the lunar terrain were of great benefit in selecting landing sites for Apollo. Photographic coverage was over a much wider area than the Ranger photographs.

BACKGROUND OF THE LUNAR ORBITER PROGRAM

With the advent of space flight by the Soviet Union in 1957 and by the United States in 1958 came the realization that a large amount of knowledge about the moon could be gained by high-resolution photographs taken by a spacecraft

© Springer Nature Switzerland AG 2018
T. Lund, *Early Exploration of the Moon*, Springer Praxis Books,
https://doi.org/10.1007/978-3-030-02071-2_3

Fig. 3.1 Engineering model of Lunar Orbiter (NASA photograph)

orbiting the moon. In addition to the photographs, information about the gravitational field, cosmic radiation, and particles in the vicinity of the moon would tell us much about the moon and perhaps about the primordial earth itself.

NASA asked the Jet Propulsion Laboratory (JPL) to perform a study of the benefits and requirements of a multiphase program to explore the moon. The findings of the study group, which was headed by Albert Hibbs, were submitted in a report to NASA in April 1959. One of the recommendations of the report was to place a satellite around the moon to take high-resolution photographs.

Responsibility and management of the new Lunar Orbiter program was given to the NASA Langley Research Center.

President Kennedy's stirring speech to a joint session of Congress on 25 May 1961 was a game changer for the US space program. The speech contained the following declaration:

"I believe that this nation should commit itself to achieving the goal, before this decade is out, of landing a man on the moon and returning him safely to the earth. No single space project in this period will be more impressive to mankind, or more important for the long-range exploration of space; and none will be so difficult or expensive to accomplish."

The Congress and the nation agreed to the challenge, and one of the greatest engineering achievements of all time, named Apollo, went forward.

The goal of the Ranger program was altered to derive the most benefit for the upcoming Apollo program to land men on the moon. The Lunar Orbiter program was seen as providing much needed high-resolution photographs of the terrain at potential Apollo landing sites.

NASA Langley Research Center worked diligently to define the requirements for the Lunar Obiter. They released a request for proposal (RFP) for the Orbiter to potential contractors in August 1963. Five aerospace companies submitted proposals in response. The winner of the competition was announced in December 1963 to be the Boeing Company.

An incentive contract amounting to $80 million was signed by Boeing and Langley Research Center representatives in April 1964. The Lunar Orbiter program included spacecraft for five flights. The launch vehicle would be the Atlas-Agena D.

Key NASA individuals for the Lunar Orbiter program were Oran Nicks of NASA Headquarters and Clifford Nelson and James Martin of NASA Langley. Nelson was the Project Manager for the Lunar Orbiter and Martin was the Business Manager.

The Boeing Lunar Orbiter Program Office was led by Robert Helberg, who was an able administrator with an engineering background. Other key managers on the Boeing Lunar Orbiter team were George Hage, Chief Engineer, and Carl Krafft, Business Manager.

OVERVIEW OF LUNAR ORBITER PROGRAM

The Langley Research Center requirements for the Lunar Orbiter included determining the altitude of the spacecraft for each picture taken and the location of the photographed area within the lunar coordinate system. As secondary to photographs, the spacecraft was also required to determine properties of the moon's gravitational field and measure micrometeoroid flux, energetic particles, and gamma radiation. The photograph was to have sufficient resolution to show cones 50 cm (20 inches) in height with a base of 2 meters (79 inches) and show slopes not less than 7 degrees on an area 7 meters (23 feet) square.

Boeing planned a spacecraft stabilized in three axes and a photography system by Eastman Kodak similar to the Eastman Kodak system already in use by the US Air Force on an Agena spacecraft.

A total of eight Lunar Orbiter spacecraft were built by Boeing. Three of these were used for ground testing, and five were flight models sent to orbit the moon. A photograph of an engineering model of the Lunar Orbiter spacecraft was given on the second page of this chapter.

The spacecraft was 2.08 meters (6.83 feet) high and 3.76 meters (12.4 feet) across the solar panels. The extent from the edge of the high-gain antenna on the left to the end of the omnidirectional antenna on the right side of the photograph was 5.21 meters (17.1 feet). The spacecraft weighed 390 kg (853 pounds).

The spacecraft was stabilized in three axes by inputs from a sun tracker and from a star tracker that locked on to Canopus. A three-axis inertial system provided stabilizing inputs when the spacecraft was maneuvering.

The spacecraft hosted a capable camera system and three scientific experiments. Of these instruments, the camera system was deemed the most important. The camera system, built by Eastman Kodak, was rather ingenious. It used very fine grain film to capture images from two lenses simultaneously on different portions of the same roll film. The film was developed and fixed and stored on a reel.

The images were later read out by a fine scanning spot of light, and the resulting video data was transmitted to the ground. One of the two lenses had a focal length of 610 mm, and it provided high-resolution photographs. The other lens had a focal length of 80 mm that provided wider angle, medium-resolution photographs. Lenses for two cameras can be seen in the photograph of the engineering model of Lunar Orbiter on the second page of this chapter. The lenses are located near the center of the spacecraft with a door partially obscuring them.

The other experiments aboard the spacecraft involved gathering selenodetic information and measuring radiation and micrometeoroid flux near the moon. The word selenodetic, in effect, means a description of the moon. The word was derived from the ancient Greek word for the moon goddess, Selene.

Selenodetic information would be obtained by tracking the spacecraft as it orbited the moon. The spacecraft carried a transponder that allowed accurate measurements of range and range rate from the earth to the spacecraft. This data was used to determine variations in the gravitational field of the moon.

Micrometeoroid flux was measured by 20 pressurized semicylinders 188 mm (7.4 inches) long and 37 mm (1.46 inches) high arrayed around the spacecraft outside of the thermal blanket. When a micrometeoroid struck one of the thin-walled semicylinders, gas pressure would be lost and an electrical switch would close and signify the event.

Radiation levels were measured by two cesium iodide radiation detectors. One detector was shielded by a cover with 0.2 grams of aluminum per square centimeter, and the other was shielded by 2.0 grams of aluminum per square centimeter.

The spacecraft were launched from earth by Atlas/Agena D launch vehicles. The Agena D second stage was a standardized version of Agena B that was used to launch the Ranger spacecraft.

The first Lunar Explorer spacecraft, Lunar Orbiter 1, was launched from Cape Canaveral on 10 August 1966. The launch and midcourse correction went well, and the spacecraft was placed in a near equatorial orbit of the moon. A total of 42 high-resolution and 187 medium-resolution pictures were returned to earth.

Lunar Orbiter 2 was launched on 6 November 1966, and Lunar Orbiter 3 was launched on 5 February 1967. Both missions were successful in returning high-quality photographs. Lunar Orbiter 2 returned 602 high-resolution and 208 medium-resolution photographs, and Lunar Orbiter 3 returned 427 high-resolution and 149 medium-resolution photographs. These first three Lunar Orbiters were designated as Apollo landing site survey missions, and they were flown in near equatorial orbits with perilune altitudes of about 40 km over the sites of interest.

Lunar Orbiters 4 and 5 were placed in near polar orbits, and they were able to photograph nearly all the surface of the moon including the backside. Lunar Orbiter 4 was launched on 4 May 1967, and Lunar Orbiter 5 was launched on 1 August 1967. Lunar Orbiter 4 returned 419 high-resolution and 127 medium-resolution photographs. Lunar Orbiter 5 returned 633 high-resolution and 211 medium-resolution photographs.

LAUNCH AND INSERTION INTO LUNAR ORBIT

The Lunar Orbiter spacecraft were launched toward the moon from Cape Canaveral in Florida by Atlas SLV-3/Agena D launch vehicles. The spacecraft, with solar arrays folded up and the antennas folded away, was fit into the nose cone attached to the Agena D upper stage. Events during the flight of Lunar Orbiter 4 will be described as an example. A photograph of the launch of Lunar Orbiter 4 in May 1967 is shown on the next page (Fig. 3.2).

The Lunar Orbiter was lofted into space by a SLV-3 model of Atlas, which was a standardized version of the LV-3 used for Ranger launch, and the Agena was a D model, which was a standardized version similar the Agena B. Agena D incorporated improvements to the nozzle of the rocket engine.

At the time of cutoff of all Atlas engines, 5 minutes and 10 seconds after liftoff, the launch vehicle was traveling about 5,646 meters per second (18,517 feet per second) in an elliptical earth orbit with semimajor axis of 4,422 km (2,748 miles) and semiminor axis of 3,873 km (2,407 miles).

After separation from Atlas, the Agena fired its engine for 2 minutes and 32 seconds to establish a high parking orbit around the earth. After coasting in the parking orbit for 20 minutes and 43 seconds, the Agena was commanded to fire again for 1 minute and 28 seconds to insert Agena with Lunar Orbiter attached into a translunar trajectory. After entering the translunar trajectory, the Lunar Orbiter was separated from Agena.

Following separation from Agena, the Lunar Orbiter unfolded its solar arrays and deployed the omnidirectional and high-gain antennas. It then maneuvered to align the longitudinal axis with the sun and rolled to lock to the star Canopus with the star tracker.

Fig. 3.2 Launch of Lunar Orbiter 4 by Atlas/Agena D

Thus stabilized in space, the Lunar Orbiter cruised toward rendezvous with the moon. The spacecraft trajectory was accurately determined from earth, and parameters were established for a midcourse correction to allow the spacecraft to be put into a polar orbit around the moon.

The midcourse correction was made 17 hours and 35 minutes after launch. The spacecraft was maneuvered into the proper orientation for the velocity correction by the rocket burn. A 78 degree roll was performed followed by a 67 degree pitch. A velocity change of 60.85 meters per second was then commanded from earth. The rocket burn was successful, and the correction was sufficiently accurate that a second midcourse correction was not required.

As the Lunar Orbiter 4 approached the moon, 88 hours and 44 minutes after launch, a deboost operation was performed to allow the spacecraft to be captured by the gravitational field of the moon and to inject the spacecraft into a lunar polar orbit. This operation required maneuvering the spacecraft to the proper orientation by rolling -29.47 degrees and pitching -96.13 degrees. A velocity decrease of 659.62 meters per second was then commanded. That velocity increment was accomplished by a burn of the rocket engine for about 502 seconds.

The deboost operation placed the spacecraft in an elliptical orbit about the moon with perilune of 2,706 km (1,683 miles), apolune of 6,114 km (4,047 miles), and inclination from the lunar equator of 85.5 degrees. This near polar orbit allowed nearly the entire moon to be photographed as the moon rotated under the plane of the orbiting spacecraft.

DETAILS OF LUNAR ORBITER

A drawing of the Lunar Orbiter spacecraft is shown below (Fig. 3.3). The spacecraft was 2.08 meters (6.83 feet) high, spanned 3.76 meters (12.4 feet) across the solar arrays, and 5.21 meters (17.1 feet) from the edge of the high-gain antenna to the tip of the omnidirectional antenna. Its weight, including fuel, was 380 kg (853 pounds).

The main structure of the spacecraft consisted of three decks upon which the major components were mounted. The middle deck was supported from the lower deck by a tubular truss structure. The upper deck was supported by an upper structural module located between the tanks shown in the drawing. Most of the major components are mounted to the lower deck. The components mounted to the lower deck on the opposite side of the spacecraft include two batteries, command decoder, multiplexer encoder, and the traveling wave tube amplifier.

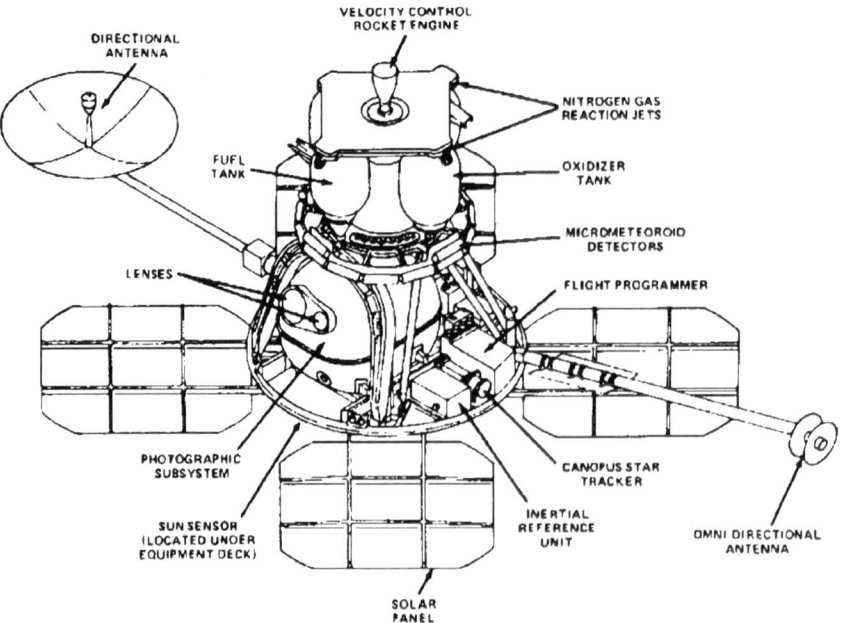

Fig. 3.3 Drawing of Lunar Orbiter (NASA graphic)

A simplified block diagram of the Lunar Orbiter spacecraft is shown below (Fig. 3.4).

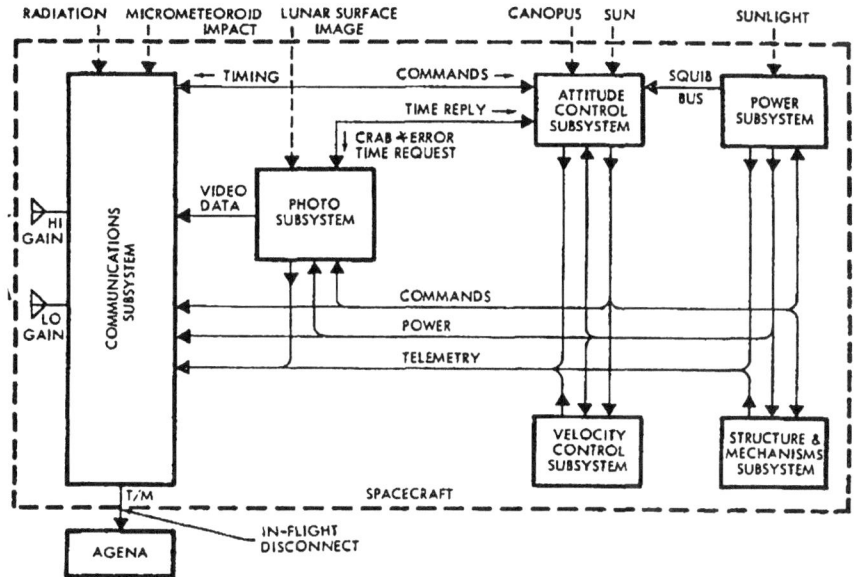

Fig. 3.4 Simplified block diagram of Lunar Orbiter (NASA graphic)

Photographic Subsystem

The unique photography system was film based. The use of fine grain film was the only practical way at the time of obtaining high-resolution photographs from a spacecraft orbiting the moon. It took advantage of the fact that an extremely large amount of information can be stored on a small piece of photographic film. The film used was Eastman Kodak SO-243 roll film 70 mm wide that was designed for aerial photography.

One edge of the film was preexposed with frame number, a nine-level gray scale, and a resolution chart. The film had extremely fine grain and that allowed very fine resolution in the photographs. As a consequence of the fine grain, the film was not highly sensitive. In common terms, it was low-speed film. This was an advantage for use in space because it was relatively insensitive to cosmic radiation.

A diagram of the photographic subsystem is given on the next page (Fig. 3.5). The photographic subsystem contained two lenses with shutters that operated simultaneously to project images on separate areas of the film. One lens had a 610 mm focal length, and it provided high-resolution photographs, and the other lens had a focal length of 80 mm, and it provided medium-resolution photographs. A focal plane shutter was used with the 610 mm lens, and a between-the-lens shutter was used with the 80 mm lens. Both lenses had an f-number of 5.6. Shutter speeds of 0.01, 0.02, or 0.04 could be selected by command from the ground.

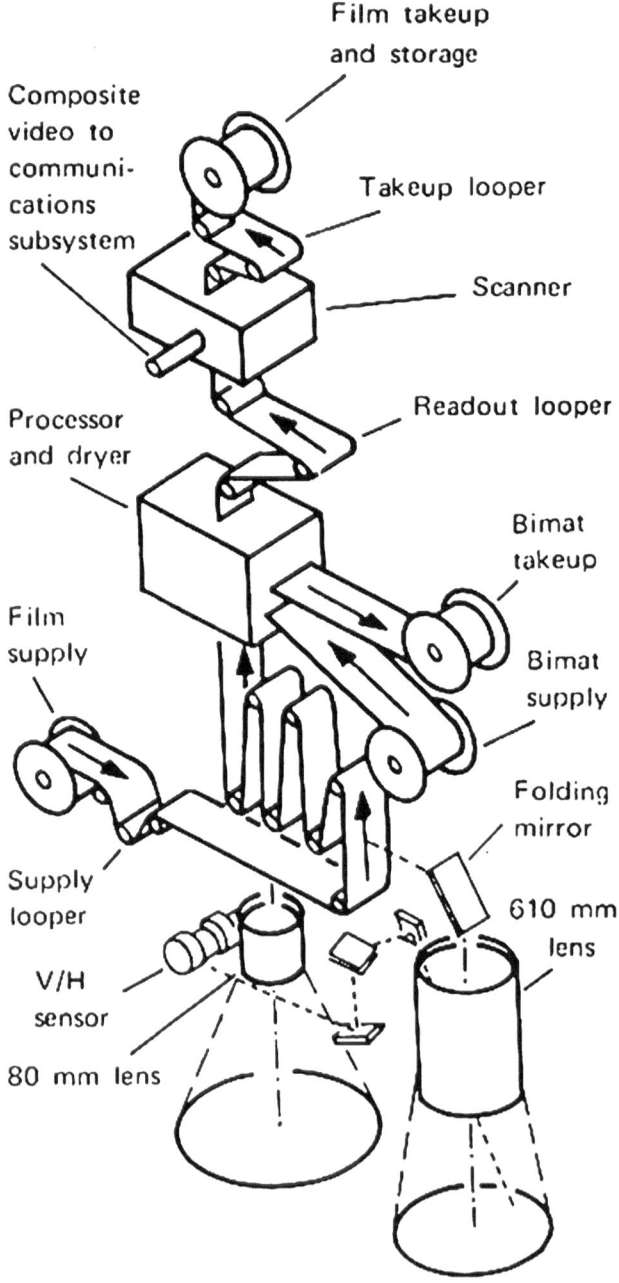

Fig. 3.5 Diagram of photographic subsystem (NASA graphic)

The image size on the film was 55 mm by 219 mm for the high-resolution lens and 55 mm by 60 mm for the medium-resolution lens. At an altitude of 46 km, the equivalent size of the area on the moon that was imaged was 4.2 km by 16.6 km for the high-resolution lens and 31.6 km by 37.4 km for the medium-resolution lens. The high-resolution image was positioned at the center of the photographs. At an altitude for photography of 46 km, the equivalent resolution was 0.98 meters for the high-resolution photographs and 7.6 meters for the medium-resolution photographs.

NASA document SP-206, *Lunar Orbiter Photographic Atlas of the Moon*, states that the angular resolution of the recovered line pairs from the photographs was 4.4 seconds of arc for the high-resolution photographs and 34 seconds of arc for the medium resolution. The resulting resolutions on the surface from an altitude of 46 km are 0.98 meters and 7.6 meters, respectively.

A diagram from SP-206 that shows the image format on the film and fields of view of the two lenses is shown below (Fig. 3.6). The rectangles in the upper left of the figure show the fields of view, and the detail in the upper right shows the information prerecorded on the edge of the film.

There was a potential problem with image smearing due to image motion, while the shutter was open at the relatively low altitudes and high velocities of the missions of the first three Lunar Orbiter spacecraft. This smearing was minimized by moving the film at the same rate as the image was moving. The film was moved by clamping it to a platen that was moved by a mechanical connection to a velocity over height (V/H) sensor. The V/H sensor was an optical tracker that looked through the 610 mm lens and physically moved to track the moving image.

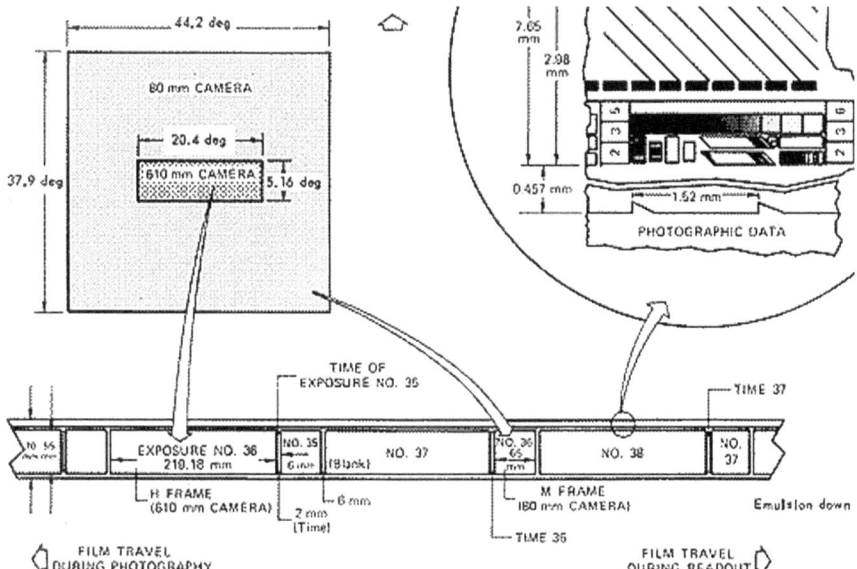

Fig. 3.6 Film format and fields of view of the two lenses (adapted from NASA graphic)

After exposure, the film was released from the platen and moved to a camera storage looping system that consisted of a multi-roller looping system that held up to 20 feet of film. This was long enough to store a sequence of several pairs of photographs. Automatic sequences of 1, 4, 8, or 16 photographs could be commanded from earth. The multiple sequences resulted in overlapping photographs in the direction of travel.

When the camera storage looper was full, or whenever operators on the ground wanted to process the film, the firm was moved through the processor-dryer where it was pressed against Kodak Bimat transfer film. The Bimat film was coated with a gelatin layer containing a film processing solution. The two films were pressed tightly together around a drum for about 3.4 minutes to allow the processing solution to develop and fix the image on the camera film. The temperature during the developing process was held closely at about 30 degrees C.

After developing and fixing, the Bimat film was separated from the camera film and wound onto a take-up reel. The developed camera film was moved to a dryer drum where it was dried at a temperature of about 35 degrees C for 11 minutes. After drying the film was moved through the scanner and onto a take-up and storage spool.

After the photographs have all been taken for a mission, the exposed film was read out, and the resulting video data was transmitted to earth. The first step in this process was to cut the Bimat film so that the exposed film could be moved backward through the scanner. The film supply spool was used as a take-up reel during the readout process.

The film was moved back into the scanner for readout. The exposed film was scanned by a spot of light only 6.5 microns in diameter (a micron is a millionth of a meter). The scanning spot was generated by a scanning electron beam striking a phosphor-coated rotating drum. The drum rotated to avoid a heat buildup from the electron beam. The resulting spot of light was focused by a scanner lens to a bright spot 6.5 microns in diameter on the surface of the film.

The scanning electron beam caused the spot to move 2.67 mm in a direction parallel to film travel. At the end of the travel, the electron beam was blanked during the retrace. The scanner lens moved vertically during the scanning process to cover the 55 mm width of the image on the film. When the vertical scan had reached the bottom of the image, the film was advanced by 2.54 mm, and the direction of motion of the scanner lens was reversed and the raster scan proceeded upward to the top of the film.

Each "framlet" so scanned consisted of a raster of 16,359 lines covering an area 2.67 mm by 55 mm on the film. The total distance covered on the film by the high-resolution and medium-resolution images plus spaces between the images was about 198 mm. Covering this distance required 117 framlets. The time required to scan and read out one framlet was 22.02 seconds and the time to scan and read out both the high-resolution and medium-resolution images was about 43 minutes.

The scanning spot of light shining through the film was modulated by the variation in darkness of the image, and this variation was picked up by a photo-multiplier tube. The photomultiplier tube converted the modulated spot of light into a varying voltage that became the video signal. The rather slow scanning of the film resulted in a video bandwidth of about 230 kHz that was readily accommodated by the telecommunications subsystem and transmitted to earth.

Communications Subsystem

The communications subsystem conveyed commands from the earth to the spacecraft, and it transmitted photograph video and engineering data back to the earth. It also allowed precise range and range rate measurements to be made from earth to the spacecraft.

A simplified block diagram of the communications subsystem is shown below (Fig. 3.7).

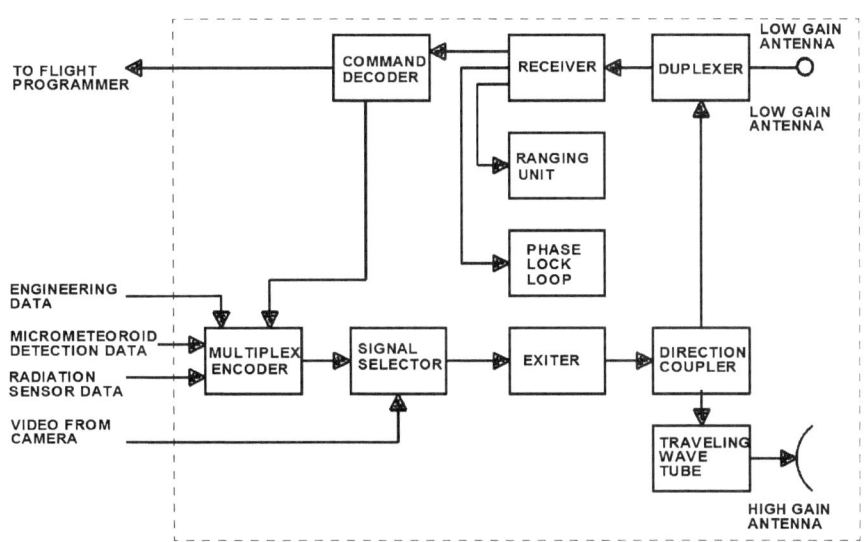

Fig. 3.7 Simplified block diagram of communications subsystem (adapted from NASA graphic)

The low-gain, omnidirectional antenna was used to receive commands from earth and to transmit telemetry data. The high-gain antenna was primarily used to transmit video from the camera and telemetry of spacecraft parameters. The low-gain antenna was a biconical discone type mounted on the end of a 2.08 meter (82 inch) long boom to minimize the influence of the spacecraft on the antenna pattern.

The high-gain antenna was a 0.91 meter (36 inch) parabolic reflector type that provided 23.5 dB (factor of 234) gain over an isotropic radiator. The beamwidth of the antenna was 10 degrees. The earth subtends an angle of about 1.8 degrees when viewed from the moon so the entire earth was readily covered by the 10 degree beamwidth.

The high-gain antenna was mounted to 1.3 meter long boom that was attached to a positioner. The positioner could rotate the antenna 360 degrees around the axis of the boom in 1 degree increments. This rotation allowed the antenna to track the earth as the spacecraft orbited the moon and the moon orbited the earth.

Commands from the earth were intercepted by the low-gain antenna and passed through a duplexer to the receiver. The duplexer allowed receiving and transmitting from the same antenna. The uplink communications frequency from the ground was 2,116.38 MHz. Transmission from the spacecraft was at a frequency of 2,298.33 MHz.

The data output of the receiver was applied to a command decoder that decoded the uplink information into a series of commands for the spacecraft. A portion of the receiver output was applied to a phase lock loop that phase locked to the carrier frequency. The phase locked signal was translated to 2,298.33 MHz for transmission back to earth. This information was used by the ground stations on earth to measure two-way Doppler shift on the communications signal and that Doppler information was used to develop very accurate range rate measurements to the spacecraft.

Range to the spacecraft was accurately determined by transmitting binary phase modulated pseudo random code on the uplink carrier. The code was demodulated by the communications subsystem and retransmitted back to earth. The time delay between the pseudo random code sequence transmitted and that received back from the spacecraft was measured and this allowed accurate determination of range to the spacecraft.

The multiplexer encoder converted analog and digital input data into a 50 bit per second serial non-return-to-zero (NRZ) data stream. Each data frame consisted of 128 9 bit words made up from the outputs of the scientific instruments and spacecraft engineering data. The scientific instruments data included that from the 20 micrometeoroid detectors and from the two radiation measurement instruments.

The 50 bps telemetry data stream was applied to the signal selector along with the video signal from the photographic scanner. The signal selector received commands from the earth to select telemetry alone or telemetry plus video to be applied to the exciter. The output of the exciter was at a power level of about 0.5 watts at 2,298.33 MHz. The exciter output was applied to a directional coupler that directed 0.48 watts to the diplexer and low-gain antenna and 0.02 watts to the traveling wave tube amplifier.

The traveling wave tube (TWT) amplifier amplified the video plus telemetering signal from the exciter and applied the amplified signal to the high-gain antenna for transmission to earth. The TWT amplifier had a RF gain of 27 dB (factor of 500) and an output power of about 10 watts at 2,298.33 MHz.

Power Subsystem

The primary power source for the spacecraft was the solar panels. A battery in the spacecraft supplied power before the solar panels were deployed and during the time that the spacecraft was maneuvering, and the panels were not in full view of the sun. Each of the four solar arrays contained 2,714 solar cells that occupied an area of 12.3 of square feet (1.14 square meters).

The solar arrays on Lunar Orbiter 4 generated total of 393 watts during the cislunar cruise phase with the solar panels normal to the sun. The spacecraft required about 115 watts of power during this phase, and the rest of the power was dissipated in resistor elements designed to dissipate heat. A shunt regulator was used to control the amount of power sent to the load resistors and to maintain the output voltage at about 30.5 volts.

The highest total spacecraft load of 260 watts occurred during the rocket engine burn because electrical power was used by the actuators that positioned the gimballed rocket engine. The total spacecraft load was 206 watts during readout of the film and with the TWT amplifier on.

The battery charging current was limited to about one ampere for Lunar Orbiters 4 and 5 because those spacecraft were in polar orbit and the solar arrays were always in the sun. Orbiters 1, 2, and 3 were flown in near equatorial orbits, and the sun was occluded for a portion of the orbit, and battery power was required during that time. The battery charging current was limited to 2.85 amperes on those spacecraft to restore charge to the battery during the sunlight periods.

Velocity Control Subsystem

The spacecraft contained a gimballed, liquid-fueled rocket engine that generated 100 pounds of thrust. The engine was fired to change velocity during midcourse correction and to inject the spacecraft into lunar orbit. It was also used to refine or change parameters of the orbit. The rocket burned Aerozine-50 fuel with nitrogen tetroxide as the oxidizer. Aerozine-50 was a 50-50 mixture of hydrazine and unsymmetrical dimethylhydrazine.

The fuel and oxidizer were each contained in two tanks with Teflon bladders. When it was time to fire the engine, pressurized nitrogen was ported into the four tanks forcing fuel and oxidizer through shutoff valves into the rocket engine. The mixture of fuel and oxidizer spontaneously combusted when they came into contact.

The rocket burn was stopped by shutting off the fuel and oxidizer. The engines were gimballed in two axes to allow maintaining proper orientation of the spacecraft during the rocket burn.

The total velocity change available from the amount of propellants carried was about 1,000 meters per second. In the case of Lunar Orbiter 4, the velocity change required during the midcourse correction was 60.8 meters per second and that was accomplished with a rocket burn of 52.7 seconds. Injection into lunar orbit required a velocity change of 659.6 meters per second and that was accomplished with a rocket burn of 501.7 seconds.

Attitude Control Subsystem

The attitude control system maintained stable orientation of the spacecraft during travel to the moon and during orbit of the moon. It also oriented the spacecraft so that the rocket burns for midcourse correction and insertion into lunar orbit took place along the proper line of direction.

The orientation was performed by nitrogen gas jets located on the upper shelf of the spacecraft. The nitrogen jets were controlled by signals from the attitude control subsystem. Nitrogen was stored in a tank pressurized at 4,000 pound per square inch at launch.

The attitude of the spacecraft was controlled by the gas jets during the travel to the moon where inputs from the sun and Canopus star trackers maintained spacecraft attitude. The longitudinal axis was aligned with the sun and the star tracker locked to Canopus.

A key element of the attitude control system was an inertial reference unit (IMU). That unit contained three Sperry SYG-1000 rate integrating gyros, one each for the pitch, yaw, and roll axes. It also contained a pulsed integrating pendulum (PIP) accelerometer aligned with the longitudinal axis of the spacecraft. The rocket motor was also aligned with the longitudinal axis.

The IMU provided an attitude reference whenever the sun or Canopus was not in view as when occluded by the moon or during spacecraft maneuvering. The PIP accelerometer provided a measure of velocity increment and that was used to mark the shutoff time for the rocket burn.

The pulsed integrating pendulum accelerometer was built by Sperry as their model 16 PIP. In that device, a pendulum mass was maintained at a null position by applying current pulses to a torquer. The pulse rate was set by a 200 Hz clock. The offset of the pendulum from null was sensed each clock period, and the direction of the current pulses through the torquer was selected to move the pendulum back toward a null. The motion of the pendulum due to acceleration was balanced by the torque pulses. Each pulse corresponded to a velocity increment of 0.1 feet per second for the accelerometer.

The SYG-1000 rate integrating gyro operated in either an angle rate mode or an angle error mode. In the angle rate mode, a torquer balanced the gyroscopic force due to angle rate, and the current required by the torquer to achieve balance was a measure of angle rate. The angle rate mode was used to provide rate data to the control loops when spacecraft attitude was stabilized by inputs from the sun sensor and Canopus sensor.

In the angle error mode, the gyro was operated open loop, and angular departure from an initial value was obtained by integrating the rate output. The angle error mode was used to remember the proper attitude for sun and Canopus alignment when either or both of these references were lost.

FLIGHT OF THE LUNAR ORBITERS

All five flights of the Lunar Orbiter spacecraft were successful and returned good quality photographs of the lunar surface. The primary mission of Lunar Orbiters 1, 2, and 3 was to photograph potential landing sites for the manned Apollo program. To that end, the spacecraft were inserted into near equatorial elliptical orbits with perilunes of about 40 km (24.9 miles) over the areas of potential landing sites. The primary mission of Lunar Orbiters 4 and 5 were to photograph nearly all of the surface area of the moon from polar orbits.

In addition to photography, the spacecraft measured cosmic radiation and frequency of micrometeoroid strikes in the vicinity of the moon. Data about variations in the gravitational field of the moon was obtained by closely tracking the spacecraft from earth.

Flight of Lunar Orbiter 1

Lunar Orbiter 1 was launched from Cape Canaveral on 10 August 1966. The launch vehicle was an Atlas SLV-3 with an Agena D as the upper stage. The launch and staging went well, and the Agena fired to put the Agena and the Orbiter into a parking orbit. After 33 minutes in earth orbit, the Agena fired again and successfully put the Agena and Lunar Orbiter into a trajectory to intercept the moon.

The Lunar Orbiter separated from Agena and deployed its solar arrays and the omnidirectional antenna and the high-gain antenna. The spacecraft is acquired and aligned with the sun and began searching for the star Canopus. The Canopus tracker could not achieve lock because of much higher apparent brightness of the star than planned. Controllers on earth arranged to have the star tracker lock on to the moon for the roll reference instead. With the spacecraft attitude thus stabilized, it cruised toward the moon.

A midcourse correction was made 28 hours and 34 minutes after liftoff. The correction was good, and a second correction was not required. Overheating of the spacecraft became a problem, and to help this situation, the spacecraft was oriented 36 degrees off of the sunline for several hours as it traveled toward the moon.

A deboost maneuver was performed to slow the spacecraft as it approached the moon. This allowed the gravitational field of the moon capture the spacecraft and pull it into lunar orbit. The initial elliptical orbit had an apolune of 1,867 km and perilune of 189 km. It was inclined 12 degrees from the lunar equator.

The first photographs were taken on 18 August at an altitude of 246 km (153 miles) above Mare Smythii. A sequence of 16 high-resolution photographs and 4 medium-resolution photographs was taken. The film was developed, fixed, and dried. The medium-resolution photographs received back at earth were of excellent quality, but the high-resolution photographs were badly smeared. It was determined that the V/H sensor, which should have moved the film to accommodate the moving image, was out of synchronization with the shutter of the high-resolution lens.

The synchronization problem between the V/H sensor and shutter of the high-resolution lens could not be corrected in flight, and so all of the high-resolution photographs taken at low altitudes during the mission were smeared.

The perilune of the orbit was lowered to 58 km to conduct the planned photography. A photograph taken with the medium-resolution lens from an altitude of 51 km (32 miles) is shown next page. The spacecraft was nearly over the equator at latitude of 0.58 degrees and longitude of -36.4 degrees when the picture was taken. The photograph spans a width on the lunar surface of about 35 km (21.8 miles), and the diameter of the large crater is about 8 km (5 miles) (Fig. 3.8).

The planned photographic mission was completed on 29 August. A total of 205 frames had been exposed, 38 frames during the initial orbit and 167 during lower orbits. The photographs included 9 potential Apollo landing sites, 11 areas of the far side of the moon, and 2 pictures of the earth.

Flight of Lunar Orbiter 2

Lunar Orbiter 2 was launched from Cape Canaveral on 6 November 1966. The launch, insertion into earth orbit, and insertion into the translunar trajectory went well. The spacecraft deployed its solar arrays and antennas, achieved lock on the sun and the star Canopus, and thus stabilized cruised toward the moon.

A midcourse correction was made by a brief burn of the rocket engine about 44 hours after launch. The spacecraft was inserted into lunar orbit about 92 hours after launch when the rocket engine slowed the spacecraft to allow it to be captured by the gravitational field of the moon. The initial orbit had an apolune of 1,850 km, perilune of 196 km, and inclination of 12 degrees to the equator.

The spacecraft was tracked by the deep space stations on earth to determine the variation in the gravitational field of the moon as the spacecraft traveled in the

Fig. 3.8 Medium-resolution photograph of lunar surface taken by Lunar Orbiter 1 (NASA photograph)

initial orbit. The parameters of the orbit were changed after 33 orbits to lower the perilune to 49 km for the photography portion of the mission.

Photography began on 18 November, and a total of 20 sites were photographed as planned. Synchronization of the V/H sensor, which had been a problem with the Lunar Orbiter 1, worked well, and both the medium-resolution and high-resolution photographs were very good quality.

A photograph was taken of the crater Copernicus although it was not in the original planned sites. That photograph of opportunity is shown on the next page (Fig. 3.9). The news media referred to the Copernicus photograph as one of the great pictures of the century. The photograph was taken by the 610 mm lens at an oblique angle when the spacecraft was about 240 km (149 miles) from the crater.

The crater itself is about 100 km (60 miles) wide. The photograph shows hills about 300 meters (984 feet) high rising from the crater floor.

Lunar Orbiter 2 returned 602 high-resolution photographs of the lunar surface and 208 medium-resolution photographs.

Flight of Lunar Orbiter 3

Lunar Orbiter 3 was launched from Cape Canaveral on 5 February 1967. The intent of the mission was to provide further photographic coverage of potential Apollo landing sites that looked promising in photographs taken by Lunar Orbiters 1 and 2. The mission would also add to knowledge about variations in the gravitational field of the moon, again of importance for Apollo.

Fig. 3.9 Central portion of Copernicus crater taken by high-resolution lens of Lunar Orbiter 2 (NASA photograph)

The launch, insertion into earth orbit, and insertion to a cislunar trajectory were successful. Lunar Orbiter 3 separated from Agena and deployed its solar panels and antennas. It aligned with the sun, and about 7 hours after launch, it became fully stabilized by acquiring the star Canopus. An accurate midcourse correction was made about 37 hours after launch by a brief firing of the rocket engine. The spacecraft reached the vicinity of the moon on 8 February. It fired the rocket engine for about 9 minutes to slow the spacecraft so that it could be captured by the moon's gravity and put into orbit about the moon.

The initial orbit of Orbiter 3 had an apolune of 1,802 km, perilune of 210 km, and inclination of about 21 degrees to the lunar equator. The spacecraft was tracked from earth for 25 orbits over a period of about 4 days to obtain data on variations in the gravitational field of the moon. The perilune of the orbit was then lowered to 55 km (34 miles) for photography.

Lunar Orbiter 3 took its first photographs on 15 February while over the south-eastern region of Mare Tranquillitatis. A total of 211 frames of photographs had been taken by 23 February and readout of the film and transmission to earth began. The photographs were very high quality and generally better than those taken by Orbiters 1 and 2. Unfortunately, after reading out 139 frames of film, the film advance motor failed, and the remaining 72 exposed frames could not be read out.

Mare Tranquillitatis was selected as the landing site for the first manned landing on the moon in Apollo 11. Photographs from the Lunar Orbiter contributed to the landing site selection. A photograph of a portion of Mare Tranquillitatis taken with the medium-resolution lens at an altitude of 50 km is shown on the next page (Fig. 3.10). The extent of the photograph is about 34.2 by 40.7 km on the surface. The center of the photograph is at latitude 0.51 degrees and longitude of 24.21 degrees in lunar coordinates. The landing site for Apollo 11 was at latitude 0.674 degrees and longitude of 23.473 degrees which puts it slightly off of the left side of the photograph and 4.98 km above the center of the photograph.

A medium-resolution photograph of the far side of the moon taken by Lunar Orbiter 3 is shown on the following page (Fig 3.11). Tsiolkovsky crater is the prominent feature near the center of the photograph. The photograph was taken during the high portion of the orbit with the spacecraft at an altitude of 1,463 km (910 miles). The spacecraft was oriented so that the principle point (center of the medium-resolution photograph) was at longitude of 127 degrees and latitude of -24 degrees (Fig. 3.10).

In total, Lunar Orbiter 3 returned 427 high-resolution photographs of the moon and 149 medium-resolution photographs.

Flight of Lunar Orbiter 4

Lunar Orbiter 4 was launched from Cape Canaveral on 4 May 1967. The launch, insertion into earth orbit, and insertion into the translunar trajectory all went well.

Fig. 3.10 Photograph of Mare Tranquillitatis by Lunar Orbiter 3 (NASA photograph)

The spacecraft deployed its solar arrays and antennas, achieved lock on the sun and the star Canopus, and cruised toward the moon.

A midcourse correction was made by a brief burn of the rocket engine about 17 hours and 35 minutes after launch. About 88 hours and 44 minutes after launch, the rocket engine was fired again to slow the spacecraft to allow it to be captured by the gravitational field of the moon and be inserted into a near polar lunar orbit.

The initial orbit had apolune of 6,111 km, perilune of 2,706 km, and inclination of 85.5 degrees to the lunar equator. The intent of the mission was to map the lunar surface by a series of photographs taken by the 610 mm lens. The resulting vertical strips of photographs as the moon rotated under the plane of the orbit could be assembled into a contiguous map by processing back on earth.

Fig. 3.11 Photograph of far side of moon with Tsiolkovsky crater near the center of the picture taken by Lunar Orbiter 3 (NASA photo)

Photography began during the sixth orbit on 11 May. Two contiguous high-resolution photographs that were taken during the mapping sequence are shown next page. The photographs were taken from an altitude of 2,724 km (1,693 miles) with the spacecraft at latitude of 14.4 degrees and longitude of 82.7 degrees in the lunar coordinate system. The width of the strip of photographs is about 249 km (155 miles) (Fig. 3.12).

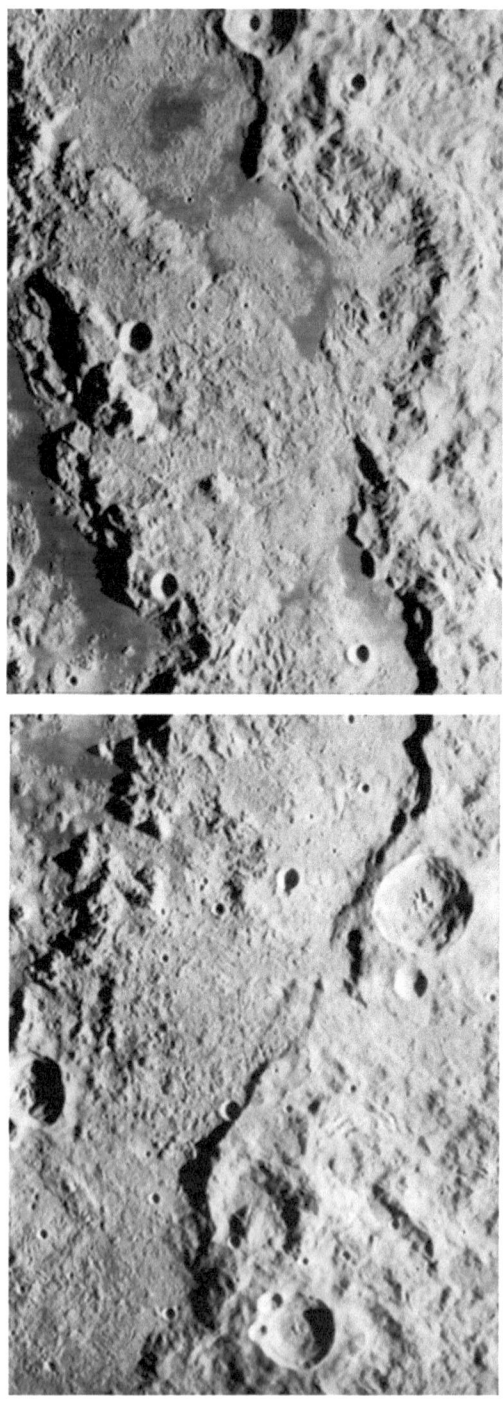

Fig. 3.12 Strip of high-resolution photographs taken by Lunar Orbiter 4
(NASA photograph)

The mapping photography began on the sixth orbit on 11 May and ended on the 35th orbit on 25 May. NASA TM X-3487 indicates that about 99% of the nearside of the moon was mapped by Lunar Orbiter 4 during that time. The resolution of the images exceeded ten times the best earth-based telescope photography.

Flight of Lunar Orbiter 5

Lunar Orbiter 5 was launched from Cape Canaveral on 1 August 1967. The launch, insertion into the cislunar trajectory, midcourse correction, and insertion into lunar orbit all went well. Lunar Orbiter 5 marked the fifth time in a row that all of these complex operations were completed successfully, and it speaks to the emerging maturity of the US Space program.

The spacecraft entered the initial orbit of the moon on 5 August with apolune of 6,023 km, perilune of 195 km, and inclination of 85 degrees from the lunar equator. The perilune was lowered to 100 km while maintaining apolune of 6,023 meters and a series of sites were photographed on both the nearside and the far side of the moon.

In all, Lunar Orbiter 5 photographed 5 Apollo sites, 36 sites of scientific interest, 23 sites on the far side of the moon. In total, Lunar Orbiter 5 returned 633 high-resolution photographs of the moon and 211 medium-resolution photographs.

The combined imaging by the five Lunar Orbiters resulted in photographs of 99% of the near- and far side of the moon with resolution in selected areas as low as one meter. In total, 2,180 high-resolution and 882 medium-resolution photographs of the moon were received back at earth.

Bibliography

NASA Technical Memorandum NASA TM X-3487, *Destination Moon: A History of the Lunar Orbiter Program*, 1977
NASA-NSSDCA/COSPAR ID:
    1966-073A, *Lunar Orbiter 1*
    1966-100A, *Lunar Orbiter 2*
    1967-008A, *Lunar Orbiter 3*
    1967-041A, *Lunar Orbiter 4*
    1967-075A, *Lunar Orbiter 5*
NASA report NASA CR-1054, *Lunar Orbiter IV*, June 1968
NASA report NASA SP-206, *Lunar Orbiter Photographic Atlas of the Moon*, 1971
The Lunar Orbiter, Boeing report for NASA, April 1966

4

The Surveyor Lunar Landing Mission

The Surveyor spacecraft was a pioneering lunar lander and a pathfinder for the manned Apollo lunar landers that followed. A photograph of an engineering test model of the spacecraft is shown next page. The model rests on the beach near the Hughes Aircraft Factory in Los Angeles. Surveyor was developed and built by Hughes (Fig. 4.1).

The very capable Surveyor spacecraft was relatively compact with three landing legs having a footprint spanning about 14 feet. The vertical mast supported a solar panel and the high-gain array antenna. The height of the spacecraft was about 10 feet.

OVERVIEW OF THE SURVEYOR PROGRAM

The Surveyor project was managed by the Jet Propulsion Laboratory (JPL) for the NASA Headquarters Office of Space Science and Applications. The JPL project manager for Surveyor was Walker Giberson. He reported to Clifford Cummings, Lunar Program Director. JPL had management responsibility for the Surveyor spacecraft system, tracking and data system, and mission operations system. The Lewis Research Center of NASA was responsible for the Atlas/Centaur launch vehicle system.

Hughes Aircraft Company designed and built the Surveyor spacecraft under contract with NASA, but the program was overseen by JPL. Hughes was selected in January 1961 after competitive bidding. Leo Stooman was the manager of the Surveyor project for Hughes, and Shel Shallon was Surveyor project scientist. The author made periodic presentations to Shallon.

© Springer Nature Switzerland AG 2018   50
T. Lund, *Early Exploration of the Moon*, Springer Praxis Books,
https://doi.org/10.1007/978-3-030-02071-2_4

Fig. 4.1 Engineering model of Surveyor spacecraft (NASA photograph)

A key objective of the Surveyor program was to verify that a spacecraft could be soft-landed on the moon and very near a preselected site. The Surveyor landings were designed to pave the way for the ambitious Apollo program that landed astronauts on the moon a few years later. In addition to proving soft-landing techniques and verifying that the lunar surface could indeed support a spacecraft, the experiments conducted provided valuable scientific data on properties of lunar soil.

Here on earth, we received a vivid appreciation of the moon from thousands of close-up and panoramic television pictures that were taken around the landing sites by the Surveyor spacecraft. One such close-up picture showing a footpad of Surveyor 1 and penetration of the pad into the lunar soil upon landing is given on the next page (Fig. 4.2).

A total of seven Surveyor missions were flown to the moon between May 1966 and January 1968. Five of these missions, Surveyor 1, 3, 5, 6, and 7, resulted in soft landings and transmission of valuable scientific data back to earth. Two missions were not successful: one due to failure of a vernier rocket engine and the other due to permanent loss of communication with the spacecraft during firing of the retrorocket engine. Most of the landing sites chosen were near potential landing sites for Apollo. Indeed, Apollo 12 landed within 600 feet of Surveyor 3.

Fig. 4.2 Footpad of Surveyor 1 imprinting lunar soil (NASA photograph)

The Surveyor spacecraft was lofted to the moon by an Atlas/Centaur launch vehicle. The spacecraft separated from the launch vehicle, cruised to the moon, and made a direct descent to the lunar surface.

A retrorocket was used to slow the spacecraft as it approached the surface. Firing of the retrorocket was initiated by a signal from an altitude marking radar at a slant range of about 60 miles. After burnout and jettisoning of the retrorocket, three vernier engines thrusting under control of the flight control subsystem brought the spacecraft to a soft landing on the moon. Key input to the flight control subsystem to orchestrate the landing was range to the surface and three-axis velocity data provided by the landing radar.

Oran Nicks, who headed the Lunar and Planetary Program at NASA, recounted in later writing that the most significant challenge for Surveyor was the landing. He writes that "The three major elements of the landing system were (1) a high-performance solid rocket motor to provide the bulk of the velocity reduction on approach, (2) liquid propelled vernier engines capable not only of varying thrust but also of swiveling to allow attitude orientation, and (3) the landing radar system, which sensed distances and vertical and lateral motions with respect to the moon."

The landing sites for the five Surveyor spacecraft are shown in a NASA graphic of the moon shown on the next page (Fig. 4.3). Four of the spacecraft were landed within an area planned for Apollo landings to reconnoiter that area. Apollo landings had been planned for a region of the moon bounded by ±5 degrees of latitude and ±45 degrees of longitude in the lunar coordinate system. The last landing, that of Surveyor 7, was in a scientifically interesting area in the highlands of the moon.

Once on the surface of the moon, a television camera on the spacecraft captured close-up pictures of the landing pads as well as panoramic pictures of surrounding terrain. The pictures of the landing pads and readings of strain gauges on the landing legs gave good assurance that the lunar surface could indeed support the weight of a large spacecraft such as the Apollo Lunar Module and not sink through the fluff as some had worried.

The very large number of pictures taken by the television camera gave the world a good visual impression of the moon. The five Surveyor spacecraft that soft-landed on the moon transmitted a total of about 87,000 pictures back to earth.

A soil sampler scoop included in the Surveyor 3 and Surveyor 7 spacecraft was used to dig trenches in the lunar surface to gain further insight into the properties of the lunar soil.

The total cost of the Surveyor program was about $469 million. It was a relatively austere program compared with Apollo, and there were many challenges during its development.

The cost did not reflect countless free hours of overtime put in by engineers and others on the program. The author was one such engineer. The effort was compensated by a certificate of appreciation received at the end of the program. That certificate contained a picture of the shadow of Surveyor 1 on the lunar surface. The certificate said:

> ***Let it be recorded that:***
> *When future generations look back on man's conquest of space, the soft landing of an instrumented spacecraft on the lunar surface will mark a most significant milestone…advancing man's technological capabilities, and providing the world its first closeup look at a celestial body, and*

> ***T. J. Lund***
> *as a member of the Surveyor team shared in this exciting venture and contributed to the successful achievement of the program's goals… paving the way for man's journey to the planets.*

Fig. 4.3 Landing sites for Surveyor 1, 3, 5, 6, and 7 spacecraft (NASA graphic)

FLIGHT OF THE SURVEYORS

Launch and Midcourse Correction

The Surveyor spacecraft was lofted toward the moon from Cape Kennedy, Florida by an Atlas/Centaur two-stage launch vehicle. The Surveyor spacecraft, with legs folded, was contained in the nose cone attached to the Centaur second stage. A photograph of the launch of Surveyor 1 in May of 1966 is shown on the next page (Fig. 4.4). The Centaur stage was the upper portion of the vehicle extending down to where the fairing strip on the side of the vehicle ends. The launch vehicle was modified to provide slightly greater performance for the launch of Surveyor 5, 6, and 7 spacecraft, and that configuration will be described here. The later Atlas/Centaur launch vehicles were 117 feet long to the tip of the nose cone and 10 feet in diameter up to the nose cone. The weight of the overall vehicle, including the Surveyor spacecraft, was 325,000 pounds.

Fig. 4.4 Launch of Surveyor 1 in May 1966 (NASA photograph)

The Atlas first stage was powered by five engines: two booster engines, one sustainer engine, and two vernier engines. The engines all burned kerosene and liquid oxygen. The two booster engines provided a thrust of 168,000 pounds each, the sustainer engine provided 58,000 pounds of thrust, and the two vernier engines provided 670 pounds each. The total thrust was 395,000 pounds at liftoff, just comfortably above the 325,000 pound weight of the vehicle. The booster engines and associated fuel pumps were jettisoned after about 2.6 minutes of flight and the sustainer and vernier engines of Atlas continued to burn until commanded to shut down.

The Atlas stage included an autopilot containing gyros for attitude reference and a programmer that initiated flight sequencing during the initial flight of the launch vehicle. Control of the vehicle was achieved by coordinated adjustment of the angles of each of the five gimbaled engines. The two booster engines were used for pitch, yaw, and roll control, while the vernier engines assisted in roll control. The sustainer engine was maintained in the centered position during boost. After jettisoning the two booster engines, the Atlas autopilot received steering commands from the inertial guidance system located in the Centaur stage, and the gimbaled sustainer engine was positioned for control.

The Centaur second stage contained two engines that burned liquid hydrogen for fuel and liquid oxygen as oxidizer. Each engine generated 15,000 pounds of thrust. The engines were gimbaled to allow control of the Centaur in pitch, yaw, and roll. The Centaur also contained 14 constant-thrust, fixed orientation, hydrogen peroxide reaction engines for control of the vehicle after the main engines were shut down. Six of these engines were grouped in two clusters of three located 180 degrees apart near the periphery of the vehicle. Each group contained one 6 pound thrust engine for pitch control and two 3.5 pound engines for yaw and roll control. The engines were individually controlled. In addition, Centaur had four engines with 50 pound thrust and four engines with 3 pound thrust mounted with their thrust axes parallel to the vehicle axis. Those engines provided acceleration to position the propellants prior to a burn, and they were used to achieve separation from the spacecraft.

A brief timeline of the initial flight and midcourse events of Surveyor 7 is given in the following paragraphs to give the reader an appreciation of the technology and the choreographing of thrust, attitude, and velocity that positioned the Surveyor spacecraft at a spot above the moon that would result in a soft landing at the selected site.

Shortly after liftoff, the launch vehicle was commanded to roll from an azimuth of 105 degrees to an azimuth of 102.9 degrees, and then a pitch-over maneuver was commanded that lasted until booster engine cutoff. A total pitch change of 72 degrees was made. The vehicle reached a speed of Mach 1 about 64 seconds after liftoff.

Booster engine cutoff was commanded about 2 minutes and 32 seconds after liftoff when the Centaur inertial guidance system detected acceleration of 5.7 g. The two booster engines were jettisoned after cutoff, and the sustainer and vernier engines continued to fire. The Centaur guidance system controlled angular position of the sustainer engine during firing of the sustainer and vernier engines of the Atlas after jettisoning of the booster engines. An additional 8 degrees of pitch-over was made for a total of 80 degrees since launch.

The clamshell nose fairing was jettisoned 3 minutes and 46 seconds after launch exposing the Surveyor spacecraft. The Atlas sustainer and Vernier engines were shut down about 4 minutes and 9 seconds after launch. Separation of the Centaur stage from Atlas occurred 2 seconds later.

The Centaur engines were then fired until the inertial guidance system determined that the vehicle had attained the desired parking orbit. In the case of Surveyor 7, the burn time was about 5 minutes and 33 seconds. The Centaur guidance system commanded additional pitch-over during the burn until a total of 98 degrees of pitch had been achieved since vertical launch. A parking orbit at a height above earth of about 90 nautical miles was achieved 9 minutes and 53 seconds after launch. The coast time in the parking orbit was 22 minutes and 26 seconds before the Centaur engine was restarted. The engine was fired for 116 seconds to put the vehicle into a lunar transfer trajectory.

The Surveyor 7 spacecraft was separated from Centaur while enroute to the moon 35 minutes and 15 seconds after launch. The landing legs and the omnidirectional antenna masts of the Surveyor spacecraft had been extended before separation.

The plane of the solar panel was positioned normal to the longitudinal or Z-axis of the spacecraft just after separation of the Surveyor spacecraft from the Centaur. The sun acquisition sequence was begun by first rolling the spacecraft and then yawing it to approximate alignment with the sun. Lock-on of the sun sensor, which was aligned with the Z-axis of the spacecraft, occurred 44 minutes and 29 seconds after launch. The spacecraft was then maintained with its Z-axis aligned with the sun by corrective signals from the sun sensor.

After alignment with the sun, a three-axis attitude reference for Surveyor 7 was obtained by tracking the star Canopus by a star tracker. Canopus was ideally suited for this operation since it is the second brightest star seen from earth and it was nearly perpendicular to a line to the sun.

The spacecraft was commanded from earth to rotate about the Z-axis until a series of stars ending with Canopus appeared in the field of view of the star tracker. Steering signals from the star tracker controlled roll of the spacecraft to maintain alignment of the –X-axis of the spacecraft with the projection of Canopus on the X–Y plane of the spacecraft. Thus aligned with the sun and Canopus, the spacecraft cruised toward the moon.

Alignment of the spacecraft in pitch, yaw, and roll was maintained by a control system that used nitrogen gas feeding three pairs of electrically controlled gas jets. The pairs of jets were mounted on the legs of the spacecraft near the landing pads. Midcourse corrections to the trajectory had been planned to adjust the landing site to the preplanned site.   Preparation for a midcourse correction was made at about 16 hours and 16 minutes after launch by orienting the spacecraft so that thrust from the vernier engines would bring about the desired velocity correction. After proper orientation of the spacecraft, the three vernier engines were commanded to fire. The burn time was 11.4 seconds, and the change in velocity was 11.13 meters per second. The spacecraft was commanded to resume cruise alignment with the sun and Canopus after the midcourse correction.

Tracking the Spacecraft

An important element in the choreography to transfer a spacecraft from earth to a preselected landing site on the moon was tracking of the spacecraft to determine its orbit. The orbit was accurately determined by tracking facilities of the Deep Space Network that included Deep Space Stations (DSS) in several locations around the earth. The stations that were designated prime during the Surveyor 7 mission were:

- DSS 11 at Goldstone Deep Space Communications Complex near Barstow, California
- DSS 42 at Tidbinbilla, Australia (near Canberra)
- DSS 61 at Robledo, Spain (near Madrid)

In addition, DSS 71 at Cape Kennedy, DSS 14 and DSS 12 at Goldstone, and DSS 51 near Johannesburg, South Africa, assisted and provided backup for the mission. The very capable DSS 14 station at Goldstone with an antenna diameter of 210 feet (64 meters) was used to back up DSS 11 at Goldstone and to gather telemetry data during the landing.

The Deep Space Stations, except for DSS 71 at Cape Kennedy, used 85-feet (26 meter)-diameter parabolic antennas to track the spacecraft. Accurate determination of spacecraft track was achieved through antenna pointing angles and Doppler shift information. The Doppler shift of the transmitted signal could be determined to a fraction of a Hertz (cycle per second), and that gave a very accurate measure of range rate from that station. Initial calculation of the translunar orbit used antenna tracking angles along with two-way Doppler shift information from two DSS facilities. As the spacecraft traveled further from earth, small pointing errors of the antenna became significant, and more accurate calculation of orbit was obtained by using Doppler data alone.

The Surveyor spacecraft contained an S-band transponder function that received the transmitted signal from the DSS, shifted it in frequency, and retransmitted it. The signal from the transponder was received by the DSS antenna and the two-way Doppler shift extracted. The orbit was calculated from two-way Doppler information gathered from two DSS stations, taking advantage of the very long baseline distance between stations.

Transit to the Moon and Terminal Descent

Following the midcourse correction at 17 hours after launch, Surveyor 7 entered a 48-hour coast phase enroute to the moon. Trajectory measurements made during this time confirmed that the landing point would be close to the desired site and a second midcourse correction would not be needed. The deep space network requested and received a wide range of engineering data from the spacecraft several times during the coast phase. Toward the end of the coast phase, heaters in the temperature-controlled compartments were turned on for a time, and various other equipment were turned on to warm up.

Events during the terminal descent are recounted below. The event times will be referred to as time after launch from earth and given in a short form as hours/minutes/seconds. During the last 3 minutes of the descent, event times will be given in seconds before touchdown in keeping with plotted range and velocity data during the final descent.

At 65:52:55, or about 40 minutes before ignition of the retrorocket engine, the Z-axis of the spacecraft was aligned with the velocity vector. After alignment with the velocity vector, a command was sent to roll the spacecraft by –16.5 degrees to achieve the desired orientation of the spacecraft upon landing.

The altitude marking radar (AMR) generated a mark signifying that the slant range had reached 60 miles from the lunar surface at 66:32:10. The three vernier engines were brought up to stabilize the spacecraft, and then the main retro engine was fired. The spacecraft velocity was about 8,900 feet per second just before firing the retro engine. The engine operated for 43 seconds before it burned out.

The radar altimeter and Doppler velocity sensor (RADVS) was activated just after retrorocket ignition at 66:32:15. All three velocity sensor beams achieve lock, and the reliable operate signal was given 32 seconds later at 66:32:47. The radar altimeter achieved lock on the lunar surface at 66:33:15.

Following burnout of the retrorocket, the engine case was ejected at 66:33:09. The case apparently passed through velocity sensor beam 1 causing loss of lock, but the signal from the lunar surface was reacquired a few seconds later.

Velocity measurements at the time of lock to the return signals from the moon indicated velocity along the longitudinal or Z-axis (V_Z) was above the upper readout limit of the telemetry of 800 feet per second (fps).   The velocity along the X-axis (V_X) was about 0, and the velocity along the Y-axis (V_Y) was about –4 fps. As the retrorocket continued to burn, the velocity along the Z-axis decreased and fell within the upper limit of 800 fps about 159 seconds before touchdown. By this time, V_X had increased to –50 fps, and V_Y had increased to −80 fps, apparently due to lateral velocity induced by burning of the retrorocket.

Velocity information from the RADVS was fed to the flight control subsystem. RADVS control was initiated about 143 seconds before touchdown. At that time,

RADVS indicated that V_Z was 452 fps, V_X was –60 fps, and V_Y was –125 fps. The slant range measurement was greater than the upper readout limit of 40,000 feet. The slant range was estimated to have been about 41,500 feet at the time. The flight path vector was about 18.1 degrees from the local vertical.

Using velocity data from the RADVS, the flight control subsystem reduced the lateral velocity along the X-axis to essentially 0 within 3 seconds. The velocity component along the Y-axis was reduced to less than −10 feet per second within 3 seconds and to essentially 0 within 13 seconds. These actions resulted in alignment of the longitudinal axis of the spacecraft with the velocity vector and alignment of the thrust axis of the vernier engines with the velocity vector as well.

With the lateral velocity components nulled and thrust applied along the longitudinal axis, the spacecraft performed a "gravity turn," and soon the longitudinal axis of the spacecraft was aligned with the local gravity vector.

The preprogrammed descent trajectory of slant range vs. longitudinal velocity is shown on the next page (Fig. 4.5). It consisted of four joined straight line segments. The thrust of the vernier engines was increased to slow the spacecraft and cause it to follow the preprogrammed trajectory. The trajectory of the Surveyor as measured by RADVS is also plotted on the chart and appears as the curved line below the straight line segments. The chart shows that the spacecraft followed the four-segment preprogramed descent trajectory quite well.

Plots of the velocity components, V_X, V_Y, and V_Z, and slant range as recorded from the telemetry data as the spacecraft descended are given on the following pages. The plots were generated by JPL.

As the spacecraft continued to descend, the flight control subsystem controlled the thrust of the vernier engines to maintain a velocity along the Z-axis between 460 and 480 fps with the lateral velocities nulled until the slant range decreased to about 20,000 feet. The spacecraft intercepted Segment 1 of the descent trajectory at that range. The longitudinal velocity was 464 feet per second at the time. Intercept occurred about 95 seconds before touchdown. The spacecraft then followed the preprogrammed descent path.

The radar altimeter generated a mark, fittingly called the 1000 foot mark, at a slant range (altitude) of 1000 feet. A mode change was made in the radar altimeter at the 1000 foot mark by increasing the scale factor of the range data by a factor of 10. This was done by increasing the slope of the FM waveform by a factor of 10, and it resulted in higher accuracy of the range measurement. The velocity along the longitudinal axis was about 110 feet per second at the 1000 foot mark, and the lateral velocity components remained nulled. The spacecraft continued to follow the preprogrammed descent trajectory using velocity and range information from the RADVS until the range had decreased to 50 feet and the velocity had decreased to 10 feet per second. The thrust was then adjusted to maintain a vertical velocity of 5 feet per second from 40 feet altitude to the 14 foot mark.

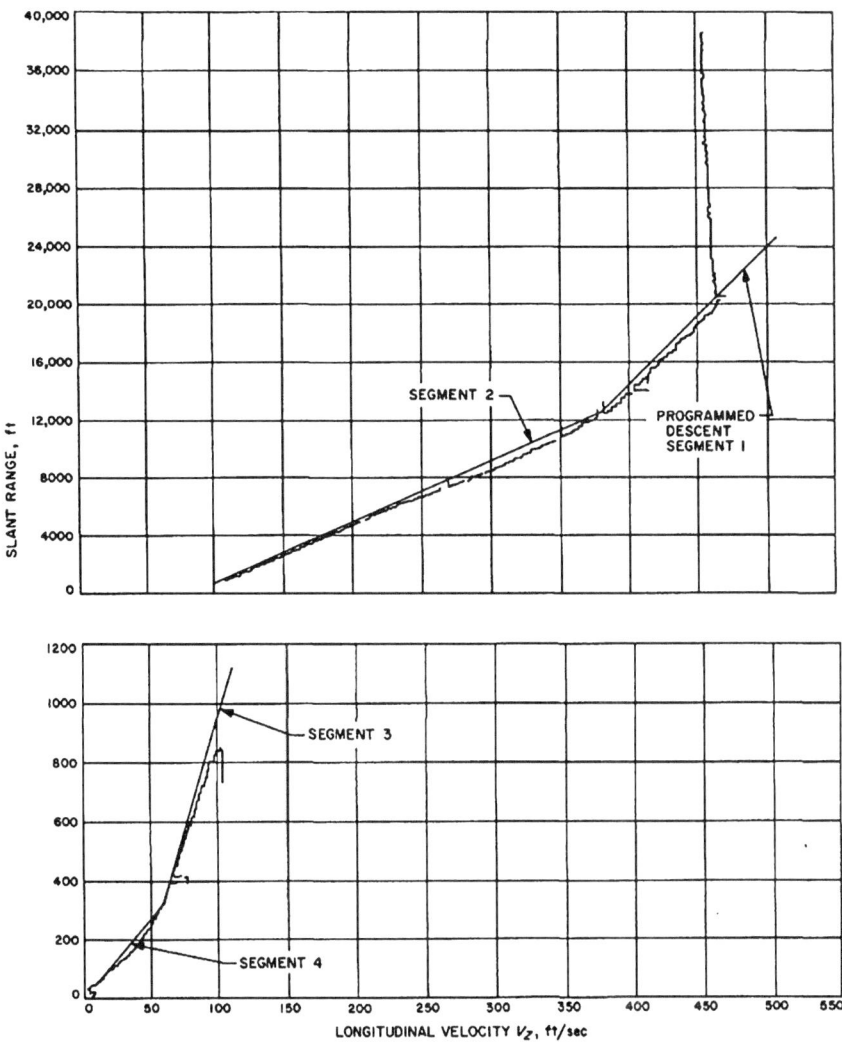

Fig. 4.5 Preprogrammed descent trajectory (straight lines) and trajectory of Surveyor 7 (curved path) (NASA graphic)

The vernier engines were shut down when the 14 foot mark was received from the RADVS and the spacecraft fell under lunar gravity to the surface. The vertical velocity of Surveyor 7 had increased to 12.4 feet per second by the time it landed. Touchdown occurred 66 hours, 35 minutes, and 35 seconds after launch from earth.

The 3 foot pads touched down within 0.05 seconds of each other. The Z-axis of the spacecraft was 3.17 degrees from the vertical after landing, and the X-axis

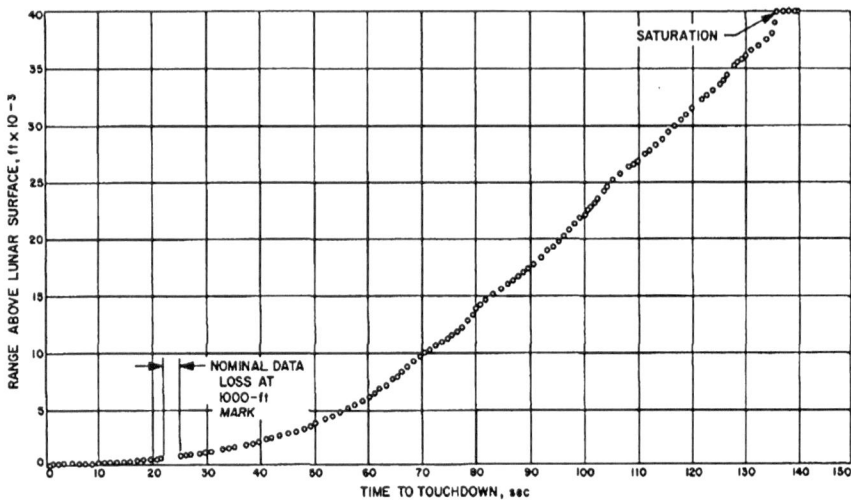

Fig. 4.6 Slant range vs. time during terminal descent (NASA graphic)

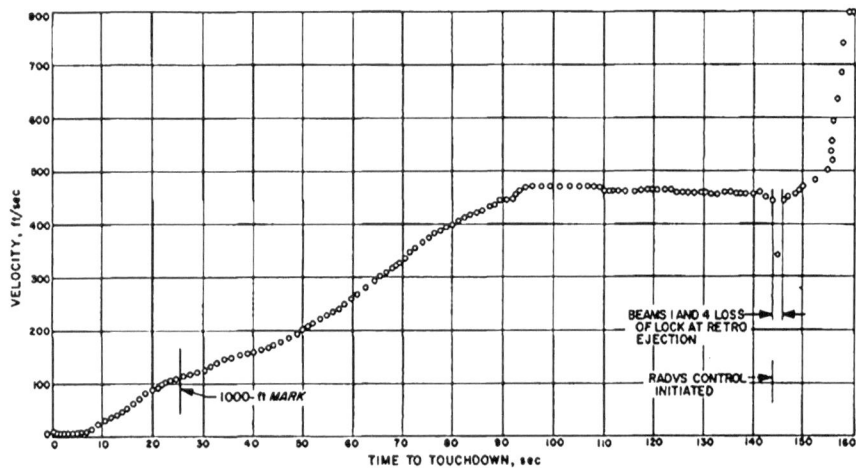

Fig. 4.7 Velocity along Z-axis during terminal descent (NASA graphic)

of the spacecraft was 290.2 degrees clockwise from lunar north. The lunar coordinates of the landing were 40.92 degrees south and 11.45 degrees west.

The slant range, Z-axis velocity, and Y- and X-axis velocities as measured by the landing radar during the terminal descent were shown above and on the next pages (Figs. 4.6, 4.7, 4.8, and 4.9). It is interesting to see how rapidly the sizable Y-axis and X-axis velocities (lateral velocities) were nulled to zero after the

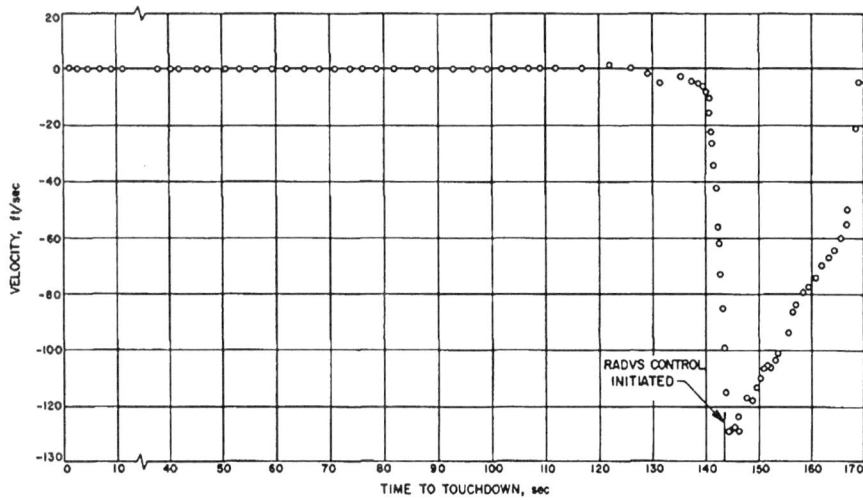

Fig. 4.8 Velocity along Y-axis during terminal descent (NASA graphic)

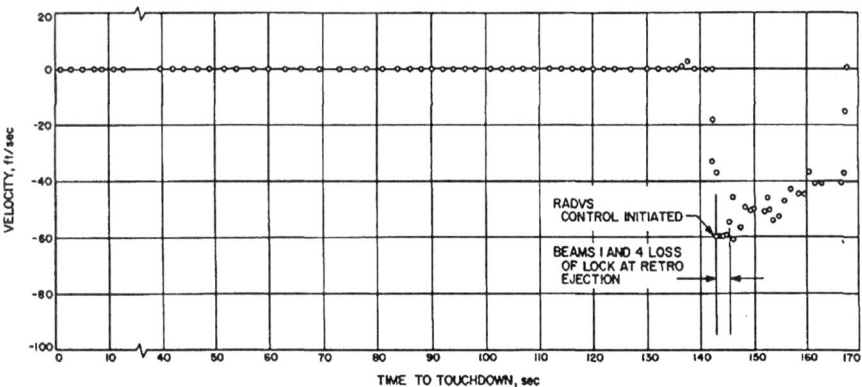

Fig. 4.9 Velocity along X-axis during terminal descent (NASA graphic)

landing radar began providing data. Surveyor 7 made a perfect three-point landing on the moon at a velocity of 12.4 feet per second. Remarkable, since the spacecraft had been hurdling towards the moon at a velocity of about 8,900 feet per second just 200 seconds earlier.

MECHANICAL CONFIGURATION OF SURVEYOR

A photograph of an engineering model of the Surveyor spacecraft sitting on the beach near the Hughes Factory in Los Angles is shown below (Fig. 4.10).

Fig. 4.10 Engineering model of Surveyor on the beach (NASA photograph)

The prominent rectangular plate on top of the mast on the left side is a solar array, and the plate on the right side is a high-gain planar array antenna. The physical size of the spacecraft can be visualized by noting that the footpads on the three landing legs would fit within a circle 14 feet in diameter, and the height of the spacecraft to the top of the mast holding the solar panel and antenna array was about 10 feet.

The frame of the spacecraft was formed from aluminum tubing with braces and mounting provisions for components attached to the tubing. The weight of the efficient frame, including mounting provisions for components was about 83 pounds. The relatively large solar panel was about 44 by 49 inches in size, and the antenna array was 39 by 39 inches in size. The landing legs were hinged where they attached to the structure to allow folding up inside of the nose cone attached to the Centaur launch vehicle. The leg struts contained hydraulic shock absorbers to absorb the kinetic energy of the spacecraft at touchdown. Three crushable honeycomb blocks under the frame protected the spacecraft should the landing legs deflect the full amount at landing.

A side view drawing of the spacecraft showing its mechanical configuration is given on the next page (Fig. 4.11). Labels on the drawings identify major components.

Two omnidirectional antennas were attached to booms that were tilted up inside the Centaur nose cone during launch and then deployed just before Surveyor separated from Centaur.

The high-gain planar array antenna and solar panel were supported by the antenna/solar panel positioner that is shown on the sketch of a side view of the spacecraft. The mast rotated in the roll axis of the spacecraft and the planar array antenna could be positioned toward the earth, and the solar panel could be positioned toward the sun. The positioner was adjusted periodically during the lunar day to keep the solar array pointed toward the sun and the planar array pointed toward the earth.

There were three thermally controlled compartments (A, B, and C), and various electronic subsystems were mounted to thermal trays within the compartments. The trays acted to distribute heat generated by the electronics throughout the insulated compartment. Compartment C was added for Surveyor 5, 6, and 7 to contained equipment for the alpha-scattering instrument that was incorporated in those spacecraft.

The thermal compartments contained a heater and a thermal control system to maintain temperatures between 40°F and 125°F for compartment A and between 0°F and +125°F for compartment B. The heaters for the three thermal compartments were commanded ON a few hours before the terminal descent phase to bring the equipment up to temperature after the long cruise from earth. The thermal control system included a series of thermally activated mechanical switches. Those switches controlled the conduction path between the top of each compartment and an external radiating surface to keep the temperatures within the compartment within assigned limits.

PROPULSION

The spacecraft was decelerated from a velocity of about 8,900 feet per second to about 460 feet per second by a retrorocket that thrusted against the Z-axis of the spacecraft. Three throttleable vernier rocket engines aligned the spacecraft with the velocity vector and further decelerated the spacecraft to a soft landing.

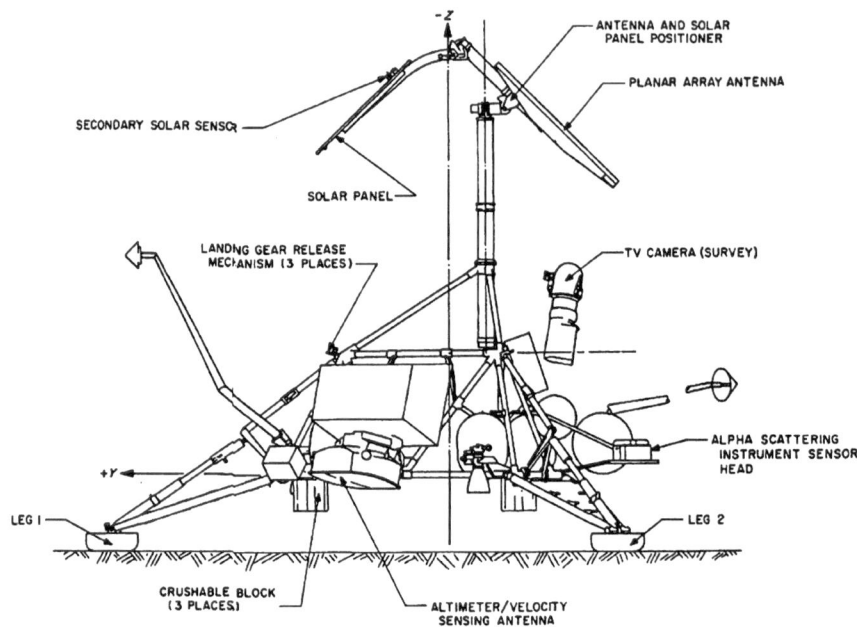

Fig. 4.11 Drawing of side view of Surveyor 7 (NASA graphic)

Retrorocket Motor

The solid fuel retrorocket was developed by the Thiokol Corporation. The retro-rocket was mounted at the center on the underside of the spacecraft. The mounts were fitted with explosive fasteners that allowed separation of the retrorocket from the spacecraft after burnout. The altitude marking radar was mounted at the exit of the nozzle of the motor by clips that allowed the radar to be jettisoned by gas pressure when the rocket was ignited.

The rocket case was spherical and 36.8 inches in diameter. The nozzle was inset into the case to shorten the overall dimension along the Z-axis. Scaling from drawings, the nozzle was inset about 15 inches into the case, and it extended about 15 inches beyond the case. The diameter at the nozzle exit was about 25 inches. The solid rocket fuel was carboxyl/terminated polybutadiene that was molded on the inside of the spherical case. The propellant weighed 1,300 pounds, and the total weight of the rocket assembly was about 1,445 pounds. The retrorocket generated about 900 pounds of thrust.

Nominally for the Surveyor missions, the velocity of the spacecraft was about 8,950 feet per second when the retrorocket was ignited at 60 miles slant range. It slowed the spacecraft to about 400 feet per second by the time it burned out. The firing time was about 43 seconds.

Vernier Engines

The thrust of three vernier engines allowed rather precise control of velocity and attitude of the spacecraft. The engines performed the midcourse velocity correction that was discussed previously. The throttleable engines, along with range and velocity measurements by the radar altimeter and Doppler velocity sensor, allowed smooth control of the spacecraft from its still high velocity after burnout of the retrorocket to a preprogrammed velocity vs. altitude trajectory ending with a soft landing. A photograph of one of the vernier engines, produced by the Thiokol Chemical Corporation, is shown below (Fig. 4.12).

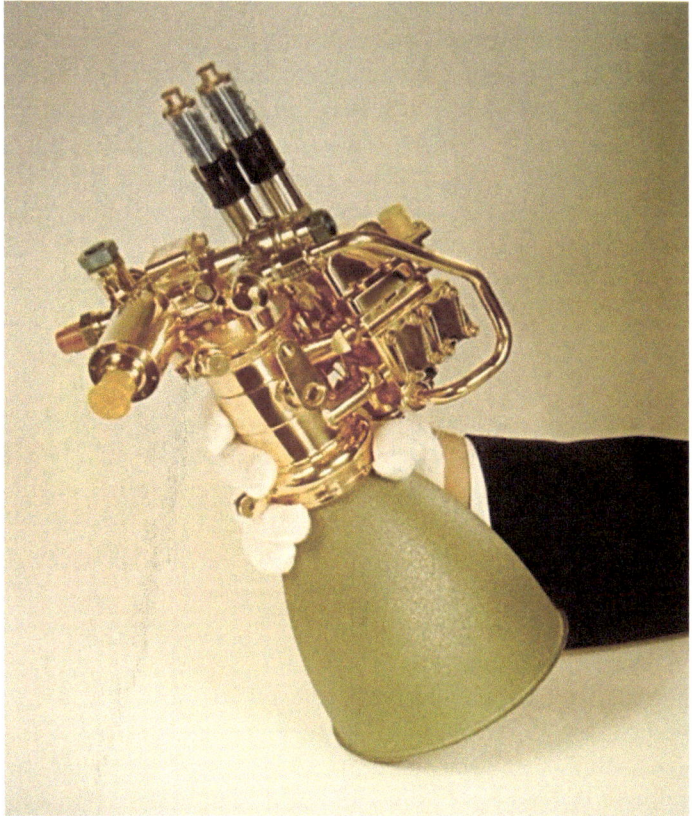

Fig. 4.12 Surveyor vernier engine (Thiokol photograph)

The three vernier engines were mounted near the attach points of the three landing legs on the bottom of the spacecraft and thrusted downward. The engines were numbered to correspond to landing leg numbering. Vernier engines 2 and 3 were mounted with their thrust axis parallel to the spacecraft Z-axis. Vernier engine 1 was gimballed to allow its thrust axis to be deflected up to ±6 degrees in the XY plane of the spacecraft to allow inducing roll of the spacecraft about the Z-axis.

The thrust of each engine could be controlled over a range of 30 to 104 pounds by throttle valves. Strain gauges attached to the mounting brackets for the engines provided measurements of the thrust produced by each engine. The strain gauge readings were telemetered back to earth. The attitude of the spacecraft could be adjusted in pitch and yaw by differential control of thrust of the engines. Roll was induced by deflecting the thrust axis of vernier engine 1.

Fuel for the engines was monomethyl hydrazine monohydrate, and the oxidizer was a combination of nitrogen tetroxide and nitric oxide. The fuel and oxidizer spontaneously combusted when they came in contact. The thrust of each engine was controlled by a throttle valve and a shutoff valve fitted in the fuel and oxidizer lines running to each engine.

Each engine had its own fuel tank and oxidizer tank. The fuel and oxidizer tanks were spherical with diameter of about 9.6 inches. The tanks contained Teflon bladders that were moved by pressurized helium to force fuel and oxidizer out of the tanks and into the engine. Helium used to pressurize the bladders was contained in a spherical tank about 14 inches in diameter. The pressure was typically regulated to be about 730 pounds per square inch (psi) when the valve from the helium tank was opened.

The Surveyor 7 spacecraft carried 73.7 pounds of useable fuel and 108 pounds of oxidizer for a total of 181.7 pounds of useable propellant. After landing it was estimated that 12.3 pounds of useable fuel and 17.4 pounds of useable oxidizer remained.

After burnout of the retrorocket, the three vernier engines were throttled to maintain velocity of the spacecraft at about 470 feet per second. The engines were differentially throttled to keep the Z-axis of the spacecraft aligned with the velocity vector. The preplanned landing sequence directed the spacecraft to descend at about 470 feet per second until it reached a slant range of 20,000 feet. At that slant range, it intercepted Segment 1 of the preplanned velocity vs. altitude descent trajectory. The preplanned trajectory and a plot of the trajectory flown by Surveyor 7 were given earlier in this chapter.

Slant range and velocity of the spacecraft relative to the lunar surface were measured by the RADVS and provided to the flight control subsystem. The thrust of the vernier engines was controlled by a closed servo loop that compared the spacecraft velocity to that of the preprogrammed value at the measured range. The thrust of the engines was continuously adjusted to match the spacecraft velocity to that in the descent profile.

A simple, elegant procedure was used to align the spacecraft with the local vertical for landing. Thrusting against the velocity vector caused the spacecraft to perform a "gravity turn" where the pull of gravity of the moon caused the flight path to become aligned with the gravity vector.

FUNCTIONAL DESCRIPTION OF SPACECRAFT

A functional block diagram of the Surveyor spacecraft is shown on the pages that follow. The diagram was obtained from NASA Technical Report 32-1262 "Surveyor VI Mission Report." The NASA block diagram was broken up into two sections by the author to make elements of the diagram larger and easier to read (Figs. 4.13 and 4.14)".

A brief description of major subsystems is given in the following paragraphs.

Telecommunications

Operation of the telecommunications subsystem was vital to the Surveyor mission. Most operations of the spacecraft and its instruments were totally dependent on commands sent from earth. The telecommunications subsystem received and decoded commands from earth, transmitted engineering data and television images to earth, and, in the transponder mode, provided coherent return of received signals to allow two-way Doppler measurements from earth for trajectory determination.

Reliability was enhanced by using dual redundant transmitters, receivers, command decoders, and omnidirectional antennas.

The telecommunications system operated in S-band using an uplink frequency of 2,113 MHz and downlink frequency of 2,295 MHz. Diplexers allowed simultaneous transmission and reception from the conical omnidirectional antennas.

The spacecraft contained a steerable, high-gain planar array antenna that was used to transmit wideband, 600-line television signals to earth. The planar array antenna had a gain of 27 dB (a factor of 500 above an isotropic radiator) and a beamwidth of 6 degrees. The earth subtends an angle of about 1.8 degrees when viewed from the moon. The entire earth was therefore covered within the beamwidth of the high-gain array antenna.

The conical omnidirectional antennas had a cardioid antenna pattern with the null of the cardioid facing the spacecraft. By switching between the two isotropic antennas, an antenna gain of 0 dB (equal to that of an isotropic radiator) could be obtained over 40% of spherical coverage, and a gain of −2 dB could be maintained over 95% of spherical coverage. Each receiver was connected to one omnidirectional antenna. Either transmitter could be switched to any of the three antennas.

Transmitter and Receiver

A brief description of the transmitter and receiver functions is given in the following paragraphs. The receivers were on continuously, so they were always able to receive commands from earth. The transmitters could be commanded ON or OFF individually.

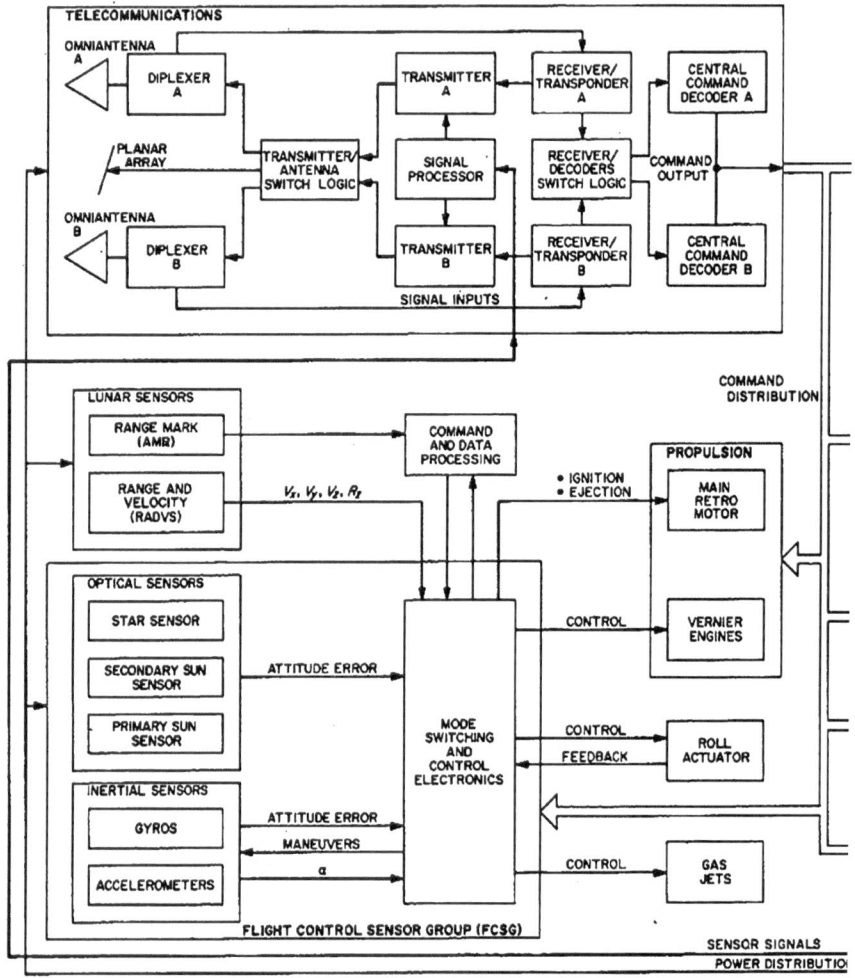

Fig. 4.13 Functional block diagram of Surveyor 6 spacecraft (adapted from NASA graphic)

Transmitter

The output power of each of the two transmitters could be commanded to be either 0.1 watts or 10 watts, and they could transmit either narrowband engineering data or wideband scientific data and television images. The transmitters could also operate in a coherent mode with input signals from a phase-locked oscillator in the receiver and that allowed two-way Doppler tracking during transit to the moon.

The transmitter contained two voltage controlled crystal oscillators operating at 19.125 MHz. One of these oscillators was frequency modulated by wideband

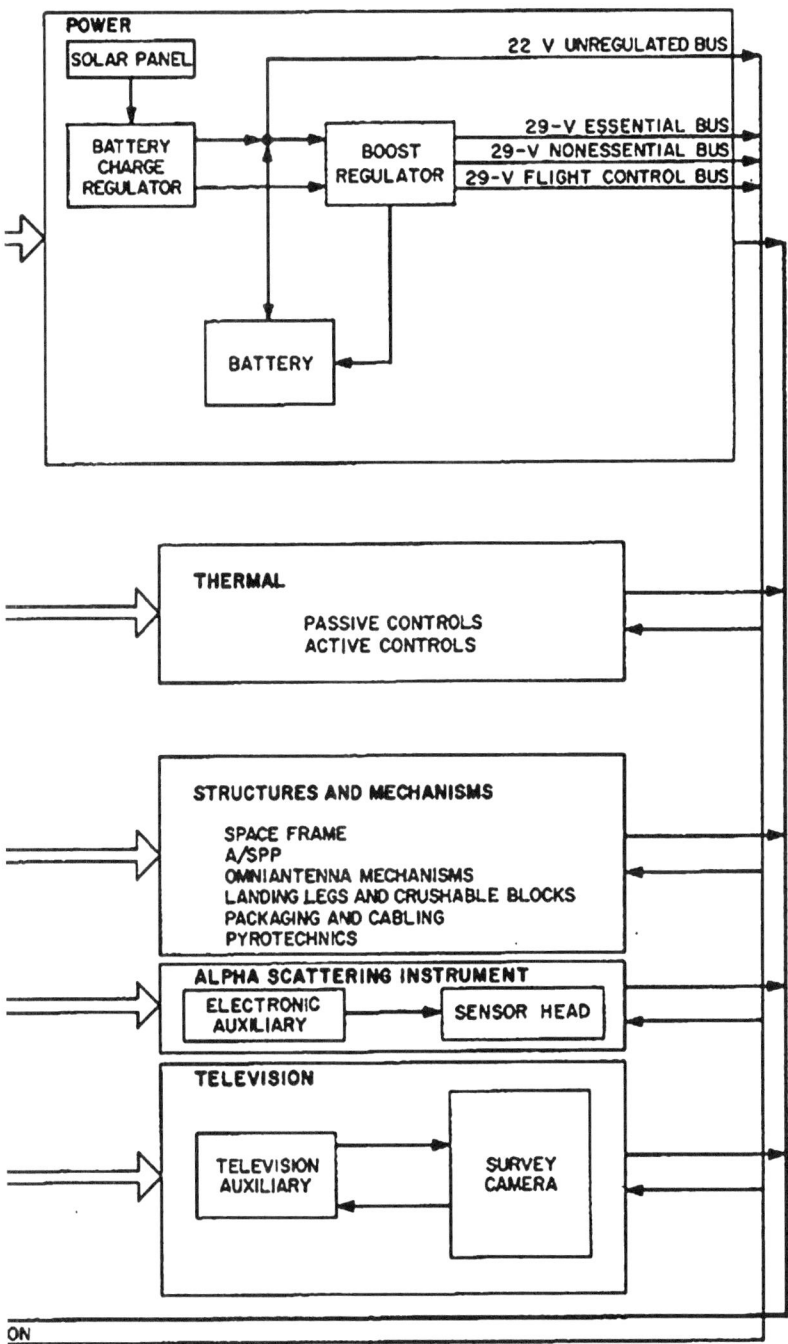

Fig. 4.14   Functional block diagram of Surveyor 6 spacecraft (continued)

data from the television camera. The other oscillator was either frequency modulated by narrowband frequency-modulated data channels or phase modulated from phase-modulated data channels.

The modulated signals from the oscillators were fed through a selector switch to frequency multiplier stages that had a total multiplication ratio of 120. The nominal frequency at the output of the multiplier was 2,295 MHz. When in the transponder mode, the selector switch at the input to the transmitter frequency multiplier connected the frequency multiplier to a voltage controlled oscillator that was part of a phase-locked loop in the receiver. The nominal frequency of that voltage controlled oscillator was also 19.125 MHz.

Each transmitter contained a traveling wave tube (TWT) amplifier that delivered an output power of 10 watts at 2,295 MHz. The TWT could be turned off and bypassed and that lowered the output power to 0.1 watts and substantially lowered the power consumed by the transmitter. The high-power mode was used to transmit 600-line television data.

Receiver

The two receivers were double superheterodyne types with different intermediate frequencies generated by two reference frequencies driving two mixers. A voltage controlled crystal oscillator operating at a nominal frequency of 19.125 MHz provided a frequency reference for the receiver. The receiver could be switched between a phase-locked loop function and an automatic frequency control function. Either loop could be selected to control the oscillator.

The output of the oscillator was applied to a multiplier/amplifier chain that multiplied the oscillator frequency by a factor of 108 to 2,065.5 MHz. This frequency was mixed with the incoming uplink signal of 2,113 MHz to give an intermediate frequency of 47.5 MHz. The intermediate frequency signal was filtered and amplified and applied to a second mixer fed from a 38.25 MHz reference obtained by multiplying the 19.125 MHz oscillator frequency by a factor of 2. The resulting 9.25 MHz second intermediate frequency was amplified and applied to a limiter followed by a discriminator.

The discriminator output yielded a direct current component that was used to tune the voltage controlled oscillator when the receiver was switched to the frequency control mode of operation. The discriminator output also yielded the subcarrier frequency that contained commands sent from earth. The subcarrier frequency signal was sent to the central command decoder.

During transit from earth to the moon, the receiver could be switched to a phase-locked mode of operation where the reference oscillator was controlled by the phase-locked loop filter. The resulting coherent signal from the oscillator was fed to the transmitter to allow two-way Doppler tracking of the spacecraft from earth.

Command Decoding Subsystem

The many critical operations of the spacecraft were commanded from earth via the telecommunications system. The commands were deciphered in the spacecraft by the command decoding subsystem. Commands from the earth were in the form of 24-bit Manchester coded words.

The first four bits of the 24-bit command word were used for synchronization. The following ten bits contained five bits of address that identified the subsystem to be controlled and five bits of complement of the address. The address and its complement allowed an error check to be made of the information. The last ten bits of the command word defined the instruction. Of the instruction bits, all ten were used for conveying quantitative information such as commands for orientation of the spacecraft. In the case of direct commands for a single action, such as turning on power to a subassembly, five of the ten bits were used for instruction, and five bits made up the complement of the instruction.

The central command decoder deciphered the data signals from the receiver and sent the command word to the appropriate command decoder located in the major subsystems of the spacecraft. There were eight such individual command decoders that provided command instructions to individual subsystems.

There were two redundant central command decoders. Either of the redundant central decoders could be switched to the output of either of the two redundant receivers. Thus, in case of a failure of a receiver or of a central command decoder, the commands to the spacecraft from earth could still be processed.

Signal Processing Subsystem

The signal processing subsystem assembled and processed data from engineering sensors and scientific instruments on the spacecraft and put the information in appropriate form to be telemetered back to earth. A total of 261 different parameters could be telemetered to earth.

The engineering signals to be telemetered back to earth were organized into six different groups that corresponded to the different phases of the mission. These groups were referred to as data modes. The data mode was commanded from the ground. For example, Mode 6 contained signals to be telemetered during the terminal descent phase. It consisted of 47 different analog signals and 74 different digital signals. The Mode 6 data signals were drawn from the flight control subsystem, power subsystem, altitude marking radar, radar altimeter and Doppler velocity sensor, and vernier engine parameters.

The analog engineering signals were applied to one of the two commutators that sequentially sampled each signal and held the resulting sample long enough to be processed by an analog-to-digital converter. One of the commutators, located in the engineering signal processor unit, could accept 100 analog signals, and the other located in the auxiliary engineering signal processor unit could accept 120 analog signals. The engineering signal processor unit handled signals that were fundamental to the spacecraft and the auxiliary engineering signal processor handled specialized signals.

Outputs of the engineering signal processor and auxiliary engineering signal processor were applied to the central signal processor that contained two analog-to-digital (A/D) converters, summing amplifiers, and clock timing functions.

Two A/D converters were provided for redundancy. A switch, controlled from a command from the ground, selected one or the other A/D converters. The A/D converter converted each of the commutated analog signals to a 10-bit digital word and added one bit for parity for a total of 11 bits. The 11-bit digital word was applied to a subcarrier oscillator.

The binary or discrete logic engineering signals were assembled into 10-bit digital words and applied directly to the subcarrier oscillator at prearranged time slots.

The digital words could be switched to any of the six subcarrier oscillators depending on bit rate chosen. The subcarrier oscillators were voltage controlled oscillators where the output frequency was proportional to the control voltage applied. Typically, a data rate of 550 bits per second was used in conjunction with a subcarrier oscillator operating at 3.9 kHz.

The A/D converter function also generated master timing signals for the signal processing operation including switching of the commutators. The data rate, set by the A/D converter, could be commanded from the ground to be 4400, 1100, 550, 137.5 or 17.2 bits per second. A data rate was selected that provided sufficient signal strength at the deep space instrumentation facilities (DSIF) on earth to yield an acceptably low error rate. The lower data rates required less bandwidth and resulted in higher signal-to-noise ratios.

The outputs of the subcarrier oscillators from the central signal processor and engineering signal processors were summed in two different sets of summing amplifiers. The output of one summing amplifier fed the frequency modulation inputs of transmitters A and B and the other fed the phase modulation inputs. The video signal from the camera in the 600-line mode was applied to a wideband voltage controlled oscillator in the transmitter that frequency modulated the transmitter output. The 200-line video signal from the camera was applied to a narrowband voltage controlled oscillator.

Altitude Marking Radar

The altitude marking radar (AMR) had an important singular purpose: to generate an altitude mark when the slant range along the spacecraft roll axis decreased to a preset range of about 60 miles from the lunar surface. The preset range could be adjusted to between 52 and 70 miles. The range mark initiated the terminal landing sequence.

The first step in this sequence was to ignite the three vernier engines and then about 1 second later ignite the main retrorocket engine. The vernier engines were used to maintain spacecraft attitude during burn of the retrorocket. The altitude marking radar was mounted by friction clips in the flare of the nozzle for the retrorocket and when the rocket ignited, the radar was ejected by the forceful burning gases.

The altitude marking radar was a pulse radar type. A simplified block diagram of the radar is shown below. The antenna was a parabolic dish 30 inches in diameter with a gain of 34 dB and half-power beamwidth of 3.6 degrees. The magnetron transmitter produced a peak output power of 1.5 kW at a frequency of 9.3 GHz. The pulse width was 3.5 microseconds, and the pulse repetition frequency was 350 pulses per second. The transmitter output was passed through a duplexer consisting of a circulator and transmit-receive (TR) tube. The TR tube protected the receiver from the leakage transmitted pulse through the duplexer (Fig. 4.15).

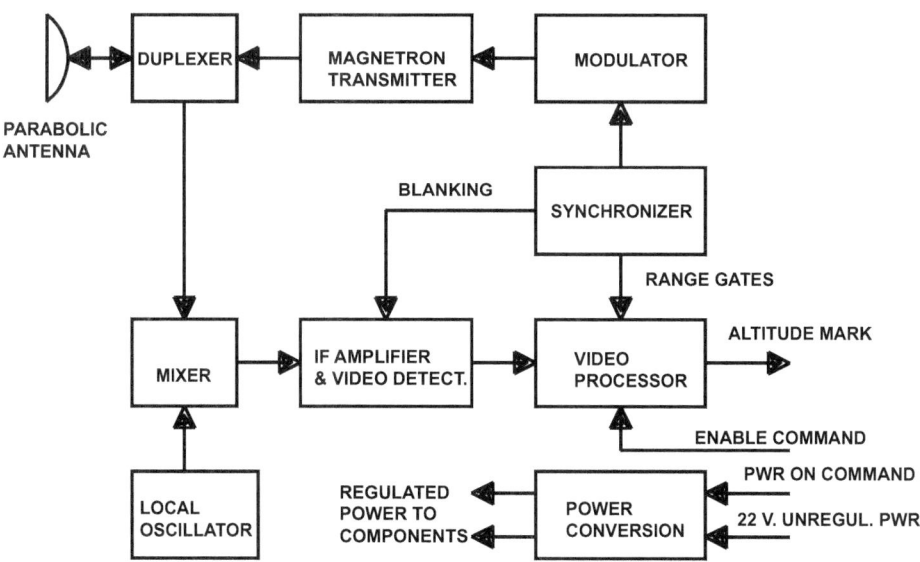

Fig. 4.15 Simplified block diagram of altitude marking radar (adapted from NASA graphics)

The magnetron was fired by a 30 volt, 3.2 microsecond pulse from the modulator. Pulsing of the modulator was controlled by the synchronizer that acted as the master timer for the radar.

The echo signal from the lunar surface was intercepted by the antenna, passed through the duplexer, and applied to a mixer. The other input to the mixer was a local oscillator signal at a frequency of 9.33 GHz. The difference signal at a frequency of 30 MHz was amplified by the intermediate frequency (IF) amplifier and rectified (detected) to form a pulsed video signal. The video signal was applied to the video processor that also received early and late range gates from the synchronizer.

The early and late gates were delayed in time from the transmitted pulse a preset amount corresponding to the time delay between the transmitted pulse and the return from the lunar surface at the desired range to generate a mark. The width of the range gates is not stated in the reports available to the author, but normally they would be about the same as the pulse width of the transmitted signal. The range gates allowed the video signal to pass during the time that the signal is within the range gate.

The two gates were adjacent in time with the echo first appearing in the late gate time and then in the early gate time as the spacecraft descended. The video signal at the output of the two range gates was summed and differenced. The logic in the video processor to generate the range mark required that the sum of the video signal in the two gates exceed a given level and the difference (early gate signal amplitude minus late gate signal amplitude) cross zero with a positive slope. An enable signal sent from earth was required to be present before the range mark signal could be sent to the flight control subsystem.

Electrical power for the altitude marking radar was derived from the 22 volt unregulated electrical bus in the spacecraft. The power conversion circuits generated regulated power at various levels required by the components of the radar. A "Power On" command sent from earth turned on electrical power to the radar.

Radar Altimeter and Doppler Velocity Sensor

The radar altimeter and Doppler velocity sensor (RADVS) provided range along the longitudinal axis (Z-axis) and three-axis velocity data to the flight control subsystem during the terminal descent to the lunar surface. A brief description of the terminal phase of the lunar landing and the use of the velocity and range data was given in the "Transient to the Moon and Terminal Descent" section of this chapter. Also given in that section are plots of range and velocity data telemetered back to earth during the descent of Surveyor 7.

The RADVS was built by the Teledyne Ryan Aeronautical Company under a contract with Hughes Aircraft Company signed in 1961. The author was intimately involved in the development of the RADVS, initially responsible for performance analysis and setting system parameters and then directing the design of the signal data converter unit and functioning as technical lead for the entire radar. Much of the information provided below was gathered from his files.

The Hughes specification for the RADVS required that velocity data be provided for slant ranges along the roll axis from 50,000 feet to 14 feet to the lunar surface. Velocity measurements were required over a velocity range of 3,000 to zero feet per second. The required accuracy of the velocity measurement was the root-sum-square of 1 fps and 2.0% of total velocity. The specification required velocity measurements and range measurements to be made for attitudes of the spacecraft from zero to 45 degrees from the local vertical.

The Hughes specification further required the radar altimeter provide range measurements for slant ranges of 40,000 to 14 feet from the lunar surface. The accuracy requirements for range data was the root-sum-square of 4 feet and 5 percent of range.

The RADVS consisted of four assemblies: velocity sensor antenna, altimeter and velocity sensor antenna, klystron power supply and modulator, and signal data converter. A photograph of these components is shown below. The klystron power supply and modulator is the box in the upper left of the photograph, and the signal data converter is the box on the lower right. The two antennas are the larger curved components in the picture (Fig. 4.16).

Fig. 4.16 Photograph of components of radar altimeter and Doppler velocity sensor (from author's files)

The antennas were shrouded parabolic types with a septum dividing each antenna into transmit and receive sections. Each antenna generated two transmit beams and two receive beams aligned with the transmit beams. The RADVS antenna beam geometry is shown below. Beams 2 and 3 emanate from the velocity sensor antenna (labeled antenna 2 in the figure) and beams 1 and 4 emanate from the altimeter and velocity sensor antenna (labeled antenna 1) (Fig. 4.17).

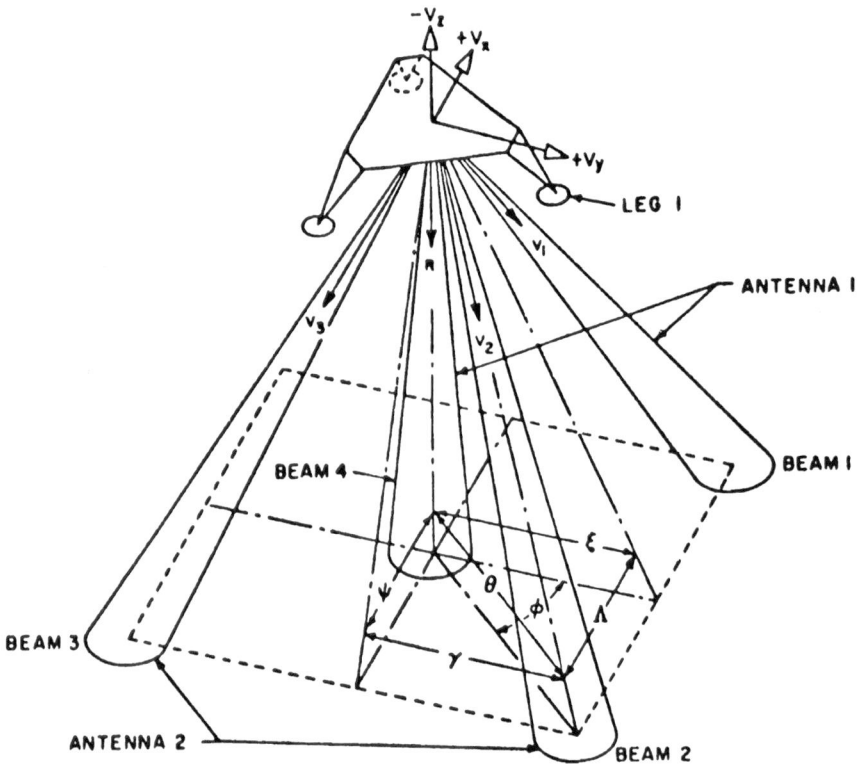

Fig. 4.17 RADVS antenna beam configuration (from author's files)

The radar altimeter antenna beam was aligned with the Z-axis of the spacecraft. The intercept of the Doppler velocity sensor beams in a plane normal to the Z-axis was a right triangle with sides parallel to the X- and Y-axes of the spacecraft. The center of each of the velocity sensor antenna beams was displaced 25 degrees from the Z-axis.

The RADVS determined the velocity along the X-, Y-, and Z-axes of the spacecraft by measuring the Doppler shift of the transmitted signal along each of the three Doppler velocity sensor antenna beams. It computed velocity along each beam from the Doppler shift and then performed a translation to velocities in the spacecraft coordinate system.

The two-way Doppler shift was related to the velocity along the beam by the expression $f_D = 2V/\lambda$ where f_D is the Doppler frequency, V is velocity, and λ is the wavelength of the transmitted signal. The continuous wave (CW) transmitted frequency was 13.3 GHz, the corresponding wavelength was 0.074 feet, and the Doppler scale factor was 27.044 Hz per foot per second of velocity.

The velocity component along the X-axis, V_X, was computed from the difference between the Doppler frequencies in beams 1 and 2. V_Y was computed from the difference between Doppler frequencies in beams 2 and 3. V_Z was computed from the sum of Doppler frequencies in beams 1 and 3.

A linear frequency-modulated signal was used by the RADVS to determine slant range to the lunar surface along the narrow radar altimeter beam. The frequency modulation was in the form of a sawtooth centered at a frequency of 12.9 GHz. The sawtooth had a repetition rate of 182 Hz, and the peak-to-peak deviation was 4.0 MHz at altitudes above 1,000 feet and 40 MHz below 1,000 feet. The difference frequency between that transmitted and received, f_{R+D}, was proportional to range time delay plus Doppler shift as expressed by $f_{R+D} = 2RS/c + 2V/\lambda$ where R is range, S is deviation rate, c is velocity of light, V is velocity component along the beam, and λ is the wavelength of the transmitted signal.

The range factor, given by 2RS/c, was equal to 1.48 Hz per foot at altitudes above 1000 feet and to 14.8 Hz per foot at altitudes below 1000 feet. The Doppler frequency component, 2V/λ, was equal to 26.23 Hz per foot per second of velocity along the antenna beam. The Doppler frequency addition to the range frequency was removed by downstream signal processing. That processing made use of velocity along the Z-axis as measured by the Doppler velocity sensor.

A simplified block diagram of the radar altimeter and Doppler velocity sensor is shown on the next page (Fig. 4.18).

Klystron Power Supply and Modulator

The transmitted signals for the Doppler velocity sensor and the radar altimeter were generated by klystron devices located in the klystron power supply and modulator. The Doppler velocity sensor transmitter consisted of a two-cavity klystron that generated 8 watts of continuous wave (CW) microwave power at a frequency of 13.3 GHz. The radar altimeter transmitter was a reflex klystron that generated 0.4 watts of frequency-modulated microwave power at a center frequency of 12.9 GHz. Reflex klystrons have the ability to generate a variable output frequency controlled by the negative voltage applied to the reflector element of the klystron. The sawtooth frequency modulation was achieved by applying a sawtooth-modulated negative voltage to the reflector element.

The reasons that the transmitter output for the velocity sensor was considerably higher than that of the radar altimeter were that the velocity sensor was required to operate at 50,000 feet rather than 40,000 feet and the velocity sensor beams

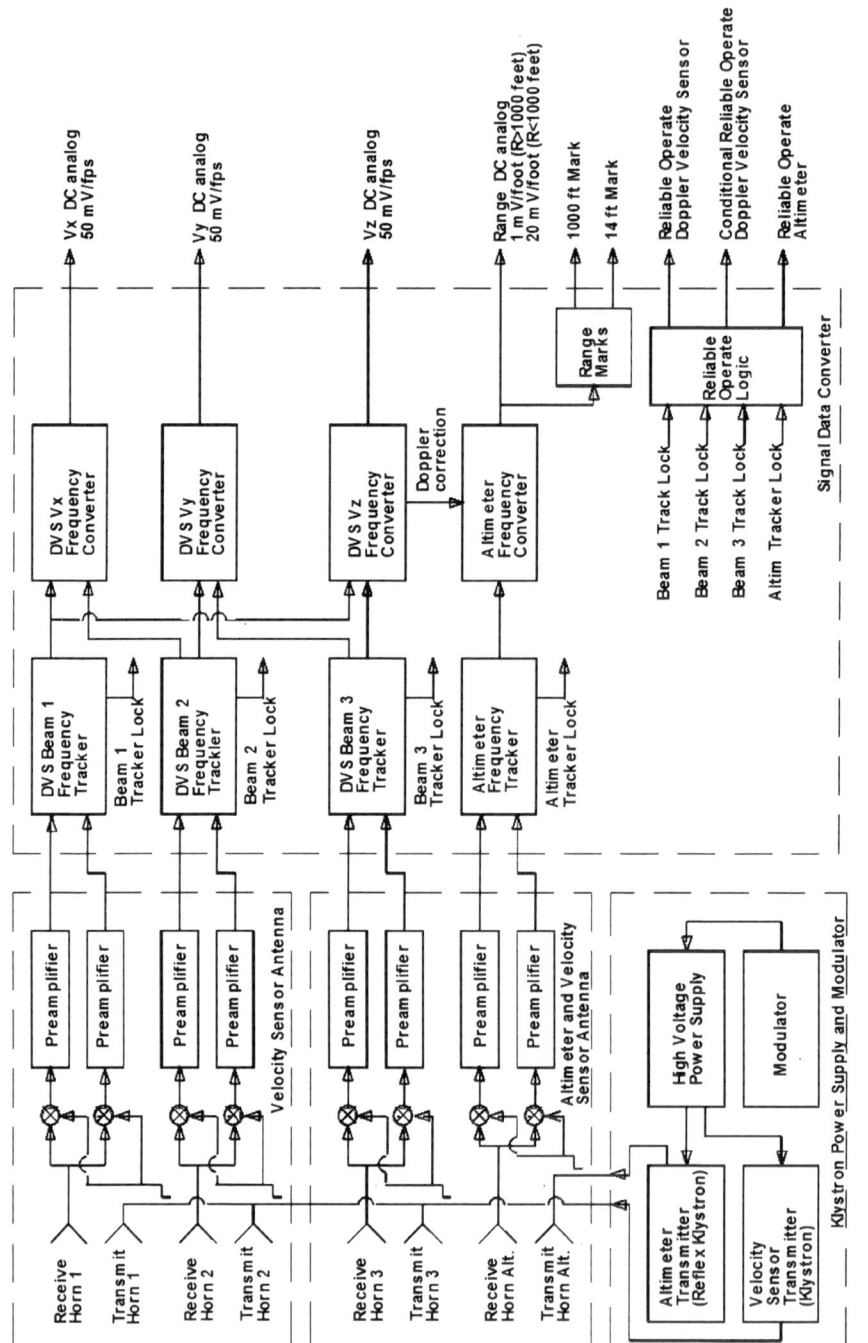

Fig. 4.18 Simplified block diagram of radar altimeter and Doppler velocity sensor (from author's files)

were displaced 25 degrees from the roll axis, while the radar altimeter beam lies on the roll axis. In the worst case at a spacecraft attitude of 45 degrees, one of the velocity sensor beams is at an angle of 70 degrees to the local vertical, and the range to the surface is about a factor of 2 longer than at 45 degrees. Another reason for needing more power is that the model used for the backscattering cross section per unit surface area of the lunar surface (related to the reflectivity) shows a value about a factor of 4 lower at 70 degrees than at 45 degrees. Lastly, the transmitter output was divided among three antenna beams in the case of the velocity sensor and only one beam in the case of the altimeter.

Antennas

The antennas were special parabolic types with a septum dividing the volume into transmit and receive sections. There were two transmit feed horns on one side of the septum and two receive feed horns on the other side. The transmit horns for the velocity sensor antenna were positioned to splay the beams about 17.4 degrees away from center. The antenna was mounted to the spacecraft to move the beams out such that the centers of the velocity sensor beams were 25 degrees from the roll axis. The feed horns and antenna alignment of the altimeter and velocity sensor antenna positioned the radar altimeter beam along the roll axis and the velocity sensor beam 25 degrees from the roll axis. The receive horns in each antenna generated antenna beams aligned with their respective transmit beams. The beamwidth of each of the antenna beams was about 4 degrees.

The Doppler-shifted echo signal intercepted by each receive antenna beam was split into two signals with one phase shifted 90 degrees from the other. These quadrature pairs of signals were applied to mixer diodes where they were mixed with a sample of the transmitted signal. The output of the mixers was a quadrature pair at Doppler frequency. Splitting into quadrature pairs allowed determining the sense of the Doppler signal. The phase relationship changes 180 degrees as the velocity along the beam changes from a closing rate to an opening rate. This property was used in the frequency trackers to avoid locking on to spurious signals such as from the retrorocket that presented an opening velocity after it was jettisoned.

The quadrature signals at the output of the mixers were applied to preamplifiers to raise the signal level to that needed by the frequency trackers. The velocity sensor preamplifiers contained sensing logic that switched the gain of the amplifiers from 90 dB to 65 dB or 40 dB as the signal level increased during descent to avoid distortion of the signal. The altimeter preamplifiers were switched in 80, 60, and 40 dB steps.

To cover the expected range of Doppler frequencies for vehicle velocities up to 3,000 feet per second, the half-power bandpass of the preamplifiers for the Doppler velocity sensor was a few hundred Hz to 82 KHz. The Doppler signal itself was a spectrum with a half-power bandwidth of up to 3,000 Hz due to the finite beamwidth of the antennas. The broadband signals at the output of the preamplifiers were applied to frequency trackers located in the signal data converter.

The radar altimeter was required to operate at slant ranges up to 40,000 feet, which, with the scale factor of 1.48 Hz per foot, yielded a range frequency of 59.2 kHz. In addition, a Doppler shift up to 20 kHz increased the altimeter signal frequency to about 80 kHz. The altimeter preamplifiers had the same bandpass as the velocity sensor preamplifiers.

Frequency Tracker

The quadrature pair of amplified Doppler signals at the input to each frequency tracker was applied to a single-sideband modulator where they were upshifted by a frequency of 600 kHz by mixing with a quadrature signals from a variable frequency oscillator (VCO). The quadrature signals were arranged to select the lower sideband of the upshifting process while providing over 30 dB rejection to the upper sideband and to the baseband Doppler signals. The action of the tracking loop was to drive the VCO frequency to 600 kHz plus Doppler plus any error frequency. The error frequency was sensed and used to control the loop to drive the average error frequency close to zero.

The frequency trackers for the Doppler velocity sensor beams and for the radar altimeter searched for the signal spectrum in the presence of wideband noise and once acquired accurately tracked the center of the spectrum.

The search program for the velocity sensor trackers was arranged to sweep an 82 kHz to 800 Hz band prior to receiving the separation signal from the spacecraft. The separation signal denoted separation of the retrorocket from the spacecraft after burnout of the retrorocket. The frequency band was searched by the tracker in 1.5 seconds with a probability of acquisition of 95% per sweep at a signal-to-noise ratio of 6 dB (factor of 4) within the tracking filter bandwidth. The search was narrowed to a frequency region 22 KHz to 800 Hz after receiving the separation signal since the vehicle velocity was much lower after the retrorocket burn and the maximum Doppler frequency was lower.

The search program for the radar altimeter tracker extended from 80 kHz to 200 Hz and that accommodated vehicle velocities up to about 800 feet per second at a range of 40,000 feet. The frequency range was searched by the tracker in 1.5 seconds.

When the frequency tracker sensed a signal-to-noise ratio of 6 dB or greater in the tracking filter bandwidth, the frequency search was stopped, and the tracking loop was activated. The tracking filter bandwidth for the velocity sensor trackers was 3,000 Hz prior to receiving the separation signal and 600 Hz after receiving the separation signal. The acquisition circuit proofed the signal for 0.1 seconds before activating the tracking loop and generating a tracker lock signal.

The output of the frequency tracker was the VCO frequency of 600 kHz plus Doppler. This signal was buffered and applied to the frequency converters in the signal data converter. Other outputs of the frequency tracker were the Tracker Lock signal and a DC analog of the signal level within the tracking filter. The signal strength output was a 0 to 5 volt DC analog voltage. The signal level information was telemetered to earth along with gain state of the preamplifiers to allow determining the reflectivity of the lunar surface at the angle of the antenna beam.

Frequency Converter

The velocity sensor frequency converters combined the frequency tracker outputs in the frequency domain to obtain frequencies proportional to the velocity components along the primary spacecraft coordinates. The V_X frequency converter developed the difference frequency between the outputs of frequency trackers for beams 1 and 2. This difference frequency was proportional to the velocity along the X-axis of the spacecraft. The difference frequency was converted to a DC analog voltage of either positive or negative polarity depending on which tracker had the higher frequency.

The V_Y frequency converter generated a DC analog of velocity along the Y-axis of the spacecraft by taking the difference between tracker outputs for beams 2 and 3. The V_Z frequency converter generated a DC analog of velocity along the Z-axis of the spacecraft by forming the sum of the tracker outputs for beams 1 and 3. The outputs of the three frequency converters were DC analog voltages with a scale factor of 50 millivolts per foot per second of velocity and with polarity denoting the sense of the velocity.

The radar altimeter frequency converter accepted the input from the altimeter frequency tracker and a signal from the V_Z frequency converter proportional to velocity along the Z-axis of the spacecraft. This velocity input allowed the Doppler portion of the range plus Doppler signal at the output of the tracker to be compensated for. The output of the range frequency converter was a DC analog voltage with a scale factor of 1.0 millivolts per foot for ranges greater than 1,000 feet and 20 millivolts per foot for ranges less than 1,000 feet. In addition, two range marks were generated. A mark labeled 1,000 foot mark was made when the radar altimeter determined that the range had decreased to 1,000 feet, and the 14 foot mark was made when the range decreased to 14 feet. These range marks consisted of 5 volt discrete signals. The 14 foot mark was used to shut off the vernier engines.

Reliable Operate Logic

The Reliable Operate Logic generated a 5 volt discrete signal called reliable operate Doppler velocity sensor when tracker lock signals from all three beams were present.

The reliable operate radar altimeter signal was generated when the radar altimeter tracker lock signal was present and the tracker lock signals for velocity sensor beams 1 and 3 were present. The latter condition assured that the velocity correction to the range signal could be made.

Cross-Coupled Sidelobe Reject Logic

Surveyor 3 performed a soft landing on the moon but then bounced and landed three times because the 14 foot mark was not generated by the radar altimeter to shut off the vernier engines and they continued to burn. The vernier engines were finally shut off by a command from earth just before the third and final landing.

The reason that the 14 foot mark was not generated involved the cross-coupled sidelobe reject logic. This conclusion was soon reached by the author during a meeting at Hughes after printouts of telemetered data became available.

Analysis performed by the author early in the program disclosed that it was possible at high attitude angles of the spacecraft for a Doppler velocity sensor tracker for one beam to lock to a sidelobe in the receive antenna pattern that lay close to the main transmit lobe of another beam. The amplitude of the sidelobe was about 25 dB (factor of 316) below the main beam. This sidelobe condition was referred to as a cross-coupled sidelobe.

Should a lock to a cross-coupled sidelobe occur, the tracker output frequency for the two beams would be the same, but the amplitude of the signal in the cross-coupled beam would be substantially lower than the main lobe signal of the other beam. The author devised logic that involved gain state of the preamplifiers and signal amplitude along with tracker output frequencies to sense that a cross-coupled lobe had been acquired. If this happened, the tracker was commanded to break lock and resume search for the true signal.

The radar altimeter required velocity input from the Doppler velocity sensor to compensate for the Doppler component of the range signal reflected from the lunar surface. If the tracker lock signals for beams 1 and 3 of the velocity sensor were not present, the reliable operate signal for the radar altimeter was inhibited, and the range mark signals were prevented from being generated.

As Surveyor 3 descended through an altitude of 37 feet about 6 seconds before touchdown, the tracker lock signal for Doppler velocity sensor beam 3 went to the unlock state, and the tracker entered the search mode. The reason for this was that conditions were satisfied to activate the cross-coupled sidelobe reject logic and the tracker was commanded to break lock and go into search even though there was an adequate signal level to track.

Telemetered data from the mission indicated that there was a sharp fade in signal strength for beam 3 relative to the signal strength in beams 1 and 2 at the time the tracker was put into search. The narrow antenna beam subtended a small area of the lunar surface at an altitude of 37 feet, and the backscatter from that particular area for a brief period of time could have been much less than experienced by the other beams and enough different to trigger the cross-coupled lobe reject logic.

A change in the RADVS was made for subsequent missions to disable the cross-coupled sidelobe reject logic once the 1,000 foot mark was generated. This avoided triggering the reject logic on potential substantial variations in signal level from beam to beam when close to the lunar surface.

The imprints of the landing pads on the lunar surface made during the extracurricular landings of Surveyor 3 were clearly shown in photographs taken by the onboard camera. In looking at these photographs later the author couldn't help but remark to himself, "I did that."

Physical Properties

Electrical power for the radar altimeter and Doppler velocity sensor was obtained from the 22 volt unregulated power bus of the spacecraft. The maximum power consumed at the extremes of the input voltage range was 590 watts.

The total weight of all components of the RADVS was 35.3 pounds. The size of each antenna was about 20.5 inches long, 15.7 inches wide, and 17.8 inches high to the top of the electronics package on top. The klystron power supply and modulator was about 7.9 inches long, 7.5 inches wide, and 6.9 inches high. The signal data converter was about 10.2 inches long, 8.8 inches wide, and 8.8 inches high.

The radar was designed in the early 1960s before the common use of electronic integrated circuits. Thus, the electronic circuits were built using discrete parts. To minimize size and weight, the electronic parts in the signal data converter were assembled into "cordwood" modules which were then installed on printed circuit boards. A photograph of a cordwood module in the author's collection is shown on the next page (Fig. 4.19). The module has been encased in clear resin for display purposes. The actual module was 1.0 inches long, 0.9 inches wide, and 0.5 inches high.

Cordwood construction was only used for a few years before integrated circuit technology became available. It was usually only used on space programs where the savings in size and weight justified the considerable extra cost of that construction.

Scientific Instruments

Main scientific instruments carried by the Surveyor spacecraft included:

- Survey camera
- Soil mechanics/surface sampler
- Alpha-scattering surface analyzer

Fig. 4.19 Cordwood module used in Surveyor RADVS (from author's collection)

All seven of the spacecraft carried a capable survey camera. In addition to the camera, Surveyors 3, 4, and 7 carried a soil mechanics/surface sampler, and Surveyors 5, 6, and 7 carried an alpha-scattering surface analyzer. Surveyors 2 and 4 malfunctioned and did not land successfully.

Survey Camera

The camera was a key instrument on the Surveyor spacecraft. No other scientific instrument could have provided as much information about the lunar surface. Indeed, it was the only scientific instrument carried on the Surveyor 1 and Surveyor 2 spacecraft. An improved version of the camera was used on the Surveyor 6 and 7 spacecraft.

In total, about 87,000 photographs were taken of the lunar surface by the five Surveyor spacecraft that landed on the moon. These photographs gave scientists on earth a very good picture of topography of the surface. The detailed photographs of the lunar terrain offered exciting new vistas to the good citizens on earth.

The survey camera was very capable for the day. It used a vidicon tube to convert an optical image to electronic form. The camera had a zoom lens that could be adjusted for a field of view from 6.4 degrees to 25.3 degrees. It could be focused from about 1.3 meters (4.2 feet) to infinity and operate on scene luminance between 0.008 foot-lamberts and 2,600 foot-lamberts. The resolution was equivalent to about 1 mm at 4 meters (0.039 inches at 13 feet). An adjustable iris allowed f-stop settings between f/22 and f/4 in ½f-stop increments. The f-stop could be automatically set by the camera from a measurement of scene luminance, or it could be commanded from earth. The camera could be operated in either a 200-line scan mode or a 600-line scan mode.

A cutaway drawing of the camera is shown on the next page (Fig. 4.20). Scaling from drawings, the camera was about 21.4 inches long and about 6.2 inches wide at the widest portion.

The camera was mounted to the spacecraft with the long axis of the camera at an angle of 16 degrees out from the Z-axis of the spacecraft with the hood end up. The photosensitive element in the camera was a vidicon tube that was mounted along the long axis so that it could view a scanning mirror. The mirror could be rotated 357 degrees in azimuth and from +40 degrees to –80 degrees in elevation with respect to the plane of the camera. That gave the camera the ability to take pictures one frame at a time over nearly 360 degrees of azimuth and over a wide range of elevation angles. It could look down to the landing pads, look out at lunar surface, or look up at the sky to image the stars.

The vidicon tube was a common device used in television cameras in the 1960s and 1970s. It was a type of cathode ray tube (CRT) with an electron scanning beam. The inside of the glass face of the tube was coated with a transparent layer of conductive film. The film was connected through an external load resistor to a positive voltage source. A photoconductive layer was placed next to the conductive film toward the inside of the tube. The image to be processed was projected onto the photoconductive layer by the camera lens. Each small element of the layer became conductive when light struck it and the degree of conductivity of each element was proportional to the intensity of the image at that point.

A charge density pattern formed on the photoconductive layer in response to the image projected onto it. The layer was scanned by an electron beam in a raster pattern and small differences in current caused by varying charge on the layer flowed through the load resistor and created a voltage proportional to the intensity of the image at each element of the layer. This signal was amplified and became the video output of the camera.

An important element of the camera was a zoom lens with a focal length between 100 mm and 25 mm, corresponding to a field of view between 6.4 degrees and 25.3 degrees. Normally, either the 100 or 25 mm focal length was selected, but the zoom could be commanded to focal lengths in-between. The aspect ratio of the image generated by the camera was 1:1. In other words, the picture was square. The zoom lens component included mechanisms to adjust focus, iris opening, and shutter time.

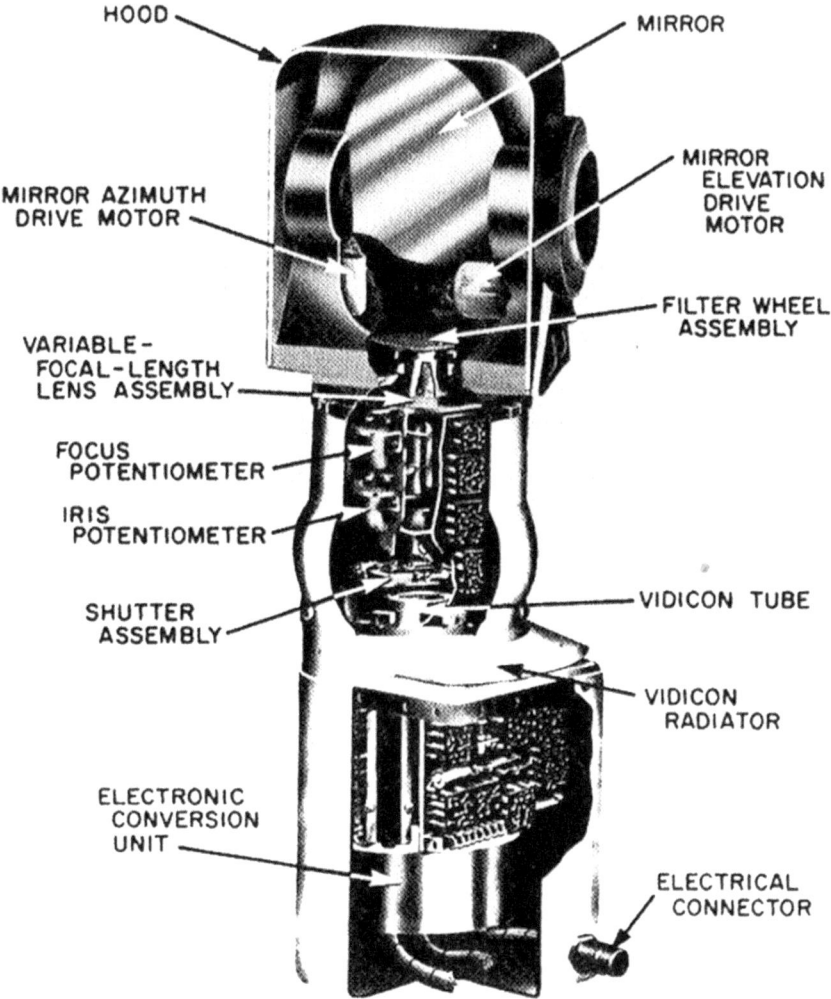

Fig. 4.20 Survey camera used on Surveyor spacecraft (NASA graphic)

The image from the mirror passed through a filter wheel before impinging on the lens. The filter wheel could be rotated by commands from the earth to present one of three different polarizing filters or a neutral density filter.

The camera was totally controlled by commands from earth except for the ability to automatically set f-stop if that option was chosen. Those commands engaged individual servomechanisms that controlled zoom, focus distance, iris opening, shutter open time, mirror azimuth angle, and mirror elevation angle. Either a 200-line scan mode or 600-line scan mode could be commanded.

A mechanical focal plane shutter was located between the camera lens and the vidicon tube. Three different shutter modes could be selected from earth. In the *normal mode*, the shutter remained open for about 150 milliseconds. In the *open shutter mode*, the shutter open time was nominally 1.2 seconds. In the *integrate exposure mode*, the shutter was opened, the vidicon was turned off, and the image was allowed to build up for several minutes on the photoconductive layer before the vidicon was turned back on to read out the image. Sixth magnitude stars (the limit of human perception) could be imaged using a 5 minute integration time.

The camera could be operated in either a 200-line scan mode or a 600-line scan mode. The higher-resolution 600-line mode was most often used. The 200-line scan mode generated an image that was read out from the vidicon in 20 seconds. The resulting low video bandwidth of 1.2 kHz allowed transmission to earth by the telecommunications system using the omnidirectional antennas. The 600-line scan mode generated a higher-resolution image that was read out from the vidicon tube in about 1 second. The video bandwidth was about 220 kHz, and that wide bandwidth required use of the high-gain array antenna to transmit the picture to earth and arrive at an adequate signal-to-noise ratio.

For photographic reference purposes, small photometric/colorimetric reference targets were mounted on the spacecraft that could be viewed by the camera. The targets were in the shape of a flat disk about 3.6 inches in diameter. The target contained nine segments of various shades of gray ranging from black to 20–30% gray and four color wedges. It also contained classic patterns of radial lines extending from the center that could be used to roughly determine the resolution of the camera.

Panoramic surveys were taken around the landing side by stepping the mirror in azimuth and elevation. The stepping enveloped angles from 132 degrees counterclockwise and 225 degrees clockwise from zero azimuth and elevation angles from 60 degrees below the plane of the camera to the horizon. Panoramic pictures were taken in both the narrow- and wide-angle modes. In the narrow-angle mode, pictures were taken each 6 degrees horizontally and every 5 degrees vertically. A total of 900 to 1000 pictures were required to cover the full panorama in the narrow-angle mode.

One such picture was a mosaic of photographs taken of the north rim of Tycho crater by Surveyor 7 shown on the next page (Fig. 4.21). The rock in the foreground is about 2 feet in extent, and the ridge on the horizon is about 8 miles away. The mosaic was made up of a wide-angle photograph and a series of narrow-angle photographs.

Soil Mechanics/Surface Sampler

The soil mechanics/surface sampler (SM/SS) was a valuable tool used to determine the physical properties of the lunar surface. It had the capability to dig into the surface, create a trench, scoop, move, and dump a sample of material and perform bearing tests of the surface. It could pick up small rocks for examination by the camera. It was used to pick up the alpha-scattering surface analyzer during the Surveyor 7 mission and move it to various locations, including over one of the trenches.

Fig. 4.21 Mosaic of photographs from Surveyor 7 (NASA image)

The soil mechanics/surface sampler was developed by the Hughes Aircraft Company. A photograph of the device at work during the Surveyor 7 mission is shown on the next page (Fig. 4.22).

The scoop could be extended out 60 inches by a pantograph-type mechanism. The mechanism operated similar to a scissor jack where the scoop end was extended or retracted by rotating a lead screw. The lead screw was rotated by an electric motor controlled by commands from the earth.

The soil mechanics/surface sampler on the Surveyor 7 spacecraft could be swung in a horizontal arc around the pivot axis up to 72 degrees clockwise and 40

Fig. 4.22 Soil mechanics surface sampler at work during Surveyor 7 mission (NASA photograph)

degrees counterclockwise from the stowed position. It could be rotated 18 degrees up from horizontal and 36 degrees down. When operating on a level surface, the scoop could engage the lunar soil from 58 inches to 23 inches from plane of the mount of the soil mechanics/surface sampler.

The SM/SS used four electric motors to perform various motions: extend/ retract, rotate in azimuth, rotate in elevation, and open and close the door of the scoop. The current used by the motors was telemetered back to earth. A calibration had been made before the flight of current vs. force on the scoop for the bearing operation and for the trenching operation.

Static bearing test operation was conducted by positioning the scoop to the desired location over the surface and then running the elevation motor to drive the scoop to a certain penetration into the surface or until the motor stalled. The door of the scoop was usually closed for these tests, and a flat surface about 1 by 2 inches in size was pressed against the lunar soil. The motor current gave an indication of the force being applied by the scoop.

Trenching was conducted by positioning the scoop to the desired location over the surface, driving the edge of the scoop into the surface, and then running the lead screw of the retract-extend mechanism to pull the scoop toward the spacecraft. The scoop door could be either open or closed for these tests. Once again, current drawn by the retract/extend motor gave a measure of the force required to pull the scoop through the soil. The width of the scoop and of the trenches was 2 inches.

Impact tests were conducted where the scoop was raised above the surface and a clutch on the elevation motor was released to let the arm with the scoop attached fall to the surface. A light spring was built into the elevation mechanism to assist the rather weak gravitation force to drive the scoop into the surface. Impact tests could be made with the scoop door open or closed. The amount of penetration of the soil was assessed by close-up photographs.

The scoop could be filled by pulling it through the surface and the material dumped at a different location. The scoop held about 6.1 cubic inches of lunar soil.

During the Surveyor 7 mission, the SM/SS conducted 16 bearing measurements and 2 impact measurements, and it dug 7 trenches. One of the trenches was dug with multiple passes of the scoop to a depth of about 9 inches. Several rocks were handled, and one was picked up and weighed. It also picked up the alpha-scattering instrument and moved it to different locations for measurements by that instrument.

Some of the conclusions drawn by principal investigators on the soil mechanics/surface sampler during the Surveyor 7 mission were:

- The lunar surface was covered by fine-grain material whose depth over rock or rock fragments varied from 1 cm to at least 15 cm (0.4 to 5.9 inches).
- The bearing capacity of the soil was about 2.1×10^5 dynes per cm^2 (3.0 psi) at a penetration of 3 cm (1.2 inches).
- The density of the surface soil was about 1.5 grams per cubic centimeter (0.054 pounds per cubic inch).
- The density of a rock that was weighed was in the range of 2.4 to 3.1 grams per cubic centimeter (0.087 to 0.11 pounds per cubic inch).

Alpha-Scattering Instrument

The alpha-scattering instrument (ASI) was used to determine the chemical composition of the upper layer of the lunar surface. Impressive for the time, the ASI allowed remote determination of the chemical composition of the moon from armchair comfort of the earth.

The instrument operated by directing alpha particles from a radioactive source against the surface and detecting the energy spectrums of alpha particles scattered nearly directly back toward the source by the nuclei of the target. The energy spectrums of both rebounding alpha particles and of protons generated were defined in terms of number of events having particular amplitudes in a given period of time. Researchers were able to determine the chemical composition of the surface by analyzing these spectrums.

There were two detectors for scattered alpha particles oriented to detect alpha particles reflected nearly straight back from the target. There were four proton detectors sited at angles to the side and spaced around the sample port. The amplitudes of the electrical signals generated by the detectors were processed by the electronic subassembly and telemetered to earth. Equipment on earth created the energy spectrums that were used by researchers to determine the chemical properties of the lunar surface.

A spectrum from sample 1 during the Surveyor 5 mission is shown on the next page (Fig. 4.23). The curve marked "sum" is the spectrum derived from the telemetered spacecraft data. The curves lying under the sum curve are the spectrums of individual elements such as oxygen (O), silicon (Si), iron (Fe), and so on that were part of a library for the instrument. The amplitudes of the energy spectrums of individual elements were adjusted so that the sum of all matched the spectrum from the moon. The relative amplitudes then gave an indication of the percentage of each element in the lunar soil.

The number of events is plotted on a logarithmic scale on the graphs. The channel numbers on the ordinate of the graphs are related to energy of the particle striking the detector in millions of electron volts.

The chemical composition determined from alpha-scattering instrument measurements of sample 1 at the Surveyor 5 landing site was reported by James Patterson et al. in the paper *Alpha-Scattering Experiments on Surveyor 7, Comparison with Surveyors 5 and 6*. Their results are given in the table on the next page (Table 4.1). The paper referenced above states that the chemical composition agrees most closely with terrestrial basalts.

The alpha-scattering instrument consisted of a sensor head that when deployed rested on the lunar surface, a deployment mechanism for the sensor head, and an electronic subassembly. A photograph of those components is shown on the following page (Fig. 4.24).

The sensor head was contained in a box about 6 inches square and 5 inches high with a 12-inch-diameter truncated circular plate attached to the open bottom. The purpose of the plate was to prevent the sensor from sinking into the surface when it was placed down. The plate had a hole 4 1/2 inches in diameter in the center that acted as the sampling port for the instrument.

The sensor head contained six small curium-242 sources of alpha particles arranged to direct the particles through the sampling port. The two alpha particle detectors were near the field of alpha particle sources and oriented to intercept some of the backscattered particles. The four proton detectors were symmetrically located around the sampling port.

The sensor head was held in a mount with clamps that could be released to deploy the instrument to the lunar surface. A nylon cord and pulley arrangement and escapement mechanism was used to lower the instrument to the surface. In the case of Surveyor 5 and Surveyor 6 missions, the test area was directly under the spot where the instrument was lowered.

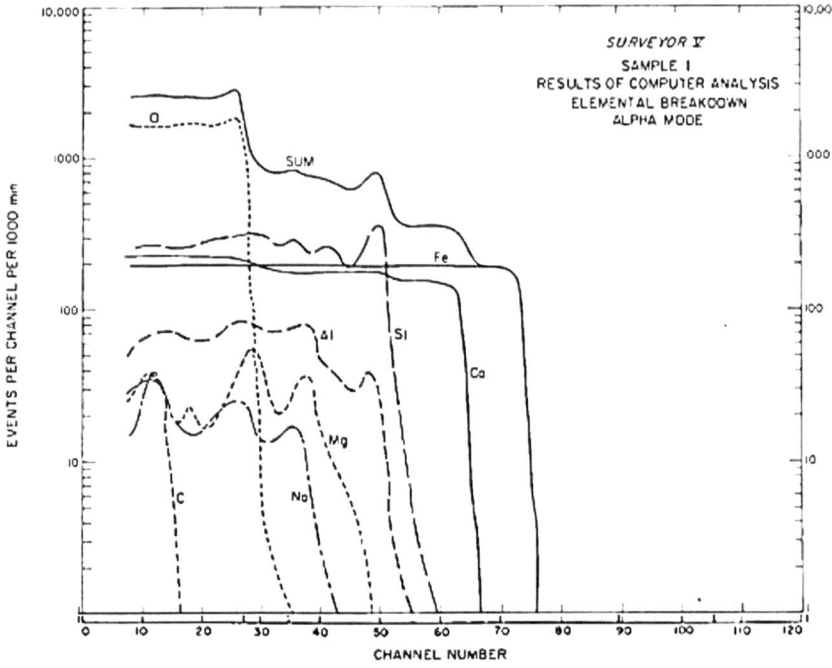

Fig. 4.23 Energy spectrum of alpha particles measured by the alpha-scattering instrument (NASA graphic)

Table 4.1 Chemical composition of lunar soil at Surveyor 5 site

Element	Atomic percent
Oxygen	62 ± 2
Sodium	0.3 ± 0.4
Magnesium	2.8 ± 1.5
Aluminum	6.2 ± 0.9
Silicon	16.3 ± 1.7
Iron	3.7 ± 0.6

In the case of Surveyor 7, a knob was installed on top of the box that could be grasped by the soil mechanics/surface sampler to allow the instrument to be placed anywhere within a 32 inch radius arc out from the mounting to the spacecraft. On Surveyor 7, the sensor head hung up and did not lower to the surface when commanded. It had to be pushed down by the soil mechanics/surface sampler to clear the jam.

Measurements were made during the Surveyor 7 mission at an undisturbed area of the surface and a trenched area and over a lunar rock. An example of chemical composition determined from measurements at the undisturbed area is given on the next page. The composition is somewhat different than measured at the Surveyor 5 site (Table 4.2).

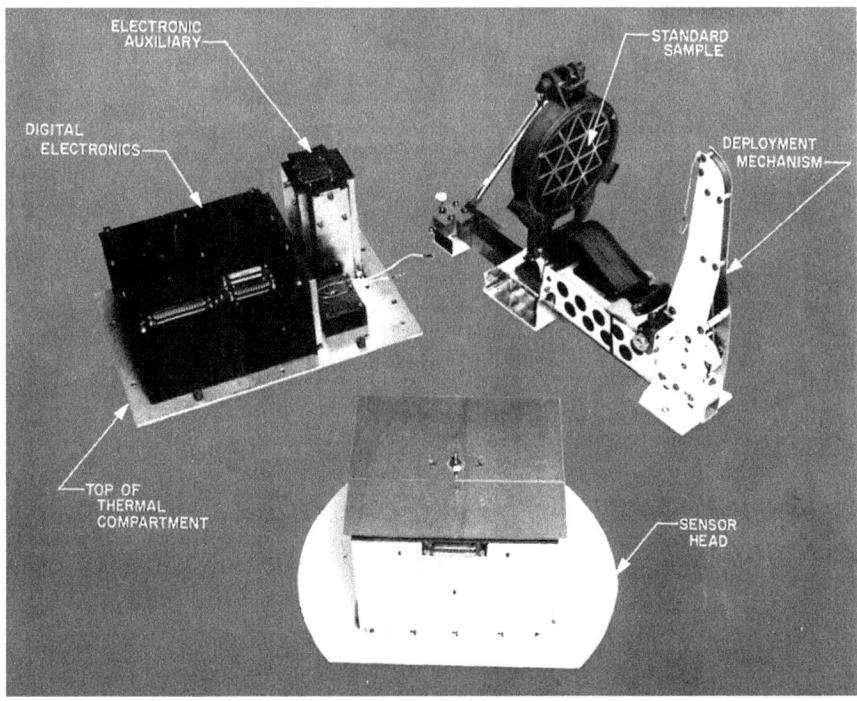

Fig. 4.24 Components of the alpha-scattering instrument (NASA photograph)

Table 4.2 Chemical composition of lunar soil at Surveyor 7 site

Element	Atomic percent
Oxygen	58 ± 5
Sodium	<3
Magnesium	4 ± 3
Aluminum	9 ± 3
Silicon	18 ± 4
Iron	2 ± 1

Electrical Power Subsystem

The Surveyor spacecraft required a significant amount of electrical power to operate its many functions. Regulated power was required for the electronic components and unregulated power served to operate heaters to warm the components.

   The sources of electrical power for the Surveyor spacecraft were a solar panel and a battery. The solar panel provided about half of the power required during transit from earth to the moon, and it provided most of the power required while on the lunar surface. It also charged the battery. The solar panel, which can be seen in photographs of the spacecraft at the beginning of this chapter, was nearly rectangular in shape with tapering sides near the attachment to the mast. Not counting the tapered form, the array was about 44 by 49 inches (1176 by 1244 mm) in size. NASA gives the area of the solar panel as 0.855 square meters (9.2 square feet). It contained 792 solar cell modules, and it could provide about 87 watts of power.

After landing, the solar panel was periodically positioned by commands from earth to keep pointed at the sun throughout the lunar day. The solar array was often pointed several degrees away from the sun to avoid overcharging the battery. It was pointed directly at the sun for several hours before sunset to fully charge the battery before the 354 hour lunar night.

The battery was a silver-zinc type with 14 cells giving a nominal output voltage of 22 volts. The energy contained in the battery at liftoff from earth was 3,608 watt-hours, corresponding to 165 ampere-hours. The battery provided about 2,000 watt-hours to the spacecraft during transit to the moon, and the solar panel provided the remaining energy required. Data from Surveyor 6 indicates that about 107 ampere-hours or 2,350 watt-hours of energy remained in the battery at the time of touchdown on the moon.

The output of the solar panel was applied to a battery charge regulator. A switch in the battery charge regulator, operated by commands from earth, allowed the output of the solar panel to be applied directly to a 30 volt preregulated bus in a boost regulator, or to a 22 volt unregulated bus, or to an OFF state where the solar panel's output was unconnected. The switch to OFF could also be made automatically by the battery logic function if the battery voltage exceeded 27.3 volts or if the battery pressure exceeded 65 psia.

The power subsystem operated most efficiently when the solar panel was connected to the 30 volt preregulated bus. If the voltage level from the solar array was lower than needed, the output of the solar panel was switched to a DC-to-DC converter that raised the voltage to the 30 volt level. A voltage regulator that followed the DC-to-DC converter provided power at a preregulated voltage level of 30 volts. This 30 volt preregulated power source was fed through two series diodes that dropped the voltage to 29 volts. This line was referred to as the 29 volt essential bus.

The nominal voltage on the unregulated bus was 22 volts, but it could vary between 17.5 and 27.5 volts. The unregulated bus provided power for those applications where closely regulated voltage was not required such as for thermal control. The preregulated bus also fed a flight control regulator and the nonessential bus regulator. The flight control regulator supplied power to the flight control subsystem. The nonessential regulator provided regulated voltage of 29 volts to those components deemed not essential at a particular time. The regulator could be turned off to disconnect the nonessential electrical load.

A shunt regulator, which was fed from the preregulated bus, allowed charging the battery when the solar panel was connected to the preregulated bus and other loads on the bus were low.

SURVEYOR ACCOMPLISHMENTS

The Surveyor program was an outstanding success! One of its main accomplishments was verifying techniques for soft landing a spacecraft on the moon. It also confirmed that the surface of the moon could support a heavy spacecraft such as the Apollo lander. Several regions of the moon that were examined by the Surveyor spacecraft were later used as landing sites for Apollo.

The five Surveyor spacecraft that landed returned 87,000 photographs of the moon ranging from close-up pictures of the soil around the landing pads to panoramic photographs of distant vistas. Detailed mechanical and chemical properties of the lunar soil were determined by the soil mechanics/surface sampler and alpha-scattering instruments.

All of this information gave good assurance that the ambitious Apollo manned lunar landings could go forward.

Bibliography

Jaffe, Leonard D., *Lunar Surface Exploration by Surveyor Spacecraft: Introduction,* Journal of Geophysical Research, Vol. 72 No.2 , 1967

Kirsten, Charles, *Surveyor Spacecraft Telecommunications*, JPL Technical Report 32-1105

Kloman, Erasmus, H., *Unmanned Space Project Management, Surveyor and Lunar Orbiter*, NASA Report SP-4901, updated July 1997

Lund, Thomas, *Radar Velocity Sensors and Altimeters for Lunar and Planetary Landing Vehicles*, First Western Space Conference, October 1970

Nicks, Oran W, *The Far Travelers: The Exploring Machines*, NASA Report SP-480, updated August 2004

Patterson, James H., et al, *Alpha Scattering Experiment on Surveyor 7, Comparison with Surveyors 5 and 6*, Journal of Geophysical Research Volume 74, Number 25, November 1969

Stokes, Lyle S., *Telecommunications from a Lunar Spacecraft*, American Institute of Aeronautics and Astronautics, Unmanned Spacecraft Meeting, 1965

Surveyor 1, NASA-NSSDC/COSPAR ID: 1966-045A

Surveyor 7, NASA-NSSDC/COSPAR ID: 1968-001A

Surveyor I Mission Report, NASA Technical Report 32-1023, September 1966

Surveyor III Mission Report, NASA Technical Report 32-1177, September 1967

Surveyor Program Results, NASA Report NASA SP-184, 1969

Surveyor V Mission Report, NASA Technical Report 32-1246, March 1968

Surveyor VI Mission Report, NASA Technical Report 32-1262, September 1968

Surveyor VII Mission Report, NASA Technical Report 32-1264, February 1969

Thurman, Sam, W., *Surveyor Spacecraft Automatic Landing System*, 27th Annual AAS Guidance and Control Conference, February 2004

5

The Apollo Manned Exploration of the Moon

The Apollo moon mission must rank as one of the most audacious scientific exploration programs attempted by man. The mission involved launching a manned spacecraft into earth orbit, then setting it on a trajectory to orbit the moon. An independent spacecraft with two astronauts aboard was detached from the orbiting vehicle and made a soft landing on the moon.

After exploring the lunar surface, the astronauts used the lower portion of the landing vehicle as a launch platform and blasted off with the upper portion to rendezvous with the main vehicle still orbiting the moon. Compared with the elaborate launch facilities and scores of people who review spacecraft health and readiness before launch at the Kennedy Space Center, the routine launch of a spacecraft in that lonely place was a triumph of human ingenuity.

The ingenuity fully unfolded toward the end of the program with serious exploration of the lunar surface by a trained geologist, Harrison Schmitt. Schmitt and the commander of the mission, Gene Cernan, spent 3 days on the moon and traveled 21 miles about the lunar surface in an electric powered, dune buggy type vehicle to gather samples and take photographs. A photograph of the Apollo 17 Lunar Module and Cernan driving the Lunar Roving Vehicle on the moon is shown on the next page (Fig. 5.1).

The Apollo program was a fulfillment of a declaration made by President Kennedy in 1961 to put a man on the moon and return him to earth before the end of the decade.

Several books describe Apollo from an historical and program point of view. It is the author's intent to describe the program from an engineering point of view and to give some reflections of a young engineer who worked and indeed lived the program.

T. Lund, *Early Exploration of the Moon*, Springer Praxis Books,
https://doi.org/10.1007/978-3-030-02071-2_5

Fig. 5.1 Apollo 17 Lunar Module and Lunar Roving Vehicle (NASA photograph)

A total of nine missions were launched toward the moon by Apollo. A summary of Apollo journeys to the moon is given in the tables on the following pages (Table 5.1). Six missions made soft landings on the moon. They were Apollo 11, 12, 14, 15, 16, and 17. Two astronauts from each mission walked on the moon and explored the terrain.

Apollo 13 suffered an explosion in an oxygen tank en route to the moon. The astronauts and mission control at Houston performed a heroic series of tasks to save the lives of the astronauts and to cause the spacecraft to loop around the moon and return safely to earth.

OVERVIEW OF THE APOLLO SPACE VEHICLE

The Apollo spacecraft was launched toward the moon by the multistage Saturn V launch vehicle. The mighty S-IC first stage, 33 feet in diameter and 130 feet long, was powered by five liquid propellant F-1 rocket engines developing a total of 7.65 million pounds of thrust. The S-II second stage was powered by five liquid propellant J-2 rocket engines, which developed a total thrust of 1.15 million

Table 5.1 Summary of Apollo journeys to the moon

Designator	Astronauts	Flight dates	Mission	Time on moon/EVA time, hours	Distance traveled, miles
Apollo 8	F. Borman J. Lovell W. Anders	21 December 1968 to 27 December	Manned mission to verify operation of the CSM in lunar orbit. A mass mockup of the LM was carried	0/0	–
Apollo 10	T. Stafford J. Young E. Cernan	18 May 1969 to 26 May	Confirm all aspects of the lunar landing mission except the actual landing	0/0	–
Apollo 11	N. Armstrong M. Collins E. Aldrin	16 July 1969 to 24 July	Land astronauts on the moon in Mare Tranquillitatis. Explore, gather samples, and photograph the lunar surface. Return astronauts safely to earth	21.6/2.5	0.63

Designator	Astronauts	Flight dates	Mission	Time on moon/EVA time, hours	Distance traveled, miles
Apollo 12	C. Conrad R. Gordon A. Beam	14 November 1969 to 24 November	Land men on the moon at a precise location (535 feet from Surveyor 3 in Oceanus Procellarum). Explore and return astronauts safely to earth	31.52 / 7.8	1.44
Apollo 13	J. Lovell J. Swigert F. Haise	11 April 1970 to 17 April	An explosion in an oxygen tank aborted the mission. The crew used the LM as a lifeboat to loop around the moon and then reactivated the Command Module to return safely to earth	0/0	–
Apollo 14	A. Shepard S. Roosa E. Mitchell	31 January 1971 to 9 February	Explore the Fra Mauro formation formed from ejecta from a massive impact that created the Imbrium Basin. Return crew safely to earth	31.5/9.4	2.5

(continued)

Table 5.1 (continued)

Designator	Astronauts	Flight dates	Mission	Time on moon/EVA time, hours	Distance traveled, miles
Apollo 15	D. Scott A. Worden J. Irwin	26 July 1971 to 7 August	Explore the Hadley-Apennine region. Expand exploration and sample gathering by using the Lunar Roving Vehicle	68.9/19.1	17.3
Apollo 16	J. Young T. Mattingly C. Duke	16 April 1972 to 27 April	Explore the Descartes highlands region using the Lunar Roving Vehicle. Conduct the first exploration of the highlands	71.0/20.2	16. 5
Apollo 17	E. Cernan H. Schmitt R. Evans	7 December 1972 to 11 December	Explore Taurus-Littrow highlands region using the Lunar Roving Vehicle. One of the explorers was a trained geologist	75.0/22.1	21.6

pounds. The S-IVB third stage was powered by one J-2 engine producing 203,000 pounds of thrust and was capable of multiple starts and stops.

One witness to a launch of the Saturn V was said to have remarked: "The question was not did the Saturn V rise but did Florida sink?" The author's experience watching the launch of Apollo 15 was a powerful shaking in my chest while being enveloped by waves of awesomely loud, low frequency sound.

A photograph of the launch of Apollo 15 is shown on the below. The spacecraft tilted slightly at launch to clear the gantry (Fig. 5.2).

The main components of the Apollo spacecraft were the Command Module (CM), Service Module (SM), and the Lunar Module (LM). The Command Module and the Service Module remained joined until just before the end of the mission when the Service Module was jettisoned and the Command Module was

Fig. 5.2 Launch of Apollo 15 (NASA photograph)

positioned to enter the earth's atmosphere. The two modules together were referred to as the Command Service Module (CSM). A sketch of the spacecraft with the Lunar Module attached to the Command Module is shown below (Fig. 5.3). The legs of the Lunar Module are folded for transport in the sketch.

The Command Module and the Service Module were developed and built by North American Aviation. The Lunar Module was developed and built by Grumman Aircraft Engineering Corporation.

The Command Module (CM) was the command, control, and communications center for the mission. It consisted of an inner pressurized crew compartment and an outer heat shield. The CM contained display and control capability as well as life support systems for the crew. It had provisions for docking to the Lunar Module and allowed ingress and egress between the Command Module and Lunar Module.

The Service Module stored most of the spacecraft consumables including oxygen, hydrogen, water, and propellant. It also contained the main propulsion engine, the reaction control system engines, and the fuel cells that produced electric power and water for the spacecraft.

The Lunar Module (LM) transported two astronauts to the lunar surface and returned them to the orbiting Command Service Module. The LM consisted of an ascent stage and a descent stage joined together by fittings with explosive bolts. The decent stage contained a throttleable, restartable engine that was gimbaled to allow compensating for center of gravity change as fuel burned off.

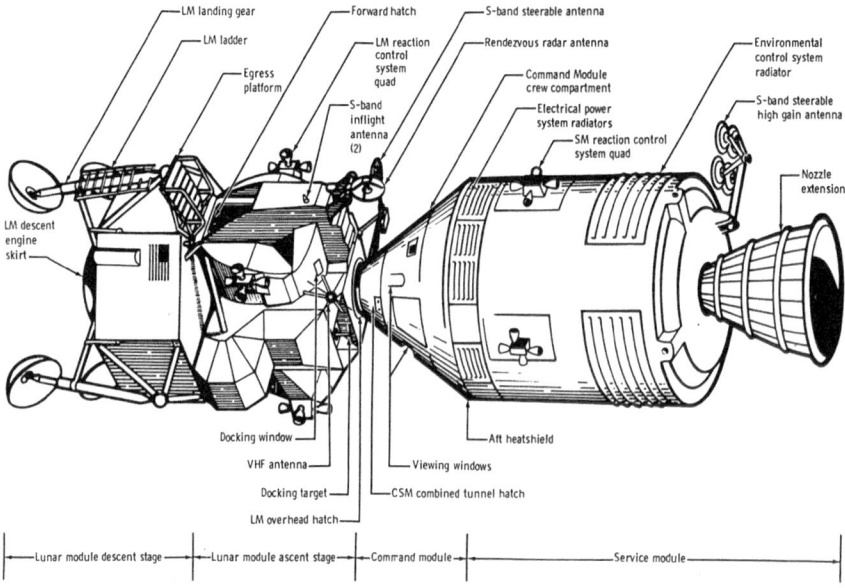

Fig. 5.3 Configuration of Apollo space vehicle with Lunar Module attached to Command Module (NASA graphic)

The ascent stage contained a restartable but non-throttleable engine. The ascent stage contained the crew compartment, display and control capability, and the life support system.

HISTORY OF EARLY MANNED SPACE FLIGHT IN THE UNITED STATES

The National Aeronautics and Space Administration (NASA), which was established in October 1958, directed both unmanned and manned space programs. Early unmanned space programs were Ranger, Lunar Explorer, and Surveyor. Those programs have been described in earlier chapters of this book. There were two manned space programs leading up to Apollo: Mercury and Gemini.

Early manned space programs were managed by the Space Task Group headed by Robert Gilruth and located at Langley Research Center in Hampton, Virginia. With the increasing importance of manned space flight programs, NASA formed the Manned Spacecraft Center (MSC), and the Space Task Group was transferred into it. The first director of MSC was Dr. Gilruth. The Manned Space Flight Center was relocated from Langley to Houston, Texas, where a large facility was being built for it.

The Mercury program was initiated in December 1958. The Mercury spacecraft was a small, single-person capsule with maneuvering capability in pitch, yaw, and roll. It contained a blunt heat shield for reentering the atmosphere. There were two manned suborbital Mercury flights and four manned orbital flights. Alan Shepard flew the first suborbital flight in May 1961, and John Glenn flew the first orbital flight in February 1962. The last Mercury flight took place in May 1963.

The Mercury spacecraft was built by McDonnell Aircraft Corporation. The spacecraft was lofted into space by a Redstone launch vehicle for the suborbital flights and by an Atlas launch vehicle for the orbital flights.

Pilots for the Mercury flights were drawn from the initial group of seven astronauts referred to as Astronaut Group 1. That group of astronauts had been selected from an elite group of military test pilots, each with thousands of hours flying high-performance aircraft. The seven were Scott Carpenter, Gordon Cooper, John Glenn, Gus Grissom, Wally Schirra, Alan Shepard, and Deke Slayton. Their selection was announced in April 1959.

In support of upcoming Gemini and Apollo programs, a second group of astronauts, Astronaut Group 2, was announced in September 1962. They were Neil Armstrong, Frank Borman, Charles Conrad, James Lovell, James McDivitt, Elliot See, Thomas Stafford, Edward White, and John Young.

The Gemini program, initiated in January 1962, was a stepping stone to Apollo. Objectives of the Gemini program were to demonstrate rendezvous and docking in space with another vehicle, demonstrate extravehicular activity (spacewalk), and demonstrate endurance of astronauts and spacecraft for at least 8 days required for a moon landing and return.

The Gemini spacecraft was larger and more capable than Mercury, and it carried two astronauts. It contained an Orbit Attitude and Maneuvering System that included 16 independent thrusters and associated pilot's controls. The system allowed attitude control in pitch, yaw, and roll as well as translation control along the spacecraft three principle axes. These controls allowed maneuvering for rendezvous and docking with another space vehicle.

The Gemini spacecraft was built by McDonnell Aircraft Corporation. It was lofted into space by a Titan II launch vehicle.

There were two unmanned Gemini flights and ten manned flights. All of these flights were successful. Seven of the manned flights demonstrated rendezvous with another space vehicle, and full docking took place on three of those flights. Extravehicular activity was demonstrated on five flights. The longest flight lasted 13 days and 18 hours. The first manned Gemini flight took place in March 1965, and the last flight took place in November 1966.

FORMULATING APOLLO

There was more purpose in the US space community and at NASA than putting men in orbit around the earth. Humans are explorers, and the moon was the logical target for a manned space mission. Landing on the moon was a well-defined goal and generally within possibility, given a little optimistic expectation of evolving space capabilities. NASA awarded three contracts to industry in October 1960 to study the feasibility of a manned lunar landing and return to earth.

The possible manned mission to the moon was given the name "Apollo" by Dr. Abe Silverstein, Director of Space Flight Programs at NASA. The mythical god Apollo was son of Zeus and god of music and god of healing. One of his important jobs was to drive his four-horse chariot carrying the sun across the sky each day. Silverstein said the image of "Apollo riding his chariot across the sun was appropriate to the grand scale of the proposed program."

Many prominent scientists favored a vigorous scientific exploration of space rather than the much more complex and expensive manned mission to the moon. However, in March 1961 the National Academy of Sciences National Research Council sent a powerfully reasoned letter to James Webb, NASA Administrator, in support of man's involvement in lunar exploration. The last paragraph in their policy position that accompanied the letter stated:

"Second, the members of the Board as individuals regard man's exploration of the Moon and planets as potentially the greatest inspirational venture of this century and one in which the entire world can share; inherent here are great and fundamental philosophical and spiritual values which find a response in man's questioning spirit and his intellectual self-realization". …

In the spring of 1961, there was much concern in the Kennedy White House that American prestige around the world was diminishing and that of the Soviet Union rising. A significant accomplishment in space was looked on as a chance to regain US standing in the world. These hopes received a setback when the Soviet Union sent Cosmonaut Yuri Gagarin on a flight to orbit the earth on 12 April 1961. Gagarin was the first human to fly in space.

President Kennedy had appointed Vice President Lyndon Johnson to be Chairman of the National Aeronautics and Space Council. There was likely considerable discussion with President Kennedy about the status of the US space programs and what significant and spectacular accomplishment in space might be pursued. The President sent a formal memo to Johnson on 20 April 1961 with a list of five questions. The memo and the first question read as follows:

Memorandum for Vice President

"In accordance with our conversation I would like for you as Chairman of the Space Council to be in charge of   making an overall survey of where we stand in space.

1. Do we have a chance of beating the Soviets by putting a laboratory in space, or by a trip round the moon, or by a rocket to land on the moon, or by a rocket to go to the moon and back with a man. Is there any other space program which promises dramatic results in which we could win?"

A memorandum responding to the President, dated 28 April 1961, was prepared by Edward Welsh, Executive Secretary of the National Aeronautics and Space Council, and signed by Vice President Johnson. A portion of the response to the President's first question reads as follows:

"As for a manned trip around the moon or a safe landing and return by a man to the moon, neither the U.S. nor the USSR has such capability at this time, so far as we know…. However, with a strong effort, the United States could conceivably be first in those two accomplishments by 1966 or 1967."

NASA released their recommendations in response to the President's request in a lengthy report on 8 May 1961. The report concluded that the best way to surpass the USSR in space was a manned lunar landing and return to earth. It also stressed the need to develop more powerful launch vehicles to support such a program.

Vice President Johnson worked the senate to drum up support for a large and very expensive manned lunar landing program. Thus, Kennedy had good assurance before his historic speech to Congress on 25 May 1961 proposing a manned lunar landing that an industry-NASA team could likely pull off it off and Congress would support the program.

A key paragraph of Kennedy's well-crafted speech to Congress dealing with space in May 1961 is given below.

"First, I believe that this nation should commit itself to achieving the goal, before this decade is out, of landing a man on the moon and returning him safely to the earth. No single space project in this period will be more impressive to mankind, or more important for the long-range exploration of space; and none will be so difficult or expensive to accomplish."

A video recording of a 9-minute section of Kennedy's speech dealing with space can be found on the John F. Kennedy Presidential Library website. Kennedy was an outstanding speaker and a charismatic personality. Even today, his speech is inspiring to the author.

NASA sought to inform industry of essentials of the Apollo program and draw on the considerable experience and expertise of industry by conducting a NASA - Industry Apollo Technical Conference. The conference, held in Washington, DC in July 1961, drew representatives from over 300 companies.

The first contract for the Apollo program was awarded to the Massachusetts Institute of Technology (MIT) Charles Draper Laboratory. The contract, which was awarded on 10 August 1961, was for guidance and control of the planned complex spacecraft. MIT was preeminent in the field of inertial guidance systems and automatic pilots at the time.

The second contract to be awarded was for the Command Module and Service Module. Requirements for these space vehicles were established by NASA from feasibility studies conducted by industry and from NASA's own studies. NASA issued request for proposals to industry on 28 July 1961. Five companies competed for the contract. North American Aviation was announced to be the winner on 28 November 1961.

The question of the best approach to send a manned spacecraft to land on the moon and return to earth was still unanswered at the time of contracting for the Command Module and Service Module. The approach selected would drive the design of all major components of the Apollo program. The difficult subject had been under consideration for a few years before the decision was made to embark on the maned lunar landing program. Early study concentrated on either a direct flight to the moon or an earth orbit mode of conducting the mission.

The direct approach would use a very powerful launch system that would send a spacecraft directly to the moon. The spacecraft would perform a soft landing on the moon. After exploration of the moon, the astronauts would launch from the moon and travel back to earth.

The earth orbit rendezvous approach would launch a propulsion unit into a parking orbit around the earth and then launch a manned spacecraft to rendezvous and dock with it. The joined unit would accelerate out of earth orbit into a translunar flight path. The propulsion unit would be jettisoned after its boost task was finished. As the manned unit approached the moon, a retrorocket would fire to slow the spacecraft for a lunar landing. The landed spacecraft would be used for the return trip to earth.

Another approach pushed strongly by John Houbolt at Langley research Center involved a lunar orbit and rendezvous. A multi-vehicle spacecraft would be sent to orbit the moon, and then a small landing spacecraft would be dispatched to the surface. After exploring the moon, the astronauts would launch from the moon and rendezvous with the orbiting spacecraft. The lander would be jettisoned, and the spacecraft would depart the moon and return to earth.

The lunar orbit approach was projected to save up to 20% of total weight to be launched and promised a significant saving in program cost. The main objection to lunar orbit was lack of an abort capability should the rendezvous fail. The prospect of astronauts being stranded in lunar orbit was sobering.

As late as December 1961, the earth orbit approach seemed to be favored at NASA Headquarters although interest in a lunar orbit approach was growing. Several meetings and presentations were held in an effort to select the best approach in early 1962. Finally, at a meeting of the Manned Space Flight Management Council on 22 June 1962, a decision was made to go with the lunar orbit rendezvous approach using the Saturn C-5 launch vehicle. The decision was formally announced by NASA Headquarters on 11 July 1962. The Saturn C-5 was renamed Saturn V in 1963.

Reasons given for deciding on the lunar orbit approach over the earth orbit approach were compelling:

- Higher probability of mission success with essentially the same mission safety
- Cost about 15% less
- Development time several months shorter
- Require the least amount of technical development

Another decided advantage of the lunar orbit and rendezvous approach was that the lander could be smaller and lighter and designed specifically to land on the moon and return to the orbiting spacecraft.

With the decision made for lunar orbit and rendezvous, NASA prepared a request for proposal for the landing spacecraft to be called the Lunar Excursion Module (LEM). The LEM would separate from the larger spacecraft that was orbiting the moon, land on the moon, and then launch from the moon to rendezvous with the orbiting spacecraft. The RFP was issued to 11 companies on 25 July 1962. The list included most of the major aircraft manufacturing companies in the United States. The winner of the competition was announced on 7 November 1962 to be Grumman Aircraft Engineering Corporation.

Thus, by the end of 1962, the direction for the Apollo mission was firm, and companies had been selected to build the major components of the Apollo spacecraft. New direction was given to North American Aviation to redesign the Command Module to reflect the lunar orbit mission approach. The contract for the revised approach was signed in August 1963.

President Kennedy was passionate about the Apollo program, and he was an inspirational speaker. He had the country largely with him. A photograph of Kennedy during his address in the Rice University football stadium is shown below (Fig. 5.4). The photograph, taken by Robert Knudsen, is displayed in the Kennedy Presidential Library.

A snippet of his speech at Rice University in September 1962 is given below.

...... *"But why, some say, the Moon? Why choose this as our goal? And they may well ask why climb the highest mountain? Why, 35 years ago, fly the Atlantic? Why does Rice play Texas?*

We choose to go to the Moon. We choose to go to the Moon in this decade and do the other things, not because they are easy, but because they are hard, because that goal will serve to organize and measure the best of our energies

Fig. 5.4 Kennedy promoting the Apollo program at Rice University in September 1962

and skills, because that challenge is one that we are willing to accept, one we are unwilling to postpone, and one which we intend to win, and the others, too.

It is for these reasons that I regard the decision last year to shift our efforts in space from low to high gear as among the most important decisions that will be made during my incumbency in the Office of the Presidency."

NASA MANAGEMENT FOR PROJECT APOLLO

James Webb was NASA Administrator from February 1961 to October 1969. He was an effective manager and guided an organization with about 33,500 people spread among 37 blocks in the NASA organization chart of 1966. Webb strove to achieve the most effective organization, and he was not adverse to change. Different organization charts were issued in November 1963, January 1966, and March 1967.

The two top executives at NASA Headquarters during much of the Apollo program were James Webb, NASA Administrator, and Robert Seamans, Deputy Administrator. Willis Shapley, Associate Deputy Administrator, functioned as chief of staff. Webb arranged to have Seamans and Shapley run NASA on a day-to-day basis while he concerned himself with policy and key problems of NASA. The Apollo program was handled by the very capable Major General Samuel Phillips, Director of the NASA Headquarters Apollo Program Office. Phillips was on loan from the US Air Force. The Apollo program accounted for 17% of the total NASA budget in 1963 and 70% of the budget at its peak in 1967.

There were 11 departments under the executive group that were run by Assistant Administrators, and there were four large organizations run by Associate Administrators. One of those large organizations was the Office of Manned Space Flight led by George Mueller. That organization directly managed the Apollo program.

There were three centers under the Office of Manned Space Flight: Marshall Space Flight Center in Huntsville directed by Wernher von Braun, Manned Spacecraft Center in Houston directed by Robert Gilruth, and the Kennedy Space Center in Florida directed by Kurt Debus. The Kennedy Space Center included the launch complex at Cape Canaveral.

An outstanding leader under Gilruth at the Manned Spaceflight Center was Joe Shea, manager of the Apollo Spacecraft Program Office. Shea was a smart, hard-driving manager who instilled systems engineering discipline into the manned spaceflight organization. He guided early development of Apollo.

Other significant people in earlier times at NASA were Hugh Dryden, Deputy Administrator until his death in 1965 and Brainerd Holmes, first leader of the Office of Manned Space Flight. All of those capable NASA leaders contributed much to the outstanding success of Apollo.

DEVELOPMENT OF HEAVY-LIFT LAUNCH VEHICLES THAT SUPPORTED APOLLO

In 1956 the US Department of Defense issued requirements for heavy-lift vehicles to place objects weighing 20,000 to 40,000 pounds into earth orbit. The large booster group of the Army Ballistic Missile Agency (ABMA) in Huntsville, Alabama, concluded in 1957 that the first stage of such a vehicle would have to have thrust of about 1.5 million pounds. ABMA was headed up by Wernher von Braun.

The Army Ballistic Missile Agency was placed under the Army Ordnance Missile Command (AOMC) in March 1958. In March 1960 the AOMC facilities in Huntsville, Alabama, were transferred to NASA and renamed the Marshall Space Flight Center (MSFC). Wernher von Braun became the first Director of MSFC.

Capable launch vehicles were also sought by the US Air Force Ballistic Missile Division. The Air Force had contracted with the Rocketdyne division of North American Aviation in 1955 to conduct preliminary design of a rocket engine that would develop 1.0 to 1.5 million pounds of thrust. The engine would burn Rocket Propellant-1 (RP-1) with liquid oxygen as oxidizer. RP-1 was refined kerosene. Rocketdyne named the new engine F-1 since it followed a smaller engine named E-1 being developed for the Army. The F-1 engine would be key to the success of Apollo. Details of its development and performance are given later in this chapter.

Another player in the heavy-lift rocket game was the Advanced Research Projects Agency (ARPA). ARPA, formed in February 1958 as an agency of the Department of Defense, was put in charge of requirements for the heavy-lift program. Since the Rocketdyne E-1 and F-1 engines were still in development, ARPA orchestrated a program to obtain heavy-lift capability as soon as possible by employing a cluster of rocket engines based on the Rocketdyne S-3D engine used on the Jupiter ballistic missile.

A contract for the modified engine, known as the H-1, was issued to Rocketdyne in September 1958 by the Department of Defense. The H-1 engine also burned RP-1 fuel with liquid oxygen as the oxidizer. Later engines developed 205,000 pounds of thrust.

The Saturn series of heavy-lift vehicles followed. What became Saturn 1 had a first stage powered by a cluster of eight H-1 engines. Upper stages were powered by RL-10 liquid hydrogen-liquid oxygen engines. The first launch of a Saturn 1 with dummy upper stages was conducted in October 1961, and it demonstrated the soundness of clustered rocket engines to obtain high thrust. The total thrust of the eight-engine first stage was about 1.7 million pounds.

The Saturn 1 launch vehicle was used by NASA to launch elements of Apollo into earth orbit as part of the methodical Apollo test program. The first four Saturn

1 vehicles were launched with dummy upper stages. The fifth launch (SA-5) was with an active S-IV upper stage with RL-10 engines that achieved earth orbit. The next four vehicles (SA-6, 7, 8, and 9) launched "boilerplate" (non-operating) versions of the Apollo Command and Service Module (CSM) into earth orbit.

The Saturn 1B launch vehicle with a more powerful upper stage, the S-IVB, was developed to launch Apollo Command and Service Modules and Lunar Excursion Modules into earth orbit. The S-IVB was powered by one J-2 engine with thrust of 200,000 pounds. The first flight, designated AS-201, was launched on 26 February 1966. It put a Block 1 Command and Service Module into a sub-orbital flight to test the engine of the Service Module and the heat shield of the Command Module. The second flight, launched in July 1966 and designated AS-203, was a test flight for the S-IVB engine with restart after a cold soak. The third flight, launched in August 1966 and designated AS-202, contained sufficient fuel in the S-IVB stage to put a Block 1 CSM into a high suborbital path with splashdown of the Command Module in the Pacific Ocean.

The Block 1 CSM reflected a design by North American Aviation based on an earth orbit and rendezvous approach that was favored early in the program. The hardware developed during that early effort served well in the Apollo test program.

Apollo mission AS-204 had been scheduled to launch the first manned Command and Service Module into earth orbit in February 1967. Tragically, during rehearsal for the launch on 27 January, a fire started in the Command Module and raged out of control in the pure oxygen environment. The three astronauts perished. The astronauts who lost their lives were Virgil "Gus" Grissom, Edward White, and Roger Chaffee.

The accident led to stepped-up safety and deliberateness in the Apollo program. The time spent redesigning the Command Module to minimize the chance of another fire allowed the development effort of several systems that had been lagging due to problems to catch up in an orderly manner.

The AS-204 launch vehicle was undamaged by the fire, and it was later designated AS-204B. AS-204B was used to launch an unmanned Lunar Module, LM-1, into earth orbit on 22 January 1968. The original name, Lunar Excursion Module, had been shortened to Lunar Module by that time. A nose cone was installed in front of the Lunar Module for the launch. The final use of the Saturn 1B launch vehicle for Apollo testing was the launch of the first manned Command and Service Module into earth orbit on 11 October 1968. The Command Module for that flight was the substantially redesigned Block 2 model.

SATURN V LAUNCH VEHICLE

The Apollo spacecraft was launched toward the moon by the colossal, multistage Saturn V launch vehicle. The most imposing element of Saturn V was the first stage where a cluster of five F-1 rocket engines generated a total thrust of about

7.6 million pounds. The Saturn V launch vehicle contained three major stages and an Instrument Unit as shown in the sketch of the vehicle below (Fig. 5.5).

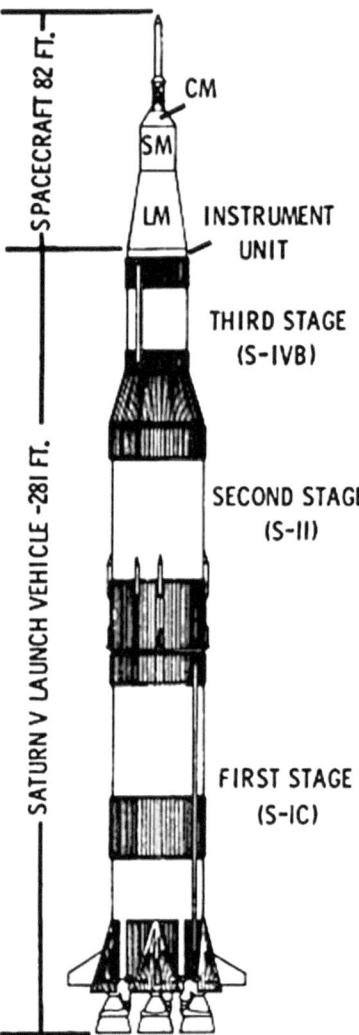

Fig. 5.5 Configuration of Saturn V launch vehicle (NASA Graphic)

The total length of the vehicle, including the spacecraft section, was 363 feet. Lying down, it would extend 63 feet beyond the confines of a 100-yard football gridiron. The diameter of the first and second stages was 33 feet. The weight of the entire vehicle including the spacecraft at the time of launch of Apollo 11 was 6,478,000 pounds. Thrust of the combined five engines of the first stage of 7.6 million pounds was comfortably above the initial weight to be lifted.

The Marshall Space Flight Center provided overall direction for the development of Saturn V. The director of MSFC was Werner von Braun. The head of the Saturn V program office at MSFC was Arthur Rudolph.

The Saturn V proved to be a very reliable launch vehicle as well as a very capable one. A total of 13 launches were made without a failure. This perfect record must have been difficult for the Russian launch vehicle managers to accept after four failures in a row of their massive N1 launch vehicle. The final launch of Saturn V was that of Skylab in May 1973 after the Apollo program had been completed.

A brief description of each of the three stages and of the Instrument Unit of Saturn V is given in the following paragraphs.

S-1C First Stage

The mighty S-1C first stage was 138 feet long and 33 feet in diameter. It weighed about 4,881,000 pounds when loaded with fuel and oxidizer. It contained five F-1 rocket engines, each having 1,522,000 pounds rated thrust. The total thrust was 7,610,000 pounds. The engines burned Rocket Propellent-1 (RP-1) fuel, which was highly refined kerosene, with liquid oxygen (LOX) as oxidizer. RP-1 fuel was contained in a 209,000 gallon tank at the bottom of the vehicle when vertical and LOX was contained in a cryogenic 334,500 gallon tank mounted above the RP-1 tank. Five insulated lines were run through the RP-1 tank to supply LOX to the five engines. Two RP-1 fuel lines were run from the RP-1 tank to each of the five engines.

Four of the five engines were mounted in a circle around the periphery of the bottom of the stage in the pattern of a cross. The fifth engine was mounted in the center. The locations of the engines can be seen in the photograph of the S-1C during assembly at the Michoud facility on the next page (Fig. 5.6).

An essential element of the first stage was the thrust structure at the engine end of the stage. That structure, which weighed about 48,000 pounds, was 33 feet in diameter and 20 feet long. It supported the entire vehicle and distributed the thrust of the engines. The four outer F-1 engines were mounted on gimbals that allowed the stage to be steered. The center engine was fixed mounted. The thrust structure contained four hold-down anchors that held the vehicle down at launch until all the engines were up to rated thrust.

The firing sequence for the engines was arranged to distribute the forces from the strong thrust of individual engines. The center engine was started first, and then a diagonal pair was started followed by the other diagonal pair. A 300 milli-second delay was programmed between engine starts.

Preliminary design of the S-1C first stage was performed by the Marshall Space Flight Center. The MSFC program manager for the S-1C was Matt Urlaub. The Boeing Company was awarded the contract in December 1961 for detail design and manufacture of the S-1C stage. Detailed design was a joint effort between

Fig. 5.6 Engines being installed on S-1C stages at Michoud (NASA photo)

MSFC and Boeing engineers who shared engineering space at MSFC. The Boeing program manager was George Stoner, head of the Saturn Booster Branch at Boeing.

The very large S-1C stages were assembled at NASA's Michoud Assembly Facility near New Orleans. Michoud was home to one of the largest manufacturing buildings in the world.

It had 2 million square feet under one roof. A photograph of S-1C stages being assembled at Michoud is shown above. Fabrication and assembly of the S-1C stage was a Boeing responsibility, but again, it was a joint effort between MSFC and Boeing personnel. The Michoud facility was managed by George Constan of MSFC. The Boeing operation at Michoud was managed by Richard Nelson.

After assembly of production models of S-1C stages, they were transported by barge about 40 miles to the Mississippi Test Facility (MSF) where the engines were test fired. After testing, the stages were transported by barge to the Kennedy Space Center. The Mississippi Test Facility is now known as the Stennis Space Center.

The first all-up test of an S-1C-T test stage with five engines firing was performed at a static test stand at the Marshall Space Flight Center in April 1965.

S-II Second Stage

The substantial S-II second stage was 81.6 feet long and 33 feet in diameter. It weighed about one million pounds when loaded with fuel and oxidizer. It contained five J-2 rocket engines, each having 230,000 pounds rated thrust. The total thrust of the stage was 1,150,000 pounds. The engines burned liquid hydrogen (LH$_2$) with liquid oxygen (LOX) as oxidizer. A photograph of the second stage taken by the author at the Johnson Space Center in Houston is shown below (Fig. 5.7).

Fig. 5.7 Engine end of second stage of Saturn V launch vehicle (author's photograph

The use of liquid hydrogen and liquid oxygen propellants yields the highest specific impulse of known rocket propellants. NASA overcame many technical challenges in handling liquid hydrogen as fuel. The Soviet Union did not develop the technology in time to effectively compete with the United States in the race to put a man on the moon.

The S-II stage was developed and built by North American Aviation in Seal Beach, California, under contract with NASA issued in September 1961. The fact that the J-2 engine was being developed by the Rocketdyne division of North American may have helped North American's cause in the competition for the S-II. The S-II program was led by Harrison Storms, head of North American's Missile Division. The MSFC program manager for the S-II stage was Roy Godfrey.

There were several technical problems in development of the stage leading to late deliveries. Welding of the large curved structures was particularly difficult. To add to the problems, there were two catastrophic failures of S-II stages during early testing. Despite the problems, S-II stages were produced in time to support the Apollo launch schedule.

The J-2 engine used in the stage was developed by the Rocketdyne division of North American Aviation under contract from NASA issued in June 1960. The first prototype J-2 engine was successfully tested in October 1962. The engine was capable of being stopped and restarted several times in flight, although this capability was not needed for the S-II. The first all-up test of an S-II-T test stage with five engines firing for full duration was performed at Rocketdyne's Santa Susana static test facility in August 1965.

Liquid oxygen was contained in a cryogenic 83,000-gallon tank located at the bottom of the S-II stage. Liquid hydrogen was contained in a cryogenic 260,000-gallon tank located above the LOX tank. The amount of propellant gave an engine burn time of about 395 seconds. The liquid oxygen (LOX) tank was ellipsoidal in shape, 22 feet high and about 33 feet in diameter. The upper bulkhead of the LOX tank served as the lower bulkhead for the liquid hydrogen (LH_2) tank.

The liquid hydrogen tank was made up of a curved bulkhead at the forward end and six cylindrical sections welded together. The cylindrical walls of the LH_2 tank served as the outer skin of the S-II stage. The length of the LH_2 tank, not counting the curved bulkheads, was about 47 feet. It was 33 feet in diameter.

A thrust structure supported the stage and the five J-2 rocket engines. The mounting arrangement of the engines was similar to that of the S-1C first stage with the outer engines gimballed and the center engine fixed.

S-IVB third stage

The S-IVB third stage was 58.3 feet long and 21.7 feet in diameter. It weighed about 265,000 pounds when loaded with fuel and oxidizer. It was powered by one J-2 rocket engine with 230,000 pounds rated thrust. A thrust structure supported the stage and the single J-2 engine. The engine was gimballed to allow controlling the thrust vector. The engine could be shut down and restarted several times. Typically for Apollo flights, the first burn on the S-IVB stage lasted about 2.8 minutes to establish a low earth orbit, and the second burn lasted about 5.2 minutes to achieve translunar injection.

A photograph of the S-IVB third stage of the Saturn V launch vehicle on display at the Johnson Space Center is shown on the next page (Fig. 5.8).

The S-IV stage was developed and built by the Douglas Aircraft Company in Huntington Beach, California, under contract with NASA issued in June 1960. It was the first of the three stages of Saturn V to be contracted for. The S-IV development effort at Douglas was led by Max Hunter. The MSFC program manager for the S-IV stage was J. McCulloch. The "hands-on" management style of MSFC irked Douglas at times, but it led to a better product. There would be several iterations of the S-IV stage reflecting changing requirements and engine availability during the course of the program.

Fig. 5.8 Engine end of third stage of Saturn V launch vehicle (author's photograph)

The fuel and oxidizer for the engine were contained in a single cryogenic tank assembly with a bulkhead separating the fuel and oxidizer. The tank held 20,150 gallons of liquid oxygen and 69,500 gallons of liquid hydrogen.

The first overall test of an S-IVB stage occurred in August 1965 with static firing of the J-2 engine for 452 seconds. The smaller size of this stage allowed it to be tested at Douglas's facilities in California and transported to the Kennedy Space Center by a special large airplane, the Super Guppy.

Instrument Unit

The instrument unit (IU), located between the third stage and the spacecraft housing, was the control and nerve center of the Saturn V launch vehicle. It controlled the firing and orientation of the rocket engines of the three stages from launch to inserting the spacecraft into earth orbit.

The Instrument Unit was a cylindrical assembly 36 inches high and 21.7 feet in diameter. It was a sturdy assembly weighing 4,500 pounds and strong enough to rigidly support the weight of the spacecraft located above it. All of the components in the IU were mounted to the inside of the cylindrical wall. A photograph of the Instrument Unit during fabrication is shown on the next page (Fig. 5.9).

The Instrument Unit was designed by engineers at the Marshall Space Flight Center. The Federal Systems Division of IBM was awarded a contract in April 1964 to fabricate, assemble, and test the Instrument Units. Those activities took place at IBM's Space System Center in Huntsville, Alabama. IBM was

Fig. 5.9 Instrument Unit during fabrication (NASA photograph)

responsible for integration of the units into the Saturn V, and they developed the software for the digital computer. The MSFC program manager for the Instrument Unit was Friedrich Duerr.

Major components within the Instrument Unit were an inertial platform, launch vehicle digital computer (LVDC), analog flight control computer, rate gyros, communication system, and batteries.

Power for the Instrumentation Unit was obtained from ground power sources until just before launch when transfer to internal battery power was made and the umbilical that supplied external power was retracted. Battery power was stored in four silver-zinc batteries, each with a capacity of 350 ampere hours at a nominal voltage of 28 volts. The total energy stored in the four batteries was 39.2 kilowatt-hours.

The inertial platform was run up and aligned in the vehicle before launch. It measured vehicle attitude and acceleration in three axes from liftoff to translunar insertion. The measurements were used by the digital computer to determine spacecraft velocity and position, in other words, to determine the state vector. The computer compared the state vector with preprogrammed values as a function of time and generated attitude correction signals. The correction signals were used by the analog flight control computer along with inputs from rate gyros to adjust the orientation of the gimballed engines in the three stages to cause the vehicle to follow a preprogrammed trajectory.

Other important components in the Instrumentation Unit were two C-band transponders and associated antennas that operated in conjunction with ground-based radar systems to allow accurate tracking of vehicle from the ground. A transponder was also included for an independent tracking system called AZUSA. The very accurate tracking data developed on the ground was sent up by a radio command link to update the guidance system.

Telemetry of critical parameters of the launch vehicle to the ground was an important function of the Instrument Unit. Several hundred measurements made throughout the launch vehicle were organized and transmitted to ground stations.

THE F-1 ENGINE

The impressive performance of the F-1 rocket engine was key to the success of the Apollo program. The importance of the engine, which was extraordinary in size and performance, cannot be overemphasized. It remains the most powerful single-chamber liquid-fueled rocket engine ever developed. A photograph of the engine is shown on the next page (Fig. 5.10).

NASA lists some pertinent characteristics of the engine as follows:

Length	19 feet
Width	12.3 feet
Nozzle exit diameter	11.6 feet
Weight	18,500 pounds
Thrust at sea level	1.52 million pounds
Flow rate, oxidizer	24,811 gallons per minute
Flow rate, fuel	15,471 gallons per minute
Mixture ratio	2.27:1 oxidizer to fuel
Combustion temperature	5,970 °F

The original requirements called for thrust of 1.5 million ±3% pounds, but that requirement was changed later to 1.522 million ±1.5% pounds. This is likely the reason that one sees both 1.5 million and 1.52 million values of thrust listed for F-1 engines.

A photograph of F-1 engines mounted in the S-1C first stage of the Saturn V is shown on the following page. Wernher von Braun is standing in the foreground. The fifth engine is out of the picture to the upper right (Fig. 5.11).

Development of the F-1 Engine

NASA issued a request for proposal to seven companies in October 1958 for a single-chamber engine capable of developing 1.0 to 1.5 million pounds of thrust. The Rocketdyne division of North American Aviation won the competition, likely

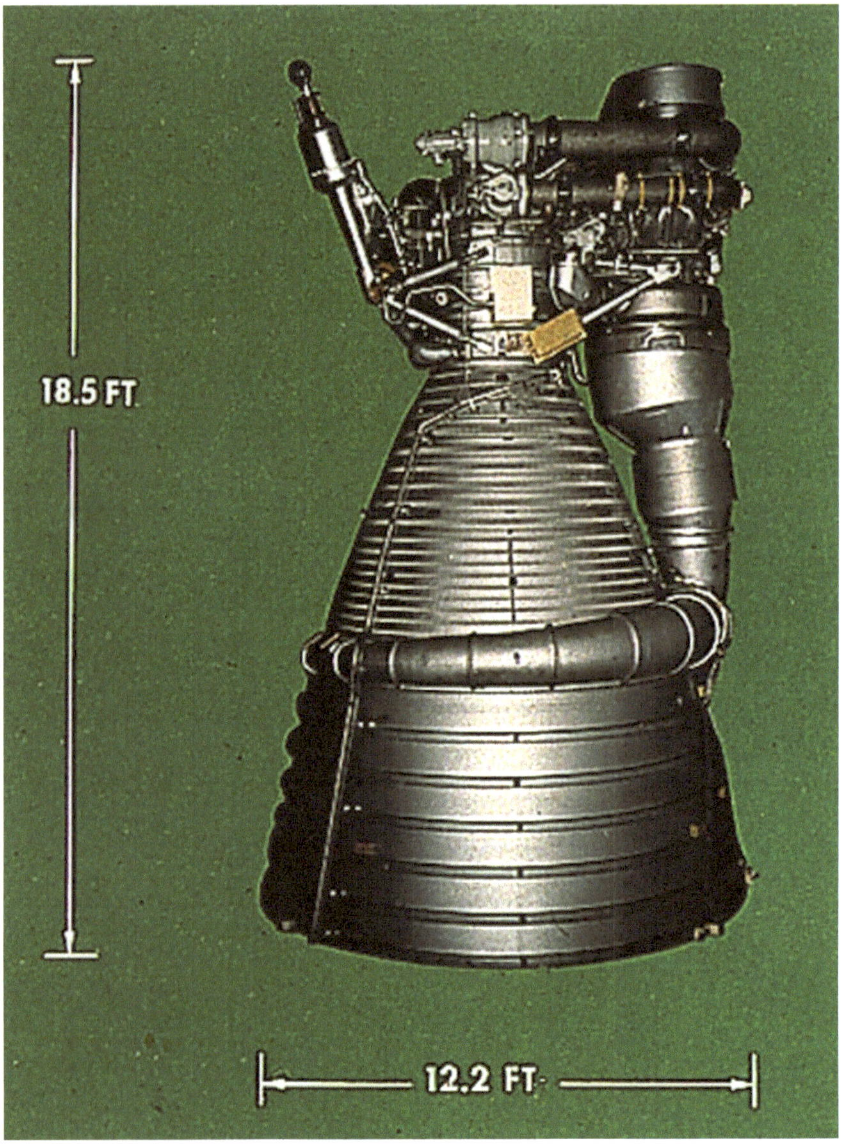

Fig. 5.10 Photograph of an F-1 engine (NASA photograph)

by virtue of their work for the Air Force on a similar engine. A contract was signed with Rocketdyne in January 1959 to develop an engine with 1.5 million pounds of thrust.

As was common with early rocket engines, combustion instability was a major problem in the development of the F-1 engine. The instability could get so severe that without an automatic engine shutdown, the engine would destroy itself. A

Fig. 5.11 F-1 engines mounted in Saturn V first stage. Wernher von Braun stands in the foreground (NASA photograph)

great deal of work and study went into the problem at both MSFC and Rocketdyne. The design of the injector assembly, which sprays fuel and oxidizer into the combustion chamber, was critical to combustion stability. Various forms of the injector assembly were analyzed and tested during the years 1961 to 1963. It was not until June 1964 that a satisfactory design evolved.

Full-thrust testing of the F-1 engine could not be conducted at Rocketdyne's Santa Susana rocket engine test site because of excessive noise at nearby housing sites. A new test facility with three static test stands for F-1 engines was constructed at the Edwards Field Laboratory in the Mojave Desert of California.

The first test of an F-1 engine at a thrust of 1.5 million pounds and for the expected flight duration of 2.5 minutes was conducted at the Edwards site in May 1962. Qualification testing of the engine was completed in November 1965. Once the engine went into production, each engine was trucked to Edwards and subjected to a 45 second calibration test and a 165 second mission duration test at full power before being delivered to NASA.

EARLY TESTING OF SATURN V

The individual stages of all of the Saturn V launch vehicles were test fired in special test stands at power levels and durations required during flight. A particularly massive test stand was used to test the very large S-IC first stage at the Mississippi Test Facility (MTF). The multiengine S-II stage was also tested at MTF. Douglas tested the S-IVB stage at their test facility in California.

After each stage had been test fired, it was refurbished and then moved to Cape Kennedy. The stages were transported to the vertical assembly building where they were assembled into the Saturn V stack.

Multistage rockets had historically been qualified for flight by first flight testing a ballasted first stage and then adding the second and third stages on later test flights. Rather than following this methodical path, the head of NASA's Manned Flight organization, George Mueller, pressed for all-up testing on the first flight. His all-up testing would include all three stages of Saturn V plus an operational Command Service Module (CSM). Not only would Saturn V be tested, but critical elements of the CSM would also be tested including reentry of the Command Module into earth's atmosphere. Mueller's reasoning prevailed, and the first flight of Saturn V with a CSM payload took place in November 1967.

That first flight of Saturn V, designated Apollo 4, was a complete success. The CSM was placed in an earth parking orbit, and then a simulated translunar injection burn of the S-IVB engine was made to place the S-IVB and CSM into an earth orbit trajectory with an apogee of 9,366 nautical miles. The CSM separated from the S-IVB stage, and the service propulsion system (SPS) engine of the CSM was fired to raise the apogee to 9,836 nautical miles. On the way back toward earth, the SPS engine was again fired to increase the spacecraft velocity to that expected on the return from the moon. The command module separated from the Service Module, entered the atmosphere, and splashed down about 9.9 nautical miles from the recovery ship. Impressive performance indeed for a first flight.

Werner Von Braun writes in *Apollo Expeditions to the Moon* that: "In retrospect it is clear that without all-up testing the first manned lunar landing could not have taken place as early as 1969. Before Mueller joined the program, it had been decided that a total of about 20 sets of Apollo spacecraft and Saturn V rockets would be needed. Clearly, at least ten unmanned flights with the huge new rocket would be required before anyone would muster the courage to launch a crew with it."

Von Braun went on to write: "Mueller changed all this and his bold telescoping of the overall plan bore magnificent fruit. With the third Saturn V ever to be launched, Frank Borman's Apollo 8 crew orbited the Moon on Christmas 1968, and the sixth Saturn V carried Neil Armstrong's Apollo 11 to the first lunar landing."

TEST MODELS OF APOLLO LEADING TO FIRST LUNAR LANDING

Apollo 5

The flight of Apollo 5 was designed to test an unmanned but operational model of the Lunar Module in space. The LM was lofted into an 88 by 120 nautical mile earth orbit by a Saturn 1B launch vehicle. A nose cone had been placed on the LM to protect it during launch.

The purpose of the flight was to verify operation of functions of the LM including firing and control of the descent and ascent rocket engines. The descent engine was fired and stopped multiple times, the last time being a simulated abort during landing where the descent and ascent engines were fired together. The ascent engine was fired twice. The LM performed well during the voyage, and the flight test was considered a success.

Apollo 6

The second launch of Saturn V carried a Command and Service Module and a mass mockup of the Lunar Module into earth orbit. The purpose of the flight was to qualify the Saturn V and the Command and Service Module for manned flight. The launch was successful, and the S-IVB third stage and the CSM were placed in a 94 nautical mile earth orbit.

However, unlike the first Saturn V flight, a few problems were encountered. First, just over 2 minutes after launch, a pogo oscillation occurred along the longitudinal axis of the vehicle. Second, several panels were lost from the Lunar Module adapter. Third, two of the five engines of the second stage shut down prematurely. Fourth, the S-IVB third stage could not be fired the second time.

Despite the problems, a successful entry into the atmosphere at near lunar return velocity was made by the Command Module. The Command Module splashed down 43 nautical miles from the intended splashdown point, but it was successfully recovered.

Post flight investigation by engineers at the Marshall Space Flight Center found the likely causes of all of the problems, and corrective measures were taken. Engineers and management at the Marshall Space Flight Center were satisfied that they understood and had remedies for the problems encountered in the flight of Apollo 6. Saturn V was certified for manned flight.

Apollo 7

Apollo 7 was the first manned Apollo mission. The purpose of the flight was to demonstrate performance of the CSM and fight crew and to qualify the CSM for manned space flight. Also to be demonstrated was the rendezvous capability of the CSM and the effectiveness of mission support on the ground. The duration of the mission was 10 days.

The crew consisted of Walter Schirra, commander; Donn Disele, CSM pilot; and Walter Cunningham, Lunar Module pilot. The titles for the crew reflected their functions during later lunar landing missions. Each crew member was assigned specific tasks in the missions leading up to the landings.

Apollo 7 was launched on 11 October 1968. Saturn 1B flawlessly inserted the CSM into a 152 by 123 nautical mile orbit around the earth. Manual control of attitude of the S-IVB stage was demonstrated before the S-IVB was separated from the CSM.

A rendezvous operation with the S-IVB was performed the next day when the S-1VB was about 80 nautical miles from the CSM. The rendezvous would simulate the condition of rendezvousing with a disabled LM. The S-IVB was tracked with the sextant in the Command Module during the rendezvous. The reaction control system thrusters and the SPS engine of the Service Module were used in steps to achieve rendezvous and station keeping with the S-IVB. The CSM closed to about 70 feet from the tumbling S-IVB and then flew around it.

A number of tests were made of CSM systems and of the thermal control system of the spacecraft. The effectiveness of a continuous slow roll of a spacecraft to maintain desired interior temperatures during the long transit to the moon was demonstrated. The SPS engine was fired eight times.

The crew performed well during the 10-day mission even though the commander came down with a head cold the second day and soon the other crew members caught it. It must have been miserable indeed to have a head cold in space.

The SPS engine was fired for final time during the 163rd orbit to set up the Command Module for reentry into the earth's atmosphere. The Command Module separated from the Service Module and made a near perfect entry into the atmosphere. The Command Module splashed down in the Atlantic Ocean 1.9 nautical miles from the target point. The mission had lasted 10 days and 20 hours.

The flight of Apollo 7 verified that the CSM was qualified for manned operation in an earth orbit environment.

Apollo 8

Apollo 8 was the first mission to send humans on a spacecraft to orbit the moon. The spacecraft for this pioneering flight was the Command Service Module, and the crew consisted of Frank Borman, commander; James Lovell, Command Module pilot; and William Anders, Lunar Module pilot. The CSM orbited the moon ten times and then returned to earth. The flight verified many of the actions and procedures necessary to place a man on the moon.

Apollo 8 was launched on 21 December 1968 by the Saturn V. The S-IVB third stage with the CSM attached was inserted into earth orbit. The S-IVB engine was fired again during the second orbit for translunar injection. The S-IVB stage was separated from the CSM after it had established coast to the moon. Two midcourse corrections were made by the SPS engine of the Service Module during the translunar flight. The spacecraft swung around the backside of the moon, and the engine was fired again for the lunar orbit insertion burn. The resulting initial orbit was 60 by 168 nautical miles, but this was circularized to about 61 nautical miles by another engine burn.

The astronauts would spend about 20 hours orbiting the moon. They described verbally the lunar surface that they passed over, and they took over 700 pictures of the moon. One of the pictures taken of earthrise during their fourth orbit became a classic. It is reproduced on the next page (Fig. 5.12).

An important goal of the mission was visual assessment of potential landing sites for later missions. Mare Tranquillitatis, the eventual landing site of Apollo 11, was examined closely and a camera was arranged to take a picture of the terrain every second as the spacecraft passed over the site.

Six television sessions were conducted during the 6-day mission. The telecasts received worldwide coverage. One of the telecasts was conducted on Christmas Eve and included reading of passages from the Book of Genesis in the *Bible* by the crew. The closing statement was made by Frank Borman: "Good night, good luck, a Merry Christmas and God bless all of you – all of you on the good Earth."

After completion of the scheduled ten orbits of the moon, the trans-earth injection burn was made by the SPS engine while on the backside of the moon. The burn placed the CSM on a near perfect trajectory to return to the earth. The Command Module was separated from the Service Module when close to the earth, and the Command Module was oriented for entry into the atmosphere. The Command Module splashed down on parachutes 2.5 nautical miles from the recovery ship. The mission had lasted 6 days.

Fig. 5.12 Earthrise taken from orbit around the moon by Apollo 8 crew (NASA photograph)

The flight of Apollo 8 was considered a success with all mission objectives and test objectives achieved.

Apollo 9

Apollo 9 continued the careful step-by-step testing to prove systems and procedures for the ultimate lunar landing. The primary purpose of the flight was to conduct tests of the Lunar Module in earth orbit by the crew. The crew members were James McDivitt, commander; David Scott, Command Module pilot; and Russell Schweickart, Lunar Module pilot.

Apollo 9 was launched by Saturn V on 3 March 1969. The S-IVB third stage, CSM, and LM were placed in a 117 by 119 nautical mile earth orbit. The CSM was separated from the LM adapter and flown out, turned around in space, and flown back to dock with the LM. The S-IVB was later jettisoned from the CSM/LM stack.

On the third day McDivitt and Schweickart entered the LM wearing spacesuits and conducted checkout of LM systems. As part of this checkout, the descent engine of the LM was fired, and the thrust level of the engine was varied throughout its range including maximum thrust. The astronauts returned to the CSM after checkout of the LM.

On the fourth day, an EVA (extravehicular activity) was scheduled that would have Schweickart travel out in space between the open side hatch of the LM and the open side hatch of the CSM. Unfortunately, Schweickart was somewhat nauseous at the time so the planned intravehicular EVA was canceled. He stood on the porch of the LM for 37 minutes while suited up in the EVA mobility unit and wearing the portable life support backpack to verify operation of that equipment in space.

On the fifth day, McDivitt and Schweickart again entered the LM and prepared for the first manned free flight of the LM. After checkout of systems, the LM was separated from the CSM. The astronauts in the LM followed a rather extensive test plan to move through the several major phases required in an actual rendezvous when returning from the moon. The Apollo 9 Mission Report shows that the largest separation of the LM from the CSM was 98 miles. The descent stage was separated from the ascent stage, and the ascent stage engine was fired to begin the rendezvous process.

Rendezvous with the CSM was carried out autonomously using range and range rate data from the Rendezvous Radar, knowledge of the state vector, programs in the LM guidance computer, and attitude and motion information from the abort guidance system. Attitude and motion information was also available from the primary guidance system, but it had been decided to use the abort guidance system for this test series.

Rendezvous and docking were successful, and the astronauts were back aboard the CSM after flying the LM for about 6 hours. They had successfully put the Lunar Module through its paces, including rendezvous with the CSM. The rest of the flight in orbit around the earth was uneventful, and the command module entered the atmosphere on 13 March 1969 and splashed down within 3 miles of the recovery ship. The mission had lasted for 10 days.

The flight was considered a success, and the LM was deemed ready for manned flight.

Apollo 10

Apollo 10 was a full-fledged dress rehearsal for the planned lunar landing in Apollo 11. As such, the manned Lunar Module of Apollo 10 descended to about 47,400 feet above the lunar surface before pulling up and rendezvousing with the CSM orbiting about 70 miles above the surface. All elements of the full Apollo mission were exercised except for the actual landing. The crew members were Tom Stafford, commander; John Young, Command Module pilot; and Gene Cernan, Lunar Module pilot.

Apollo 10 was launched on 18 May 1969 by Saturn V. The joined S-IVB third stage, CSM, and LM were put into a 110 nautical mile earth parking orbit. A translunar injection burn was made by the S-IVB, and the long coast to the moon began. The CSM separated from the LM adapter and was turned around and flown back to dock with the LM. The S-IVB was later jettisoned from the CSM/LM stack.

The CSM/LM entered an orbit around the moon. The orbit was circularized at 61 by 59 nautical miles. The next day, Stafford and Cernan entered the LM and prepared it for flight. The LM was undocked from the CSM, and it flew for a time station keeping with the CSM while all LM systems were checked out and visual verification, including landing gear deployment, was made from the CSM.

The descent engine was fired to place the LM in an orbit of 60 by 8.5 nautical miles with the low point of the orbit in the vicinity of the planned landing area for Apollo 11. During the pass over the landing area, visual assessment of the site was made, a stereoscopic series of photographs were taken, and the LM landing radar locked on the surface and provided range and velocity data. The lowest measured altitude was 47,400 feet.

Following the low pass, the descent engine was fired to raise the orbit to 190 by 12 nautical miles. Later, the descent stage was jettisoned, and the ascent stage engine and thrusters were used to rendezvous with the orbiting CSM. The rendezvous and docking were successful, and the astronauts entered the CSM. They had flown the LM for 8 hours and 10 minutes.

The trans-earth injection burn was successful, and the coast back to earth was uneventful. The Command Module splashed down 1.3 nautical miles from the target point. The flight had lasted 8 days.

The flight was considered a success. The performance of both the CSM and the LM were satisfactory except with a problem with the steerable LM communications antenna. All mission objectives were accomplished. The mission provided final evaluation necessary to proceed with a lunar landing.

READYING THE SATURN V AND APOLLO PAYLOAD FOR LAUNCH

Events leading to the launch of Apollo 11 are a good example of the steps and procedures used to prepare the Saturn V and its payload for launch.

The Saturn V launch vehicle with the Apollo 11 CSM and LM on top was assembled stage by stage within the massive vertical assembly building at the Kennedy Space Center. That building, which is still the largest single-story building in the world, is 525 feet high, 716 feet long, and 518 feet wide. It had sufficient volume to assemble four Saturn V launch vehicles and spacecraft at the same time.

Assembly was performed with the launch vehicle resting on a mobile launcher so that the mobile launcher with the Saturn V and the lunar spacecraft standing vertically could be moved from the vertical assembly building to the launch pad.

The base of the mobile launcher was 160 feet long and 135 feet wide. It was made up of two floors with a total height of 25 feet. Heavy bracing between floors made a very strong structure to support the Saturn V, Apollo spacecraft, and a gantry. The launcher had an opening 45 feet square directly under the Saturn first stage to let exhaust flame pass through and down into the flame deflector at the launch pad.

A gantry as tall as the spacecraft was mounted on the mobile launcher to stabilize the spacecraft and provide access to the Command Module. The mobile launcher was supported on 6 legs 22 feet long so that a crawler-transporter could move under it. The crawler-transporter raised a platform to lift the mobile launcher and then ponderously moved the launcher and spacecraft to the launch pad 3.5 miles away. A photograph of Apollo 11 being moved out of the vertical assembly building on 20 May 1969 is shown next page (Fig. 5.13).

The crawler-transporter was 131 feet long, 114 feet wide, and about 20 feet high. It weighed 6,000,000 pounds. Each corner was supported by twin tracked drives. The eight tracked drives each contained 57 shoes that were 7.5 feet long, 1.5 feet wide, and weighed 2,100 pounds each. The crawler-transporters were built by the Marion Power Shovel Company. Marion steam shovels had been instrumental in digging the Panama Canal. The immense size of the vehicle can be judged from the size of people in the photograph of the crawler-transporter shown on next page (Fig. 5.14).

The platform at the top of the crawler-transporter could be raised from 20 feet above ground level to 26 feet. In the retracted position, the platform would just fit under the mobile launcher. The crawler-transporter was driven into the vertical assembly building and under the mobile launcher. The platform was raised to load the mobile launcher and Saturn V with the Apollo spacecraft onto the crawler. The loaded crawler-transporter was then driven out of the building to the launch pad. The platform was stabilized to keep the top of the platform level within 0.17 degrees even when climbing up a 5-degree slope to the top of the launch pad.

A photograph of Apollo 11 and the mobile launcher being transported to the launch the following page (Fig. 5.15). After depositing the mobile launcher in a precise position at the launch pad, the crawler-transporter withdrew. It was used to move the mobile launcher back to the vicinity of the vertical assembly building after the launch. The crawler-transporter traveled about 1 mile per hour when loaded with the Saturn V and Apollo. Unloaded, it traveled at the breakneck speed of 2 miles per hour.

Fig. 5.13 Apollo 11 exiting the vertical assembly building en route to the launch pad (NASA photograph)

Fig. 5.14 Crawler/transporter unloaded (NASA photograph)

Fig. 5.15 Apollo 11 being transported to the launch pad (NASA photograph)

The author was at the Cape for the launch of Apollo 15. Aside from the launch, the thing that I most vividly remember was examining one of the two crawler-transporters during the visit.

The transporter was used for the Space Shuttle as well as Apollo. It has been recently refurbished with new engines to transport NASA's latest heavy-lift launch vehicle and the Orion manned space capsule.

SUMMARY OF THE FLIGHT OF APOLLO 11

A summary of events and maneuvers that carried the Apollo 11 astronauts away from the earth, gently deposited them on the moon, and returned them safely to earth is recounted in this section. The Apollo 11 mission is described at this point of the book so that readers may better appreciate the various subsystems aboard the Command Service Module and the Lunar Module when they are described in later chapters. The Apollo 11 astronauts were Neil Armstrong, commander; Mike Collins, Command Module pilot; and Edwin (Buzz) Aldrin, Lunar Module pilot.

Launch and Insertion into Earth Orbit

The astronauts entered the Command Module just before 7:00 am on 16 July. To get to the Command Module, they rode an elevator in the gantry of the launcher up to the 320-foot level. They then walked over swing arm Number 9 to the white room that was positioned next to the hatch of the Command Module as shown in the photograph on the next page (Fig. 5.16).

The Pad Leader in the white room for several Apollo launches, including Apollo 11, was Guenter Wendt. Wendt was a strict disciplinarian, esteemed by the astronauts.

After entering the Command Module, the astronauts and the ground crew followed the comprehensive *Apollo 11 Flight Plan* as well as detailed checklists. The checklists started with 128 switch position settings on the various instrument panels prior to liftoff and continued on with switch positions and readings to be taken throughout the mission. The final flight plan, prepared on 1 July 1969, specified the nominal launch time at 9:32 am Eastern Standard Time on 16 July 1969. Adhering closely to the schedule, the actual launch took place at exactly 9:32 am on 16 July. The flight plan gave a launch window duration of 4 hours and 24 minutes that day.

The hatch to the Command Module was closed at about 7:25 am. The countdown continued to go smoothly. The astronauts were busily engaged running through checklists on switch positions and display and indicator states in the time before launch. They were also involved in detailed check of the Emergency Detection System and check of alignment of the guidance system.

The countdown reached minus 4 seconds, and the powerful engines of the S-1C first stage ignited in sequence. Thrust built up, and once all five engines were up to full thrust, the hold-downs restraining the Saturn V were released, and Apollo 11 lifted off the mobile launcher. Liftoff occurred at 9:32 am on 16 July 1969. A photograph of the launch of Apollo 11 is shown following page (Fig. 5.17).

At 1.7 seconds after launch, a yaw maneuver of 1.25 degrees was initiated to tilt the Saturn V away from the gantry. The vehicle was returned to the vertical 9.7

Fig. 5.16 White room positioned against side hatch of Apollo 11 Command Module (NASA photograph)

seconds after launch. At 13.2 seconds after launch, a roll maneuver was initiated to roll the spacecraft such that the Y-axis moved from 90 degrees east of north toward 72 degrees east of north. The purpose of this maneuver was to achieve the desired flight path azimuth angle over the earth of 72 degrees by pitching the vehicle. A pitch maneuver was initiated at the same time as the roll maneuver, and the vehicle was slowly tilted from vertical flight to a path that would take it out over the ocean.

Fig. 5.17 Apollo 11 leaving the earth (NASA photograph cropped by author)

The spacecraft reached a velocity of Mach 1 and an altitude of 25,736 feet about 66 seconds after launch. Maximum dynamic pressure (Q) on the spacecraft of 735 pounds per square foot occurred at 83 seconds after launch at an altitude of 44,512 feet. Dynamic pressure was a concern since bending of the spacecraft could lead to disaster when the dynamic pressure was high.

Time in this account of the mission of Apollo 11 will be given as time from launch expressed as hours:minutes:seconds. Thus, 000:02:44 indicates 2 minutes and 44 seconds after launch. All reference to miles in this chapter will be nautical miles (6,076.1 feet). The velocity readings are earth-fixed values. After the spacecraft had been set on a translunar trajectory, the velocity reference was changed from earth-fixed values to space fixed values.

The center S-1C engine was cut off 2 minutes and 15 seconds after launch (000:02:15), and the outboard engines were cutoff at 000:02:42. The acceleration had reached 3.94 g at the time of cutoff of the outboard engines. The velocity at this time had reached 7,852 feet per second, the altitude was 35.7 nautical miles, and the spacecraft was 50.5 nautical miles downrange.

Separation of the S-1C stage was commanded at 000:02:42, and ignition of the S-II stage engines occurred at 000:02:44.

The escape tower was jettisoned at 000:03:18 after reaching an altitude and velocity where if abort were required the CSM could be separated from the launch vehicle and use the Service Module engine and reaction control system engines to maneuver. The Command Module would be separated from the Service Module and reenter the atmosphere and splash down as in a normal landing.

The center engine of the S-II stage was cut off at 000:07:41, and the outboard engines were cut off 000:09:08. At cutoff of the outboard engines, the velocity was 21,368 feet per second, the altitude was 101.1 nautical miles, and the space-craft was 874 miles downrange. The S-II stage was separated from the spacecraft, and the engine of the S-IVB third stage was ignited at 000:09:12. The engine was cut off at 000:11:49 after inserting the spacecraft into earth orbit. The velocity of the spacecraft was 24,244 feet per second at that time, the altitude was 103.2 miles, and the spacecraft was 1,461 miles downrange.

The orbit had apogee of 100.4 miles and perigee of 98.9 miles. After about one and a half orbits of earth, the engine of the S-IVB stage was fired again to increase the velocity to that required for translunar injection. The engine was cut off at 002:50:03 after firing for 5 minutes and 47 seconds. This burn increased the veloc-ity to 34,230 feet per second and placed the spacecraft into a translunar injection orbit. The translunar orbit was a wide reaching orbit that would take the spacecraft on a loop around the moon before returning to the vicinity of earth.

Translunar Orbit Cruise to the Moon

After entering the translunar orbit, the next step was to separate the Command Service Module from the rest of the stack and maneuver it out and around in space and come back and dock the conical end of the Command Module with the Lunar Module. Mike Collins, Command Module pilot, was at the controls of the reaction motors in the CSM for those maneuvers.

Separation of the Command Service Module from the stack occurred at about 3 hours and 17 minutes after launch. Collins then moved the CSM away from the rest of the spacecraft with the reaction control motors on the Service Module and turned the CSM around in space so that the probe on the CM Module faced the

Lunar Module. He then carefully maneuvered to dock the probe of the CM with the drogue of the LM. After docking, the LM was securely latched to the CM. Docking took place at about 3 hours and 24 minutes after launch. Collins later said that he had moved the CSM about 100 feet away from the stack before coming back in.

The Lunar Module was separated from the Instrument Unit and S-IVB stage at 4 hours and 17 minutes after launch. The joined CSM and LM continued toward the moon.

The astronauts took several star sights with the sextant to update alignment of the inertial platform. The spacecraft was then put into a passive thermal control mode during which the spacecraft attitude was set broadside to the sun, and then it was rotated about the X-axis at a rate of about three revolutions per hour to even out temperatures in the spacecraft.

A sleep period for the crew began about 14 hours after launch. This would have been about 11:30 pm Eastern Standard Time (EST). The crew was given a wake-up call next morning at about 8:22 am EST on 17 July.

The most significant event during day 2 was a midcourse correction. The correction was made at 026:44:58. It consisted of a burn of the service propulsion subsystem (SPS) engine in the Service Module for 3.13 seconds under control of the Guidance and Navigation System computer. A velocity change of 21.3 feet per second was planned and that was achieved. The velocity after the burn was 5,010 feet per second.

Color television from a hand-held TV camera was transmitted from the spacecraft back to earth later in the day. The coverage showed the earth from a distance of about 128,000 miles as well as views inside the Command Module.

The astronauts began their sleep period at about 9:42 pm EST on day 2. They received a wake-up call about 9:41 am EST on 18 July by Mission Control who had given them an extra hour of sleep since there was nothing pressing.

The most important event of the day of 18 July for the astronauts was performing a check of conditions within the LM. The first step in the operation was to pressurize the LM which had been unpressurized until then. This was done by opening a valve that let Command Module atmosphere enter the LM. Additional oxygen was ported into the Command Module to make up the difference. The hatch to the LM was opened at 55 hours and 36 minutes after launch and after the pressures in the LM and CM had equalized.

Aldrin went into the LM first and took the TV camera with him. Television images of the inside of the LM and views back into the Command Module were transmitted back to earth. Armstrong joined him a few minutes later, and the two of them began arranging and stowing things so that the LM would be ready for their scheduled detailed LM systems checkout the next day.

After finishing their work in the LM, the crew returned back to the command module. The drogue and probe were reinstalled, and the hatches were closed. The crew began their scheduled sleep period about 61 hours after launch corresponding to 10:30 pm EST on day 3.

At about 61 hours and 40 minutes after launch, the spacecraft entered into the sphere of influence of the moon's gravitational field, and a computational change-over was made at Mission Control to shift from earth referenced data to moon referenced data. The spacecraft was 186,487 miles from earth and 33,822 miles from the moon at that time.

The astronauts were allowed to sleep until 8:32 am EST on 19 July since a second midcourse correction planned for earlier in the morning was not needed. The two big events scheduled for 19 July were insertion into lunar orbit and powering up and checking out systems in the Lunar Module.

Spacecraft in Lunar Orbit

The spacecraft passed behind the limb of the moon, and communication with earth ceased at 75 hours and 41 minutes after launch. At 075:49:51 (1:12 pm EST), while still behind the moon, the SPS engine ignited to slow the spacecraft to place it in lunar orbit. The motor burned for 5 minutes and 58 seconds and slowed the velocity by 2,918 feet per second. The velocity after the burn was 5,479 feet per second. The resulting orbit about the moon had an apolune of 169.7 miles and a perilune of 60 miles.

The spacecraft was placed in a clockwise orbit around the moon so that the astronauts in the LM would have the sun at their back when landing. During the third orbit and while on the backside of the moon, a second burn of the SPS engine was made at 080:11:37 (5:43 pm EST) to circularize the orbit. The burn lasted 16.9 seconds and resulted in a velocity change of 159 feet per second. The velocity after the burn was 5,338 feet per second. The apolune of the new orbit was 66.1 miles, and the perilune was 54.5 miles. The orbit was deliberately set up with some eccentricity since analysis indicated that the orbit would gradually become circular due to the varying gravitational field of the moon during orbit.

Armstrong and Aldrin entered the LM about 081:25 (6:57 pm EST) to power-up and perform checkout of the LM systems. The checkout included data in all modes of the communication system. LM checkout was completed, and the LM was powered down at about 83:30 (9:02 pm EST). The astronauts prepared for their sleep period at about 12:05 am EST on day 4.

Descent and Landing on the Moon

The astronauts received a wake-up call at 7:04 am EST on 20 July. This was to be the big day, featuring separating the LM from the CSM, performing a descent orbit insertion burn with the LM, and then performing a powered descent to a landing on the moon. The landing was planned to take place in the morning at their landing site so that the temperature would not be excessive and the sun would be low enough to create good shadows to aid their perspective of the landing area.

Aldrin entered the LM at about 95:51 (9:23 am EST) and began power-up and checkout. Armstrong put on his pressure suit and had joined Aldrin in the LM by 10:22 am EST to continue the checkout. Aldrin went back to the Command Module at about 10:30 am EST to put on his pressure suit and then rejoined Armstrong in the LM. Continuing the checkout, the landing legs of the LM were deployed at 11:46 am EST.

All systems in the LM checked out good. The Lunar Module was separated from the CSM at 100:12 (1:44 pm EST), while Apollo 11 was on the backside of the moon. A photograph taken by Collins of the LM shortly after separation is shown below (Fig. 5.18).

Fig. 5.18 LM after separation from CSM (NASA photograph cropped by author to remove stray light streaks)

All systems in the Lunar Module continued to look good, and the descent engine was ignited again at 102:33:05 (4:05 pm EST) for the Powered Descent Initiation (PDI) burn. The burn began at 10% of full thrust and then under computer control advanced to full thrust for a time selected to allow the LM to reach the landing zone. Total burn time was 12 minutes and 36 seconds.

Powered Descent was divided up into three phases, each conducted by a different computer program (P_). The phases were braking phase (P63), approach phase (P64), and landing phase (P65). An additional program (P66) could be selected that would have spacecraft stability be controlled by the computer, but it allowed manual inputs to direct the flight path.

The LM was oriented with engine forward and windows facing the surface of the moon at the beginning of the burn. A roll of the spacecraft was initiated at 102:36:46 (4:08 pm EST) while still in the braking phase to change the orientation to windows up. The astronauts followed progress of the flight closely by selecting a display in the cockpit that showed range to the landing site, time remaining in the braking phase, and velocity of the spacecraft.

Aldrin reported lock-on of the Landing Radar at 102:38:04. The radar indicated that the altitude to the surface was about 2,800 feet less than computed by the navigation and guidance system. Radar data plots indicate slant range of 44,000 feet just after lock-on. The computer was updated with radar data, and by 102:39:24 the computer estimated altitude and radar altitude agreed.

An alarm with a code of 1202 went off at about 102:39 and continued to go off frequently for the rest of the descent. Mission control deemed this alarm not cause for abort, but it was disturbing to the crew. The alarm was caused by overload of the mission computer, which caused the computer to reinitialize and resume operation on highest priority tasks.

The descent engine was throttled down by the computer at 102:39:31, and the approach phase (computer program P64) began at 102:41:32. At 102:41:35 the computer initiated pitch of the spacecraft from 55 degrees toward a more upright position. At 102:41:51 Armstrong reported an altitude of 5,000 feet and altitude rate of 100 feet per second. At 102:41:55 he switched to the attitude hold mode to check handling qualities under manual control. He was satisfied with the handling and switched back to computer program control at 102:42:03.

At 102:42:37 Armstrong reported altitude of 1000 feet and altitude rate of 30 feet per second. He asked Aldrin for a LPD angle. Aldrin responded with 35 degrees. The LPD (Landing Point Designator) was a two-digit number generated by the computer used to determine the landing point. There was a scribed graduated line marked in degrees on the window of the Lunar Module. Sighting past the scribed line at the angle generated by the computer showed the predicted landing area. A sketch of the window with the scribed scale is shown on the next page (Fig. 5.19).

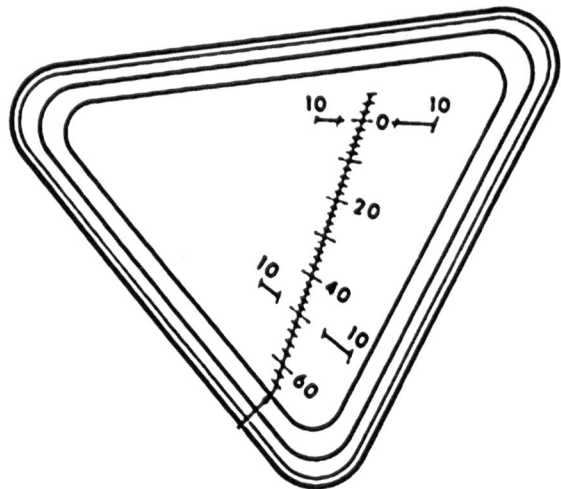

Fig. 5.19 Scribed scale on window of LM to match up with the angle of the LPD (NASA graphic)

Armstrong did not like the landing area that the computer was taking them to since it was dominated by a crater with boulders along its sides. He took over manual control of the LM at 102:43:15 by switching the PGNS to Attitude Hold and pressing the Descent Rate switch. The descent rate switch allowed incrementing the descent rate by 1 foot per second each time it was flicked. The first time it was flicked it changed the computer program to P66. Under manual control he changed the pitch angle to be closer to the vertical to decrease the descent rate and then flew the spacecraft over the crater. At 102:43:26 Aldrin reported altitude of 400 feet, altitude rate of 9 feet per second, and horizontal rate of 58 feet per second. At 102:44:02 Armstrong announced "looks like a good area here," and he pitched the spacecraft back a little to reduce the forward velocity.

Aldrin continued updating Armstrong with flight information from the computer display during the final descent. That information was given in terse form as one pilot to another. An expansion of his callouts is given below.

Paraphrased communication from Aldrin to Armstrong during landing:

102:44:07 – 250 feet altitude, 2.5 feet per second down, and 19 feet per second forward
102:44:31 – 160 feet altitude, 6.5 feet per second down
102:44:45 – 100 feet altitude, 3.5 feet per second down, 9 feet per second forward, and 5% fuel remaining.
102:45:08 – 60 feet altitude, 2.5 feet per second down, 2 feet per second forward

102:44:45 – 20 feet altitude, 0.5 feet per second down, 4 feet per second forward, and drifting to the right a little
102:45:40 – Contact light

The contact signal came from thin probes 67 inches long that hung down from the landing pads to give an alert prior to touchdown. NASA gives the landing time as 102:45:39.5 and engine cutoff by Armstrong at 102:45:41.4 (4:17 pm EST).

After a few shutdown tasks were performed, Armstrong reported his now famous words to Mission Control at 102:45:58: "Houston, Tranquility Base here. The Eagle has landed." At 4:18 pm EST Aldrin, no slouch of a pilot himself, commented: "Very smooth touchdown." Data telemetered to earth indicated that the velocity components at touchdown were 1.7 feet per second vertical, negligible velocity forward, and 2.1 feet per second to the left. For comparison, a normal walking pace is 4.4 feet per second.

The crew immediately began preparing for a contingent immediate takeoff should something be amiss in the LM or Mission Control should find a serious problem from the data being telemetered down. Mission Control verified that the LM was in good shape by 4:19 pm EST and informed the crew that they were good for stay until a prearranged time, T1.

Adhering to the checklist, the crew began a simulated countdown to launch. This procedure was set up as a practice for the actual launch from the moon but also to have things ready in case they needed to leave in a hurry. The countdown was a lengthy process and included aligning the guidance system. Alignment was done both with star sightings and using the gravity and motion parameters of the moon.

The simulated countdown was completed at 104:35:31 (6:07 pm EST), and the LM was partially powered down. The flight plan called for having a meal and then a rest period before going outside to explore the moon. The crew recommended that they skip the rest period. Mission control agreed, and so the crew began preparation for *extravehicular activity* (EVA) at 106:11 (7:43 pm EST).

Exploring the Lunar Surface

Preparation for the EVA was a lengthy process that involved donning the pressure suit, the back mounted Portable Life Support System (PLSS), and chest mounted Remote Control Unit (RCU) and hooking everything up. The last things to install were the helmet and gloves. Finally, everything checked out including the communication system that used a folding antenna that could be positioned vertically from the top of the PLSS. After venting the pressure inside the LM to the outside, the side hatch of the LM was opened at 109:07:33 (10:39:33 pm EST).

By 109:19:16 Armstrong, wearing the PLSS, had managed to squeeze out of the hatch, and he stood on the porch next to the hatch opening. He pulled a lanyard that allowed an assembly containing a camera to swing down to capture his steps down the ladder to the surface. At the bottom step of the ladder, he reported the conditions of the surface and the amount of depression of the landing pads into it. At 109:24:12 (10:56:12 pm EST), he stepped off the LM onto the moon and made his now famous remark: "That's one small step for man; one giant leap for mankind."

Armstrong made a series of observations on the lunar surface, and then Aldrin passed the Hasselbland camera down to him via the Lunar Equipment Conveyor (LEC). The LEC was essentially a long flat strap with a pulley arrangement to transfer equipment between the LM cabin and the surface.

The camera, which was a Hasselblad 500EL modified for space use, used an electric motor to advance the film and cock the shutter. It used 70 mm film that was contained in a detachable magazine holding sufficient film for 160 color pictures or 200 black and white pictures. The camera was mounted on a bracket affixed to the chest area of the space suit and that required an astronaut to turn his body to point the camera.

After taking a series of photographs, Armstrong gathered a contingency sample of lunar soil using a bag designed for that purpose and placed the bag in a special pocket in his pressure suit.

Aldrin exited the LM cabin and came down the ladder about 109:42. Armstrong on the next page (Fig. 5.20) of Aldrin on the footpad just after stepping off the bottom step. The picture shows the appreciable height of the bottom step above the footpad. The last step is at about the level of Aldrin's belt in the photograph.

Armstrong removed the TV camera from the assembly that had held it to capture video of astronauts coming down the ladder, and he changed the lens to better show the activities during their chores around the LM. He unveiled a plaque attached to one of the landing legs and read the inscription and took some video of it with the TV camera. A photograph of the plaque is shown on the following page (Fig. 5.21). The plaque reads:

HERE MEN FROM THE PLANET EARTH
FIRST SET FOOT UPON THE MOON
JULY 1969, A. D.
WE CAME IN PEACE FOR ALL MANKIND

Signatures of the three astronauts and of President Nixon are along the bottom.

The TV camera was then taken to the limit of its cable away from the LM and set up on a tripod to capture video of the activities of the astronauts. The TV video was transmitted to earth so that Mission Control and the whole world could watch the historic adventure.

Fig. 5.20 Aldrin standing on the footpad of the Lunar Module (NASA photograph)

Both Armstrong and Aldrin walked around on the surface and reported their views and observations back to Mission Control. Aldrin reported quite extensively on stability and mobility and concluded that maneuvering on the moon was not a problem. Several photographs were taken of the general area and of rocks of interest.

Mission Control asked Armstrong and Aldrin to stand together in the field of view of the TV camera for an address by President Nixon that began at 110:16:30 (11:48:30 pm EST). The first two sentences of Nixon's address were: "Hello, Neil and Buzz. I'm talking to you by telephone from the Oval Room at the White House, and this certainly has to be the most historic telephone call ever made. I just can't tell you how proud we all are of what you (the next few words were unreadable for the transcript)." Armstrong responded by saying "Thank you Mr. President. It's a great honor and privilege for us to be here representing not only the United States but men of peace of all nations, and with interest and a curiosity and a vision for the future. It's an honor for us to be able to participate here today."

Fig. 5.21 Plaque on landing leg of Apollo 11 (NASA photograph)

An important element of the Apollo program was scientific experiments designed to learn more about the moon. The experiments conducted by Apollo 11 are listed below.

- Soil mechanics investigation
- Solar wind composition experiment
- Early Apollo Scientific Experiment Package (EASEP)

    - Passive seismic experiment
    - Laser ranging reflector
    - Lunar dust detector

After becoming acclimated to moving around on the lunar surface, Aldrin raised up the doors on the side of the LM to the compartment that held the Early Apollo Scientific Experiments Package (EASEP) and other science packages. A picture taken by Armstrong of the LM on the surface of the moon and Aldrin removing the EASEP is shown on the next page (Fig. 5.22).

Fig. 5.22 Aldrin removing EASEP from the Lunar Module (NASA photograph)

The astronauts deployed and leveled the EASEP scientific instruments. Next, a strip of aluminum foil 12 x 55 inches in size was deployed in the full sunlight for the solar wind experiment. After about 77 minutes of exposure, it was retrieved and put in a plastic bag for return to earth.

Core tubes driven in by a hammer were used to obtain samples of the lunar material below the surface. They found it difficult to pound the core tubes down more than about 6 inches below the surface. The core tubes were capped and placed in a "rock box" to take back to earth.

About 20 rocks of different forms were collected and put in the two rock boxes along with lunar soil for packing material. The aluminum rock boxes, formally called Apollo Lunar Sample Return Containers, were 19 × 11.75 × 8 inches in size and had an air-tight seal when closed. A total of 47.7 pounds of lunar material gathered around the Apollo 11 landing site was returned to earth in these boxes.

A photograph taken by Armstrong of Aldrin standing near the deployed EASEP experiment package is shown below (Fig. 5.23).

Fig. 5.23 Buzz Aldrin standing near the deployed experiment package (NASA photograph)

At 111:21:16 (12:53 am EST on 21 July) Mission Control informed the crew that it was time to wrap up activities and close out the EVA. The astronauts both indicated later that they wished they could have stayed on the surface longer because there was much more that they would have liked to do. At 111:25:04 Aldrin started up the ladder of the LM. He remained on the porch to help Armstrong transfer up the two rock boxes and two film magazines for the camera. The camera itself was left on the moon to save weight in the ascent stage.

Armstrong started up the ladder of the LM at 111:37:29. He had been on the lunar surface for 2 hours and 13 minutes. Both astronauts were inside, and the hatch of the LM was closed at 111:39:13 (1:11am EST).

The cabin was repressurized and the astronauts removed their Portable Life Support System (PLSS) and had a meal. At 113:47:40 Mission Control informed the crew that liftoff time would be 124:22:02 (about 10 hours and 34 minutes from then).

A last task before their sleep period was depressurizing the cabin, opening the side hatch, and jettisoning the two PLSS units and other unneeded items to save weight in the ascent stage. Mission Control informed the crew that the seismic equipment that they had placed on the surface recorded the impact of each PLSS when it fell. To which Armstrong responded: "You can't get away with anything anymore, can you?"

They depressurized the cabin and began their sleep period at about 114:53 (4:25 am EST) on 21 July. The crew slept with their helmets and gloves on to avoid the lunar dust that had been brought into the cabin when they came in from their exploration of the lunar surface. A wake-up call was made to the LM crew by Mission Control at 121:40:36 (11:12 am EST) on 21 July.

The astronauts readied the ascent stage to launch from the moon. Aldrin aligned the inertial platform with a procedure that used the moon's gravity vector and one star sight. The abort guidance system was also aligned. The rendezvous radar was then used to get a series of range and range rate readings as CSM Columbia passed overhead. Those readings were used to refine the location of the LM on the surface and to locate the LM in reference to the CSM orbital plane. The crew then proceeded through several lunar surface checklists leading up to launch of the ascent stage.

Ascent from the Lunar Surface and Rendezvous

Explosive bolts were fired to separate the ascent stage from the descent stage, and the ascent engine was ignited at 124:22:00 (1:54 pm EST). The ascent stage rose vertically for about 10 seconds, and after the vertical velocity reached 40 feet per second, a pitchover maneuver was initiated. The pitchover resulted in a flight path 50 degrees from the local vertical so that the spacecraft could pick up the horizontal velocity needed for lunar orbit.

The vertical velocity had reached 80 feet per second at 1,000 feet altitude. At 9,000 feet altitude, the vertical velocity was 150 feet per second, and the horizontal velocity was 700 feet per second. The ascent engine cut off as designed after the stage reached orbital velocity. The crew announced cutoff at 124:29:17 after a burn of about 7 minutes. The altitude at insertion into lunar orbit was 60,300 feet, vertical velocity was 32 feet per second, and horizontal velocity was 5,537 feet per second. The orbit had an apolune of 47.3 miles and a perilune of 9.5 miles.

A personal note here: The event of ascent from the moon and achieving lunar orbit was a tremendous relief to all of us who had been involved in the Apollo program. The successful landing had been great and the fact that the landing radar had worked well was good, but the launch from the moon had been the big worry for the author. Most of the other operations during the mission had abort backups, but if the ascent engine had not fired or had been erratic, the crew would have been stranded on the moon or perished. What a calamity! There was little in congratulations at our facility until the crew was safely back on earth.

The LM crew performed alignment of the inertial platform soon after reaching lunar orbit. The LM state vector was uplinked to the computer in CSM Columbia. The CSM computer used the LM state vector, range data from the VHF ranging system, and sextant angles of sightlines to the LM tracking light to update navigation with respect to the LM. Collins was prepared to use this navigation data in the CSM computer to pilot the CSM down to rescue the LM should the LM not be able to reach the orbit of the CSM.

The Z-axis reaction control thrusters of the LM were fired to circularize the orbit. Refining orbits required some finesse that the thrusters were better suited for than firing the main ascent engine.

Several additional adjustments to the ascent stage orbit were necessary to finally rendezvous with the CSM and begin station keeping with it. A photograph of the ascent stage on its return to the CSM taken by Collins is shown on the next page (Fig. 5.24). Collins said that he liked this picture because it has the earth, moon, LM, and an edge of the CSM all in the same frame.

The ascent stage of the LM was docked with the CSM at 128:03 (5:35 pm EST) after some maneuvering by both the CSM and LM. Docking occurred 3 hours and 41 minutes after launch from the moon, while the LM was about three-fourths the way around the moon in the second orbit.

The crew of the LM and their collection of lunar soil and rocks were transferred to the CSM. The ascent stage of the LM was jettisoned from the CSM at 130:09:31 (7:41 EST).

Return to the Earth

The trans-earth injection (TEI) burn was made at 135:24 (12:56 am EST on 22 July), while the CSM was on the far side of the moon. The burn lasted 151 seconds and increased the velocity of the CSM from 5,376 feet per second to 8,589 feet per second and set the spacecraft on a path toward the earth. The coast from TEI to entering the earth's atmosphere would take 59 hours and 37 minutes. A midcourse correction burn of 10 seconds that changed the spacecraft velocity by 4.8 feet per second was made at a distance of 169,087 miles from earth.

The Command Module was separated from the Service Module, while the CSM was 1,778 miles from earth. The Command Module entered the earth's atmosphere at a velocity of 36,194 feet per second at 195:03. After the CM had been slowed by the aeroshell, the forward heat shield was jettisoned, and the

Fig. 5.24 Apollo 11 ascent stage returning to the CSM (NASA photograph)

drogue parachutes deployed. After slowing to about 180 feet per second by the drogue parachutes, the main parachutes deployed, and the Command Module floated down at about 32 feet per second and splashed down in the Pacific Ocean at 195:18 (12:35 pm EST). Splashdown occurred just 1.7 miles from the intended point and near the recovery ship, the aircraft carrier USS Hornet. The recovery area was 812 miles southwest of Hawaii.

The astronauts were retrieved from the Command Module by swimmer teams from the USS Hornet and then flown by helicopter to the Hornet. They arrived at the ship at 1:38 pm EST on 24 July. Their epic journey to the surface of the moon and return had taken 8 days.

APOLLO LANDING SITES

The US Geological Survey (USGS) has prepared detailed topographical maps of the moon. A portion of the map in the vicinity of Apollo 11 landing site in Mare Tranquillitatis is shown below (Fig. 5.25).

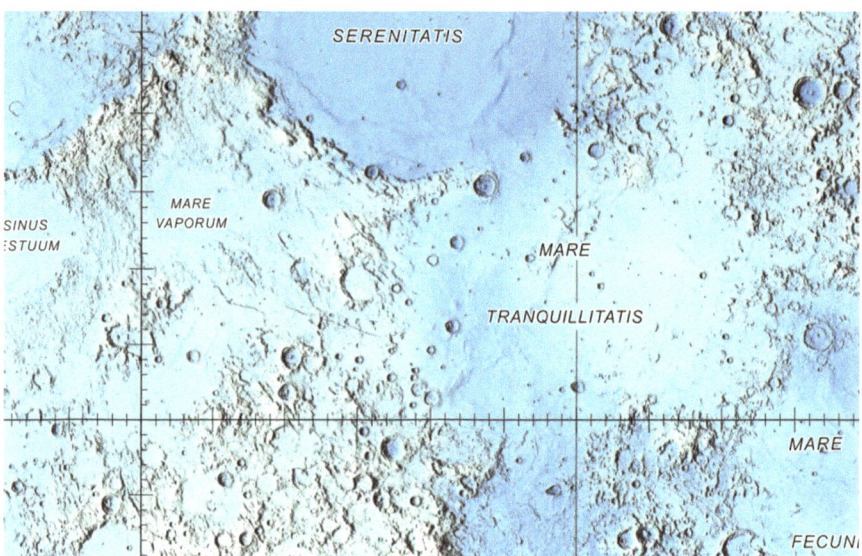

Fig. 5.25 Topographical map of moon in vicinity of Apollo 11 landing site (adapted from USGS graphics)

The landing coordinates of Apollo 11 were 0.674° N latitude and 23.473° E longitude. The vertical line on the map with tic marks is the 0° longitude line, and the horizontal line with tic marks is the 0° latitude line in the lunar coordinate system. Each tic mark presents one degree of latitude or longitude. The landing site is located 0.674 tic marks above the zero latitude line and 23.473 tic marks to the right of the zero longitude line.

The landing site for Apollo 11 was chosen in a region explored by the Surveyor and Ranger spacecraft a few years before. Apollo 11 landed about 14.5 nautical miles from Surveyor 5 and about 38 nautical miles from the impact site of Ranger 8.

A graphic of the near side of the moon showing sites of each of the six Apollo landings is shown on the next page along with a table giving landing coordinates (Fig. 5.26 and Table 5.2). The last landing, that of Apollo 17, took place in December 1972.

Fig. 5.26 Locations of Apollo landing side on the moon (NASA graphic)

Table 5.2 Apollo landing sites in lunar coordinates

Mission	Location	Latitude	Longitude
Apollo 11	Mare Tranquillitatis	0.674° N	23.473° E
Apollo 12	Oceans Procellarum	3.012° S	23.422° W
Apollo 14	Fra Mauro	3.645° S	17.471° W
Apollo 15	Hadley -Apennines	26.132° N	3.634° E
Apollo 16	Descartes	8.973° S	15.498° E
Apollo 17	Taurus-Littrow	20.191° N	30.772° E

The cost of the Apollo program was totaled up to be $25.4 billion in 1973. The equivalent cost in 2018 would be $147.3 billion. In return, the American people received the unmatched spectacle of astronauts walking and working on the moon along with a trove of scientific data. American prestige in the world soared.

COMMAND MODULE, SERVICE MODULE, AND LUNAR MODULE

The Command Module, Service Module, and Lunar Module were quite complex spacecraft embodying state-of-the-art technology of the day. The author has elected to allocate separate chapters of this book to those important modules. The Command Module is described in Chap. 6, the Service Module is described in Chap. 7, and the Lunar Module is described in Chap. 8.

Bibliography

Apollo 11 Mission Report, NASA Report NASA SP-238, 1971

Apollo 15 Mission Report, NAA Report MSC-05161, 1971

Bennett, Floyd V., *Apollo Experience report – Mission Planning for Lunar Module Descent and Ascent*, NASA Technical Note NASA TN D-6846, June 1972

Bennett, Floyd V., *Apollo Lunar Descent and Ascent Trajectories*, NASA Technical Memorandum NASA TM X-58040, January 1970

Bilstein, Roger E., *Stages to Saturn A Technological History of the Apollo/Saturn Launch Vehicles*, NASA History Office, NASA Report SP-4206, Updated August 2004

Brooks, Courtney, G., Grimwood, James, M., Swenson, Loyd, S., *Chariots for Apollo,* Dover Publications, Mineola, New York, 2009

Huntley, J. D., *U.S. Space-Launch Vehicle Technology,* University Press of Florida, 2008

Orloff, Richard, W, and Harland, David, M., *Apollo The Definitive Sourcebook*, Praxis Publishing, Chichester, UK, 2006

Orloff, Richard,W., *Apollo by the Numbers*, NASA report NASA SP-2000-4029, Revised September 2004

Saturn V Flight Manual SA 507, NASA Report MSFC-MAN-507, August 1969

Saturn V major subcontractors and NASA, Saturn V News Reference, August 1967

Seamans, Robert C., *Project Apollo The tough Decisions*, NASA report NASA SP 2005-4537, 2004

Siddiqi, Asif, A., *Deep Space Chronicle*, NASA report NASA SP-2000-4524, June 202

Woods, David, MacTaggart, Ken, and O'Brien, Frank, *The Apollo 11 Flight Journal*, NASA History Division, updated March 2016

Zupp, George A., *An Analysis and a Historical Review of the Apollo Program Lunar Module Touchdown Dynamics*, NASA Report NASA/SP-2013-605, January 2013

6

The Apollo Command Module

The Command Module and the Service Module remained joined until just before return to earth. The overall assembly was referred to as the Command Service Module (CSM).

A photograph of the Apollo 15 Command Service Module taken from the Lunar Module while in orbit around the moon is shown next page. The Service Module is the dominant cylindrical structure with the rocket motor expansion nozzle attached. The Command Module is the dark looking conical structure attached to the forward end of the Service Module (Fig. 6.1).

The Command Module was an efficient space capsule that provided living accommodations for three astronauts for several days. Along with providing comfortable accommodations for the crew, it contained controls for maneuvering the Command Service Module or the Command Module alone, firing the rocket engine of the Service Module, and provisions for communications with earth and with the Lunar Module.

A photograph of the Command Module from Apollo 9 that is prominently displayed in the San Diego Air and Space Museum is shown next page (Fig. 6.2).

Physically, the command module was 12 feet, 7 inches in diameter at the widest point of the aft heat shield and 10 feet, 7 inches high from the bottom of the heat shield to the top of the docking probe. The launch weight, including a crew of three, was about 13,000 pounds.

The Command Module and the Service Module were developed by North American Aviation in Seal Beach, California.

© Springer Nature Switzerland AG 2018   155
T. Lund, *Early Exploration of the Moon*, Springer Praxis Books,
https://doi.org/10.1007/978-3-030-02071-2_6

Fig. 6.1 Apollo 15 Command Service Module in orbit around the moon (NASA photograph)

Fig. 6.2 Command Module from Apollo 9 mission (author's photo)

COMMAND MODULE BACKGROUND

The Command Module had a turbulent beginning with initial program uncertainties and it hosted the only fatalities of the Apollo program. The uncertainty came about because the basic approach to land astronauts on the moon changed from an earth orbit rendezvous approach to a lunar orbit rendezvous approach after the initial contract to develop the Command Service Module had been awarded.

North American Aviation was announced the winner of the competition to develop the Command Service Module (CSM) in November 1961. Development of the CSM at North American was directed by Harrison Storms, president of the Space and Information Systems division. His team included John Paup, Apollo program manager, and Charles Felz, Apollo program engineer.

The NASA requirements at the time reflected an earth orbit scenario to place astronauts on the moon. That scenario would involve two or more launches to place components of the spacecraft in earth orbit where they would rendezvous and be assembled. The resulting spacecraft would depart for the moon and land astronauts on the surface. After exploration, the astronauts would launch the ascent portion of the spacecraft from the moon and return to earth. Details of the lunar landing and return to earth had not yet been worked out.

The Command Service Module (CSM) would house the astronauts and serve as the command and control center for the mission. By itself, if launched toward the moon, the CSM could circumnavigate the moon and return safely to earth.

Studies and discussions of the best approach to land astronauts on the moon continued at NASA after North American had begun designing the CSM. John Houbolt of NASA Langley was tireless in crusading for a lunar orbit rendezvous approach. Finally, NASA scientists concluded that a lunar orbit rendezvous approach had important advantages over the earth orbit approach. The decision for lunar orbit rendezvous was announced by NASA Administrator, James Webb, in July 1962.

The lunar orbit rendezvous approach involved detaching a dedicated landing spacecraft, eventually called the Lunar Module (LM), from a spacecraft assembly in orbit around the moon. The LM, carrying two astronauts, would make a soft landing on the moon. The astronauts would explore the lunar surface and then launch from the moon and rendezvous and dock with the orbiting spacecraft. The LM would be jettisoned and the orbiting spacecraft would return to earth.

NASA directed North American to change the design of the Command Module to include docking provisions and provisions to transfer astronauts between the Command Module and the Lunar Module. To take advantage of design work already done, and the fact that several Command Modules would be required for testing in earth orbit where docking would not be required, NASA decided to have several Command Modules of the original design built. They would be known as Block 1. A photograph of a Block 1 Command Module is shown on the next page (Fig. 6.3).

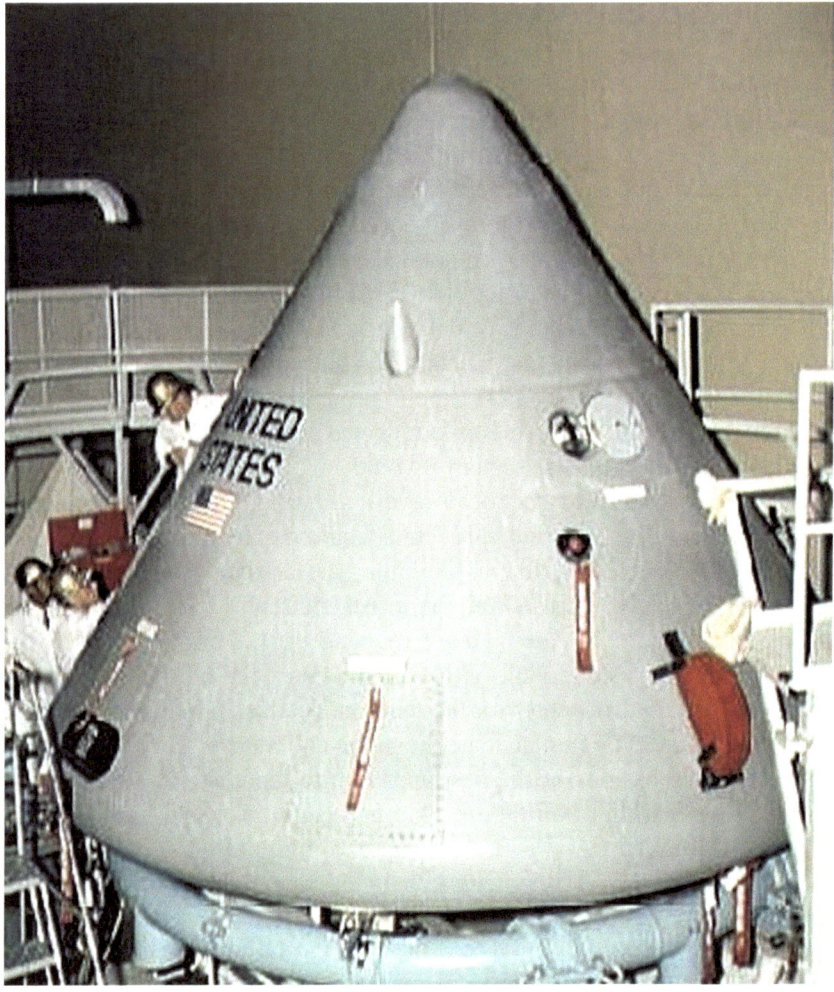

Fig. 6.3 Block 1 Model of Command Module (NASA Photograph)

The design of the Block 2 models, which incorporated docking provisions and other changes for the lunar orbit configuration, went forward concurrent with build of the Block 1 models. The final contract between NASA and North American for the Command Service Module was signed in August 1963. The contract called for 15 test models and 11 flight models. First delivery of a Block 1 flight model CSM was made in August 1966 to support mission AS-204.

Apollo mission AS-204 was scheduled for the last quarter of 1966 as a manned mission to evaluate astronaut operations and Command Service Module performance in the earth orbit. The astronauts for the mission were Gus Grissom, Edward White, and Roger Chaffee. Command Module, Block 1 Number 012, and Service Module 012 were shipped from the North American plant to Cape Kennedy in

August 1966. There were many change orders on the spacecraft that had not been accomplished before shipment to the Cape that had to be worked off. The launch of AS-204 was rescheduled for February 1967.

The great tragedy of the Apollo program happened on 27 January 1967 when a routine ground test with astronauts in the cabin of Command Module 012 went very wrong. A fire started in the Command Module and it spread uncontrollably in the 16 psi pure oxygen environment. The three astronauts perished.

Teams were set up to perform thorough investigations of various aspects of the accident. Hundreds of people were involved. Investigators took apart the fire ravaged Command Module 012 and inspected each element. They also compared it as much as they could with Command Module 014 that had been shipped to the Cape for disassembly and comparison. The exact cause of the fire was not determined, but chafing and subsequent short circuiting of an electrical cable were a likely cause. A large amount of flammable material in the cabin fed the fire.

Two substantial changes were made as a result of the fire. First, the hatch of the Command Module was changed to open outward and to be easily opened from the inside. The original hatch opened to the inside and was impossible to open with the spacecraft pressurized. Second, the atmosphere in the cabin was changed from pure oxygen to a nitrogen-oxygen mixture when on the ground. In flight, a 5 psi pure oxygen atmosphere would be used.

Increased oversite of all aspects of the Apollo program was instituted. North American increased emphasis on workmanship and inspection and assigned an assistant program manager whose only job was safety. As fallout of the catastrophe, Harrison Storms, president of North American Space and Information Systems division, was replaced by William Bergen. Joe Shea, Apollo Spacecraft Program Manager for NASA, overworked himself to a severe health problem during the accident investigation. He was replaced by George Low.

Block 1 Command Modules 002, 009, 011, 017, and 020 were flown in space unmanned as part of the test program for Apollo. Other Block 1 Command Modules were used for testing on earth, including Command Module 007 that was tested floating on the ocean with astronauts to verify post landing procedures.

Eleven Block 2 Command Modules were flown in space as part of the linked Command Module and Service Module (CSM). The first, CSM-101, flew on Apollo 7. The last, CSM-114, carried the crew of Apollo 17 on their journey to the moon.

MECHANICAL CONFIGURATION OF COMMAND MODULE

A cutaway drawing of the Command Module is shown on the next page (Fig. 6.4). The drawing depicts the three astronauts on their couches with the Command Module in the vertical launch orientation. The astronauts were seated facing along

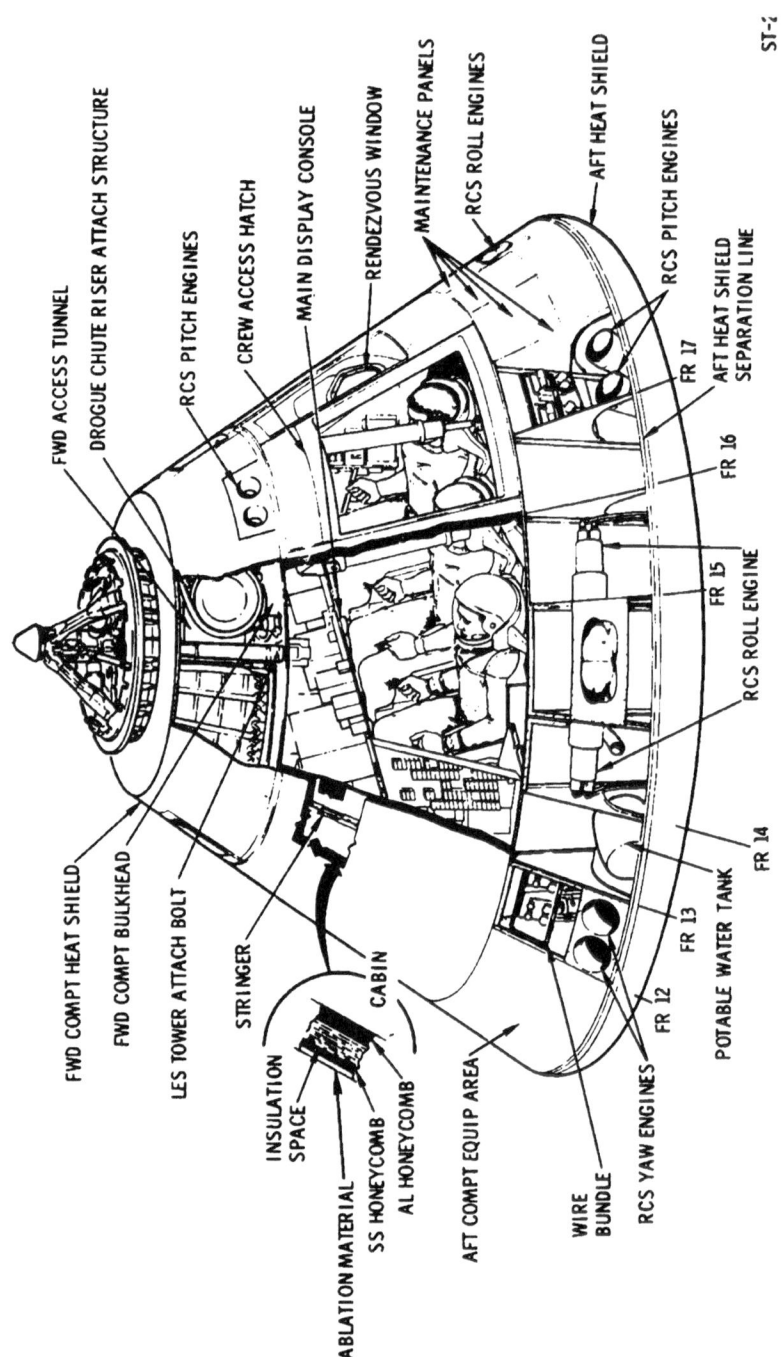

Fig. 6.4 Cutaway view of Command Module (from Apollo Training Course APC-118)

the –X-axis of the spacecraft with the main display console approximately perpendicular to their line of sight.

The outer shell of the Command Module was formed from stainless steel honeycomb faced with stainless steel sheets. The thickness of the outer shell structure varied from 2.5 inches near the blunt end to 0.5 inches near the top end.

The crew compartment was a pressurized inner shell in the center section of the module. The crew compartment shell was made up of aluminum honeycomb faced with aluminum sheets. The thickness of the inner shell varied from 1.5 inches at the lower end to 0.25 inches at the upper end. Insulation was installed between the inner and outer shells to help protect the astronauts from substantial heat generated on the outer shell during launch through the atmosphere and the extreme heat on the outer shell when reentering the earth's atmosphere.

The Command Module was protected from heat generated during passage through the atmosphere at launch by a protective fiberglass and cork cover that fit over the module. That cover was attached to the launch escape tower and it was jettisoned along with the escape tower at about 295,000 feet altitude.

The entire command module was covered with an ablative heat shield to protect it from the fierce heat generated during reentry into the earth's atmosphere. The heat shield was 2 inches thick at the blunt end, and it narrowed to ½ inch at the narrow end. The heat shield had an outer silver Mylar thermal coating and that gave the command module a polished aluminum appearance when viewed from some angles prior to return to earth.

The Command Module reentered the earth's atmosphere blunt end first and the aft heat shield took the brunt of the heat generated on reentry. The ablative material that made up the heat shield was a phenolic epoxy resin. During reentry it charred and melted and the ablative action minimized heat being absorbed by the spacecraft.

The Command Module was divided up into three sections. The forward section at the narrow end of the conical structure contained docking provisions to the Lunar Module, a tunnel allowing passage between the Command Module and the Lunar Module, and the parachutes and other elements of the earth landing system. The pressurized center section contained seats and other accommodations for three astronauts along with provisions and support equipment for the crew for 14 days. The center section also contained controls and displays for the spacecraft and most of the electronic components of the Command Module systems. The aft section contained propellants for the reaction control engines, a helium tank to pressurize the propellants, and water tanks. The large heat shield was mounted at the bottom of the aft section.

The pressurized center section had a volume of about 210 cubic feet and its dimensions allowed two astronauts to stand when the center seat was folded. A sketch of Apollo crew stations from Apollo Training Course APC-118 is given next page (Fig. 6.5). The sketch shows one astronaut standing at the optical instrument station, and it shows sleep stations under the left and right couches.

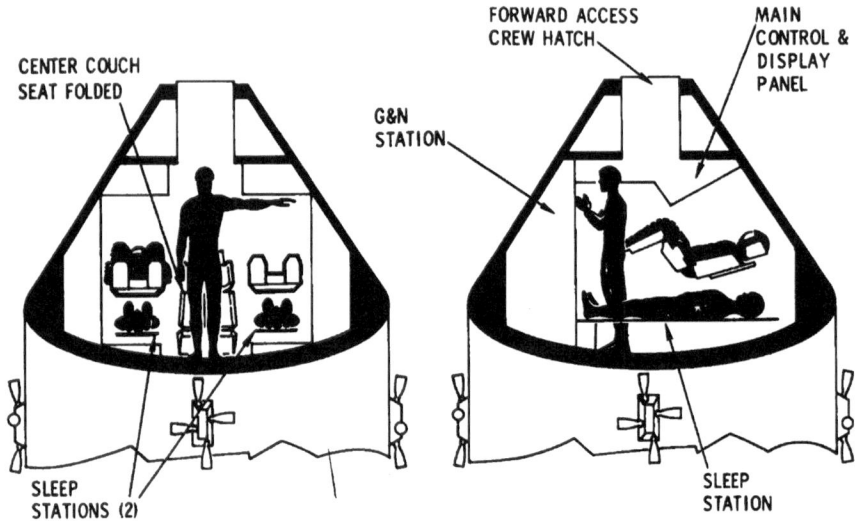

Fig. 6.5 Apollo crew stations (from Apollo Training Course APC-118)

The oxygen atmosphere in the Command Module was maintained at a pressure of 5 psi. The capsule temperature was nominally about 75°F although it could be adjusted by the crew.

There were five windows in the Command Module that are apparent in the photograph of the Apollo 17 Command Module shown on the next page (Fig. 6.6). The photograph was taken from the Lunar Module while above the moon. The two side windows, labeled 1 and 5 in the photograph, were adjacent to the outer couches for the astronauts. One window was in the entry hatch, and two, labeled 2 and 4, were arranged to allow viewing forward, and they were used during rendezvous with the Lunar Module.

The Command Module had two hatches; a side hatch and a forward hatch. The side hatch was the primary means of getting in and out of the module. It was 34 inches wide and 29 inches high and located over the center seat. The hatch opened outward. It had 12 latches around the periphery of the hatch that could be tightened or released by a handle that drove a ratcheting mechanism. It could also be opened from the outside with a special tool.

The forward hatch was mounted at the top of the docking tunnel. It was opened to transfer astronauts between the Command Module and Lunar Module. The hatch was about 30 inches in diameter with a six-point latching mechanism. A handle allowed the hatch to be opened or latched. The hatch had a pressure equalization valve so that pressure in the LM and the tunnel could be equalized before the hatch was opened.

Fig. 6.6 Photograph of Apollo 17 Command Module showing windows and docking mechanism (NASA photograph)

There were a total of 12 reaction control engines on the Command Module that were fired to maneuver the vehicle. These engines were only used after separation of the Command Module from the Service Module.

INTERIOR DETAILS OF COMMAND MODULE

A photograph of the inside of the Apollo 11 Command Module is shown on the next page. That historic Command Module is displayed in the National Air and Space Museum in Washington, DC. The excellent photograph was taken by Eric Long of NASM (Fig. 6.7).

The three seats for the astronauts and the instrument panel for the Command Module are apparent in the photograph. The mission commander sat in the left seat, the Command Module pilot sat in the center seat, and the Lunar Module pilot sat in the right seat. The Command Module pilot had sole control of the CSM when the other two astronauts were in the LM or on the lunar surface.

Fig. 6.7 Interior view of Apollo 11 Command Module (NASM photograph)

A drawing of the instrument panel in the Command Module is shown on the next page (Fig. 6.9). An expanded view of the portion of the instrument panel in front of the commander is shown on the following page (Fig. 6.10).

Key Instruments in Front of the Commander

A prominent and very valuable instrument in front of the commander was the flight director attitude indicator (FDAI) shown in the photograph below (Fig. 6.8).

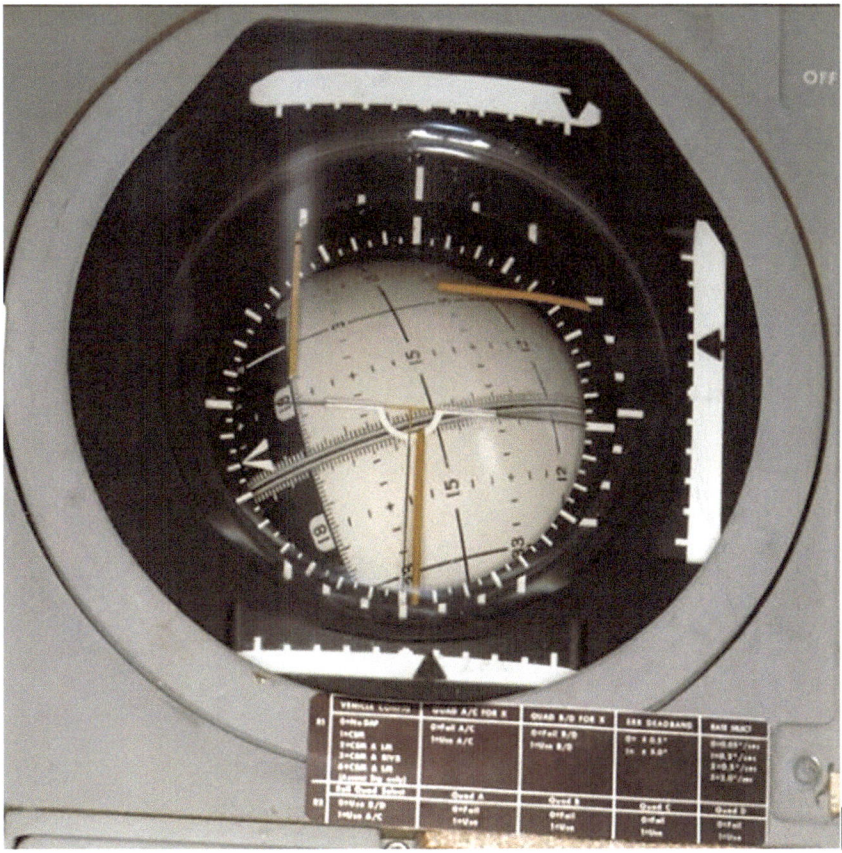

Fig. 6.8 Flight director attitude indicator in Command Module of Apollo 15 (NASA photograph cropped by author)

The FDAI was close in appearance and function to the flight director attitude indicator instrument relied on for years by the astronauts their former lives as jet fighter pilots. A second FDAI for use by the Command Module pilot was located to the right and above the one for the commander.

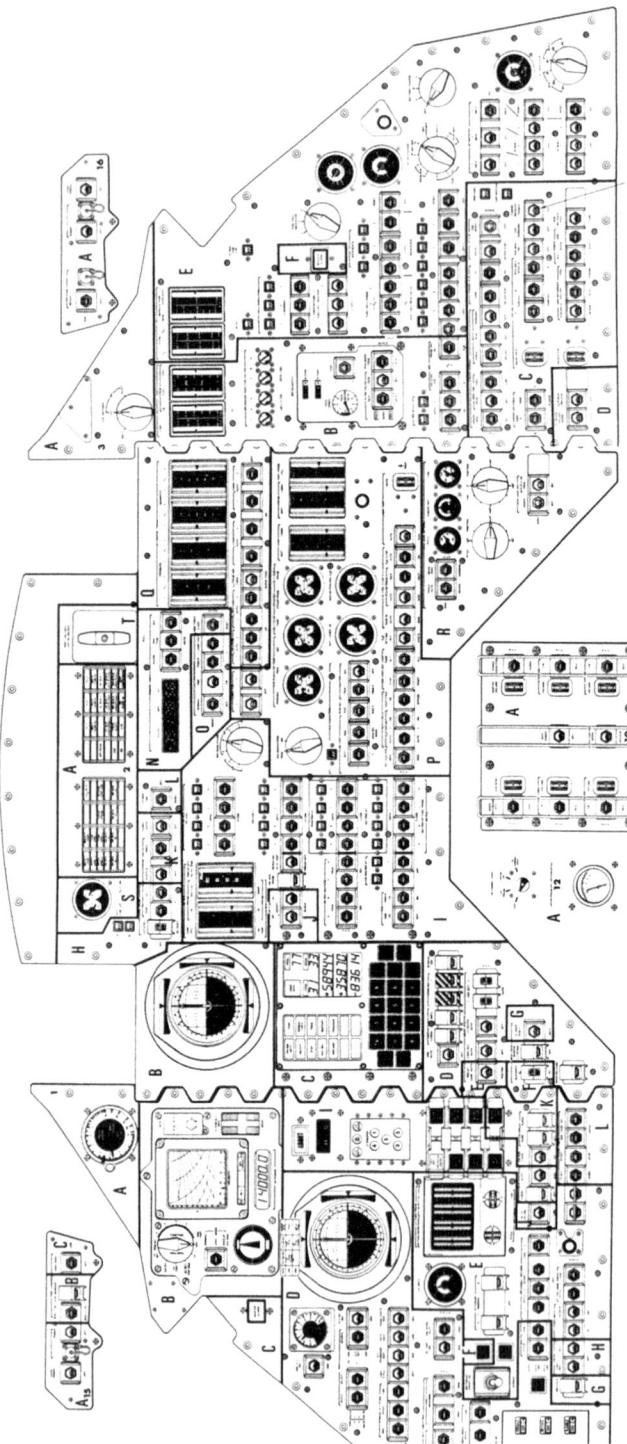

Fig. 6.9 Instrument panel of Apollo Command Module (NASA graphic)

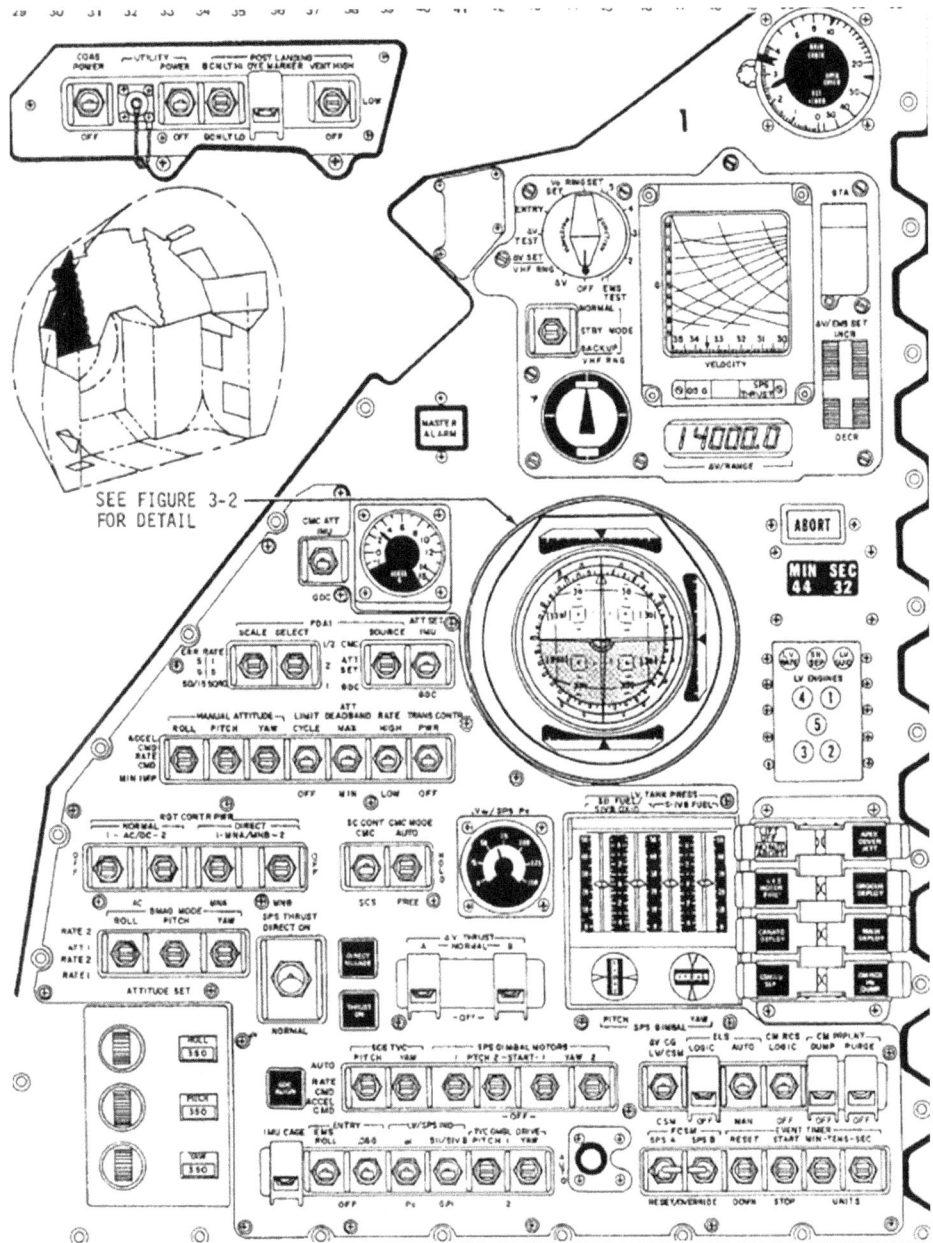

Fig. 6.10 Portion of instrument panel in front of commander (NASA graphic)

Various modes of the FDAI could be selected to display pertinent parameters depending on needs of the mission at the time. When in the ORDEAL mode in orbit around the earth or moon and in level flight with respect to the surface, the black half of the ball would be down, and the black to white boundary would be under the white wings at the center of the instrument. The lettering on the first horizontal line on the ball above center would read 3 (30 degrees).

The rotation of the ball in the vertical direction represented pitch angle of the spacecraft, and rotation in the horizontal direction represented yaw angle with respect to a reference attitude. Roll angle of the spacecraft was indicated by roll of the ball and indicated by the white arrow against the markings on the periphery of the viewing aperture.

The white scales with black arrowheads located at the top, side, and bottom of the instrument displayed attitude rates. The displacement of the arrows from the center indicated the magnitude of rate and the direction of control movement required to bring the rates back toward zero. The top scale showed roll rate, the side scale showed pitch rate, and the bottom scale showed yaw rate. Full-scale deflection of the scales in degrees per second was set by a switch located to the left of the instrument. A switch on the control panel allowed selecting the source of attitude information to be either the inertial measurement unit (IMU) or the Gyro Display Coupler (GDC).

The flight director function of the instrument was provided by the three yellow needles that are visible on the face of the instrument in the photograph. The needles indicated the magnitude of the angular error in each axis and the direction to move controls to return the error toward zero. The horizontal bar indicated displacement from the desired angle in the pitch plane, the upper vertical bar indicated displacement in the yaw plane, and the lower vertical bar indicated displacement in the roll axis. The scale factor of the flight director needles was set by a switch located to the left of the instrument.

During manual control of the spacecraft, an astronaut could set in attitudes quite accurately using the markings on the ball of the FDAI. Fine attitude control could then be exercised by using the attitude error needles to reduce the difference between spacecraft attitude and desired attitude to near zero.

Referring back to the drawing of the instrument panel in front of the commander, the round instrument at the top of the panel was a barometric altimeter that was most useful when the Command Module was descending on a parachute in the recovery process after a mission. During the descent of Apollo 11 on the parachute, Neil Armstrong communicated altitude to the recovery team as well as position of the Command Module in terms of latitude and longitude.

The Entry Monitor System panel was located just below the altimeter. Readouts from that panel provided critical information on entry dynamics during the aeroshell braking phase during entry into the atmosphere upon return to earth. The orientation of the axis of the Command Module with respect to the velocity vector was very important to a safe reentry into the atmosphere.

Normally, the Apollo guidance computer controlled the orientation of the spacecraft for the proper entry angle for safe return and to be close to the designated recovery point at sea. In case of emergency or failure of the computer, displayed information on the Entry Monitor System panel would be used along with manual control to maneuver the Command Module for a successful entry into the atmosphere. The entry phase of the mission and details of the instrumentation is discussed later in this chapter.

Another important subsystem on the CSM was the Service Propulsion System (SPS) located in the Service Module. The SPS contained a gimballed rocket engine with 20,500 pounds of thrust. The rocket engine was normally controlled by the guidance computer. In case of emergency or failure of the computer, firing of the rocket engine and adjusting gimbal angles could be performed manually. Several controls and displays associated with the rocket engine were located just below the FDAI on the instrument panel in front of the Commander. A large toggle switch labeled SPS THRUST DIRECT ON was located to the left of the two push-button switches that appear black in the drawing. The toggle switch could be moved to the ON position to energize circuitry to fire the SPS engine manually. Then pressing the THRUST ON push button fired the engine.

The quantity of propellants for the SPS engine remaining in the tanks and temperatures and pressures in the tanks were indicated by a group of displays located on the panel in front of the Lunar Module pilot.

The small analog meter located lust below and to the left of the FDAI gave readout of thrust chamber pressure of the SPS engine. These pressure readings were closely monitored by the crew. On Apollo 11, Armstrong reported reading 88 psi rather than the normal 100 psi during the first midcourse correction burn. Although of concern by Armstrong, it apparently passed muster at mission control.

Two scales on vertical indicator instrument below the FDAI indicated position of the pitch and yaw gimbals of the SPS engine. Two thumbwheels located below the vertical scales allowed manual adjustment of the pitch and yaw gimbal angles.

The astronauts were very busy during the boost phase monitoring critical systems and vehicle performance. The commander had access to much of the flight information. Some of this information is summarized in the following paragraphs.

Referring to the portion of the instrument panel in front of the commander and looking to the right of the FDAI, a red ABORT light would come on if an abort situation occurred, and abort was commanded from the ground. The next item down on the panel was an event timer that started at liftoff. The event timer could be reset and could be set to count up or count down. It was used to time events during the mission. A second timer, on the panel in front of the Command Module pilot, was referred to as the mission timer. It started counting up at liftoff and normally it was let run for the duration of the mission.

The next subpanel down from the event timer contained warning and event lights associated with the boost system. The left light in the upper row glowed red if the launch vehicle angular rates became excessive. The middle light in the upper row glowed red if there was a failure in the launch vehicle guidance system. The right light in the upper row lit up white when staging of the second stage occurred. The five lights below the upper row were arranged in a pattern similar to the first- and second-stage boost engines. The lights glowed yellow if the engines were not producing thrust. Prior to ignition they all glowed yellow. Each light went out when the engine developed 90% thrust. Two lights glowing yellow after launch could trigger an abort.

The astronauts had a front row seat to an amazing display of power and technology as the booster rockets drove the spacecraft ever further from the earth's surface. The display and keyboard (DSKY) in the next panel to the right of the commander's panel gave an unfolding account of velocity, altitude rate, and altitude. The FDAI gave attitude angles, attitude rates, and difference between attitude angles at the moment and those planned. The performance of all critical systems of the spacecraft was displayed either on gauges or by warning lights on the instrument panel. It must have been a fantastic ride!

Key Instruments in Adjacent Panel

The commander also had access to a portion of the instrument panel just to the right of the panel previously shown. An expanded view of that portion of the panel is given on the next page (Fig. 6.11).

A major display and control item on that section of panel was the display and keyboard (DSKY). It was located just below the Command Module pilot's flight director attitude indicator in the panel. The DSKY, which was accessible by both the commander and Command Module pilot, allowed the crew to communicate with the guidance computer. A second DSKY was mounted in the lower equipment bay of the Command Module next to the control panel for telescope and sextant.

An illustration showing the face of the DSKY is given on the following page. The face was about 8 inches high and 7 inches wide. The lines on the figure led to captions in the original figure. The captions were removed by the author to present a larger picture of the face of the DSKY on the page. The DSKY units were the same in the Command Module and Lunar Module except for the caution panel (Fig. 6.12).

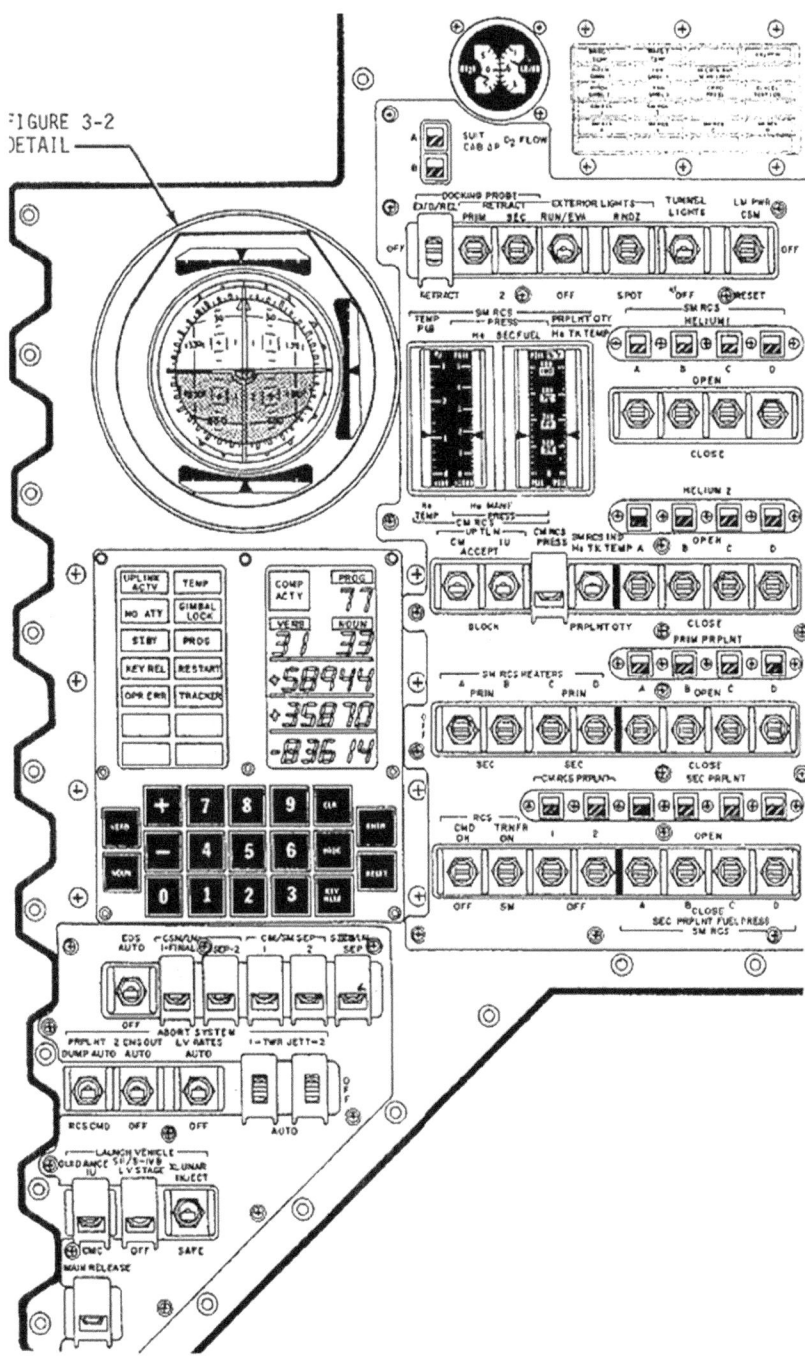

Fig. 6.11 Portion of instrument panel to the right of previous panel (NASA graphic)

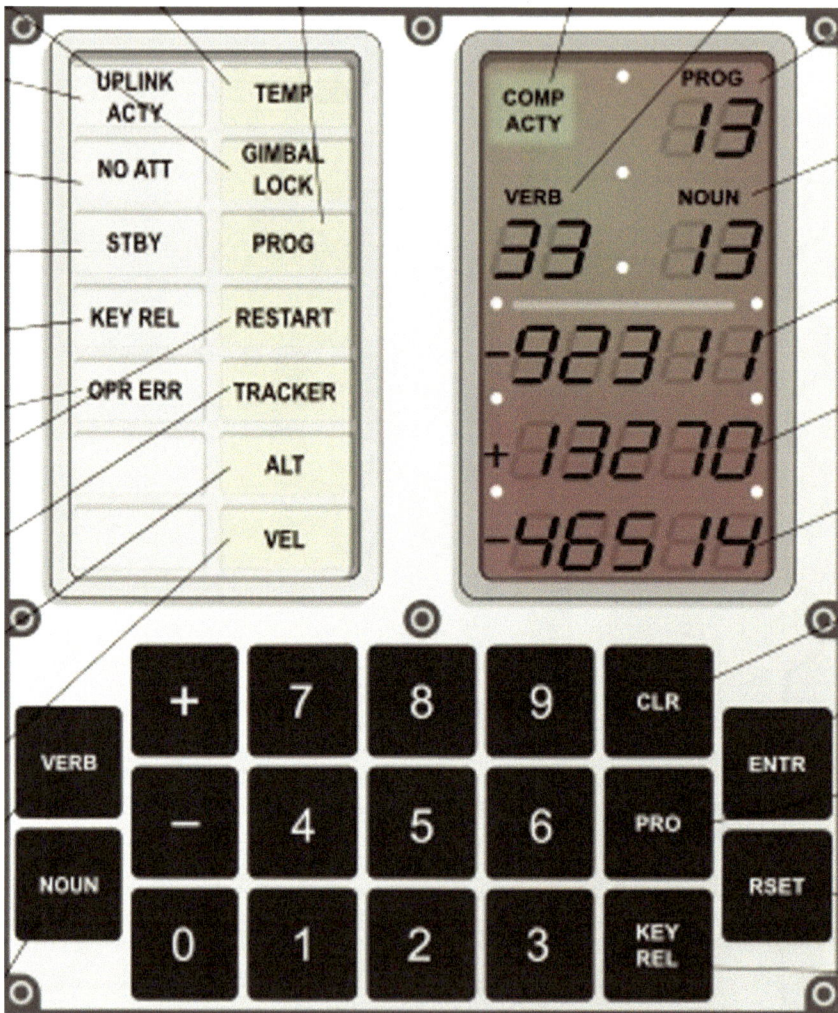

Fig. 6.12 Apollo DSKY (NASA graphic cropped by author)

The display region on the upper right side of the DSKY displayed computer activity, program number, and three lines of five-digit data plus a sign. The upper left side of the DSKY contained caution lights for ten different events for the DSKY mounted in the Command Module. Two additional caution lights were used on the Lunar Module DSKY and that DSKY is represented above. A key-board occupied the lower portion of the DSKY.

The astronauts communicated with the computer by first requesting a particular computer program by entering a two-digit number. The program selected was displayed under the PROG area of the display. The action desired from that program was then selected by pressing VERB followed by two numbers and then pressing NOUN followed by two numbers. Verbs generally requested an action. For example, keying in VERB 06 specified displaying data in decimal form (as opposed to Octal) in the five-digit display fields. Nouns generally specified what data was to be displayed. For example, keying in NOUN 62 called for displaying "inertial velocity magnitude," "altitude rate," and "altitude above pad radius" in the three five-digit display fields.

There were 44 different programs that could be requested and there were 99 different verbs and 99 different nouns to choose from. As an example, during the boost phase the astronaut would key Program P11 (Earth Orbit Insertion Monitor) into the DSKY and then key in Verb 16 and Noun 62. The three five-digit readouts on the DSKY would then show the following:

Register 1: Inertial velocity (in feet per second)

Register 2: Altitude rate (in feet per second)

Register 3: Altitude above launch pad (in nautical miles with the last digit representing 0.1 miles.

These critical parameters allowed the astronauts to follow the progress of the spacecraft as the rocket engines hurdled them away from the launch pad.

Data could also be entered into the computer from the ground. Update of the state vector was made periodically from the ground. To do this, the capsule communicator (CAPCOM) in Mission Control in Houston asked the astronauts to select Program 00 and Accept. The Accept/Block switch was located just to the right of the DSKY on the instrument panel. The switch either allowed or blocked uplinked data from the ground. Placing the switch in the ACCEPT position allowed Mission Control to have control of the computer and upload data via the telecommunications system. When the upload was complete, the astronauts regained control of the computer by turning the Accept/Block switch to BLOCK.

The state vector was very important to the mission. It contained seven numbers: three numbers representing spacecraft velocity in three axes, three numbers representing spacecraft position in three axes, and one number representing time when the data was gathered. The state vector was determined very precisely from tracking equipment on the ground. It was used to periodically update the guidance computer to correct for slight drift of the inertial platform.

The Warning/Caution panel for the Command Module DSKY had ten captions. A particular caption would light up when the computer sensed an error or that a warning was needed. The meaning of each of those captions is covered during the description of the DSKY later in this chapter.

MAJOR SUBSYSTEMS WITHIN THE COMMAND MODULE

A simplified block diagram of major subsystems within the Command Module showing data flow is given on the next page (Fig. 6.13).

Primary Guidance Navigation and Control System (PGNCS)

The Primary Guidance Navigation and Control System (PGNCS) in the Command Module performed the guidance and navigation functions of the Apollo mission remarkably well. The PGNCS was made up of the computer subsystem, inertial subsystem, and optical subsystem. Together, these subsystems allowed the state vector to be continuously determined.

The Instrumentation Laboratory of the Massachusetts Institute of Technology (MIT) was selected by NASA in August 1961 to develop the navigation and control system for the Apollo spacecraft. Three contractors were selected in May 1962 for production contracts for major subsystems of the navigation and control system. The contractors and subsystems were:

- Computer subsystem - Raytheon Company
- Inertial subsystem – AC Sparkplug Division of General Motors
- Optical subsystem - Kollsman Instrument Corporation

Computer Subsystem

The computer subsystem consisted of the Apollo Guidance Computer (AGC) and the display and keyboard (DSKY).

Apollo Guidance Computer

The Apollo Guidance Computer was crucial to the success of the Apollo mission. It was the brains and nerve center for both the Command Module and the Lunar Module. The computers used in the CM and LM were the same physically, but they had a different software that reflected the different operations conducted by the two spacecraft.

A photograph of the Apollo guidance computer and display and keyboard (DSKY) is shown next page (Fig. 6.14). The DSKY, on the right in the picture, was the astronaut's interface with the computer. It allowed them to command operations and input data as well as to read out pertinent data generated by the computer.

The computer performed many functions. It developed guidance solutions, solved flight equations, generated commands to control the spacecraft, generated

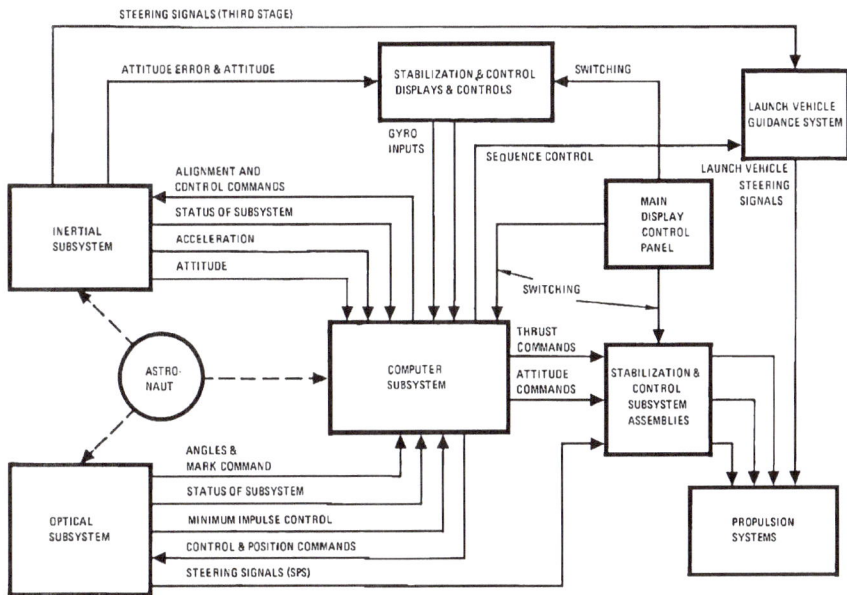

Fig. 6.13 Major elements of Command Module guidance and navigation subsystem (NASA graphic)

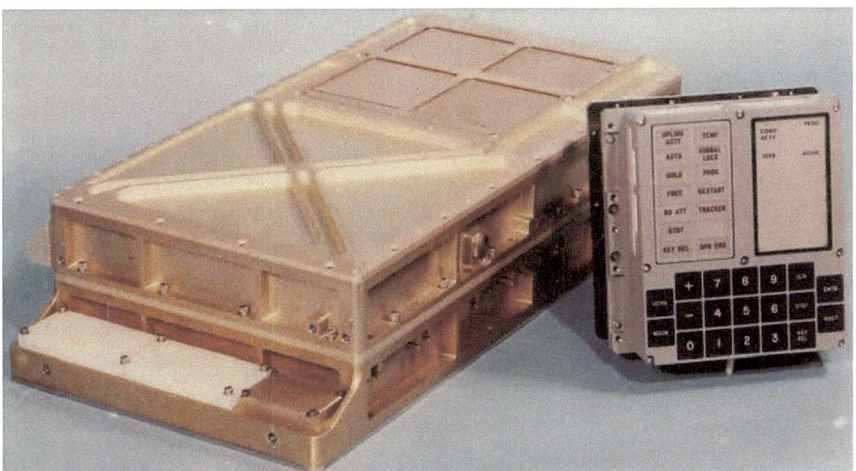

Fig. 6.14 Apollo guidance computer and DSKY (NASA photograph)

data to be displayed, and generated data used by the astronauts to align the inertial subsystem and to position the optical sextant to allow taking star sights needed for the alignment. It was referred to by some as the "fourth crew member" in the spacecraft. Readouts from the computer received close attention by the astronauts throughout the mission.

The computer was designed by the Instrumentation Laboratory of the Massachusetts Institute of Technology (MIT) and built by the Raytheon Company. Software for the computer was developed by MIT.

Physical characteristics of computer

The Apollo Guidance Computer represented state-of-the-art packaging of computer functions with reliable circuitry in a small size at the time. The computer assembly was 24 inches long, 12.5 inches wide, and 6 inches high. It weighed 70.1 pounds and consumed about 55 watts of power at 28 volts DC when operating. It consumed 15 watts during standby.

Physically, the computer was composed of an upper tray and a lower tray bolted together. The trays held a series of modules. The lower tray contained 2 power supply modules, 5 interface modules, and 24 logic modules. The upper tray contained an erasable memory module, six fixed memory modules, eight modules that controlled the memory, an oscillator module, and an alarm module.

All of the modules except for the six fixed memory modules had the same form factor although some were double-wide. Scaling from photographs, the modules were about 10 inches long and 1.6 inches high from top to the bottom of connector pins. The top frame of each module was mounted to structure of the enclosure. Electrical connections to pins at the bottom were made by wire wraps.

Most of the logic circuitry in the computer was implemented using a newly developed integrated circuit by Fairchild that packaged two three-input NOR gates in a single flat pack. Each logic module contained 120 of those integrated circuits, 60 on each side of the module.

The fixed memory modules were made up of 6 memory planes, each containing 256 cores that represented 1024 words of memory. The arrangement of six memory planes was enclosed in a sealed module. Scaling from photographs, the fixed memory modules were about 10.6 inches long, 3.3 inches wide, and 0.8 inches high. A total of 6 memory modules provided 36,864 words of fixed memory. The fixed memory modules plugged into the upper tray at the far end in the photograph of the computer unit shown previously.

The Apollo Guidance Computer was designed to withstand a harsh spacecraft environment. Part of the substantial weight of the unit was due to the strong, rugged enclosure and its internal structure. Select units underwent rigorous environmental testing and each flight unit was subjected to acceptance testing that included vibration and temperature cycling while monitoring computer operation. Vibration testing was conducted with the unit mounted to a shake table.

Computer capability

The computer contained the basic functions of a timer, central processor, memory, priority control, and input-output. The timer generated synchronization signals for the computer and other subsystems in the spacecraft. The central processor performed all required arithmetic operations and operations dictated by the program stored in memory. The memory consisted of 2,048 words of erasable memory and 36,864 words of fixed (read-only) memory. The erasable memory stored results of calculations and data supplied by other elements of the guidance and navigation system. The fixed memory stored the computer program and essential tables such as star tables.

The priority control set up priorities for operations to be performed. It made certain that highest priority tasks were performed in case of conflict and computer overload. An operation was interrupted if a higher priority task was called for. This feature may have saved the mission of Apollo 11 when request from the rendezvous radar began vying for processing time from the higher priority task of landing the Lunar Module.

The input-output functions managed and routed signals to and from other elements of the spacecraft.

The basic cycle time of the computer clock was 2.048 MHz. Cycle time for the memory was 11.7 microseconds (μsec). The time required for a single addition was 23.4 μsec and the time required for a multiplication was 46.8 μsec. There were a total of 34 normal instructions that could be written in the computer code.

The final version of the computer contained 36,864 (36k) words of fixed memory and 2,048 (2k) words of erasable memory. Those memory sizes reflect a considerable growth from initial sizing of 4k words of fixed memory and 256 bits of erasable memory. The word length was 16 bits made up of 14 bits of data, one sign bit, and one parity bit.

Obtaining the amounts of erasable and fixed memory in the computer is trivial today. However, it was a challenge in the early 1960s to develop such memory in a small physical size. The technology used for the memory functions is interesting and it is briefly described in the following paragraphs.

Erasable memory

The erasable memory was a coincident-current ferrite core type that was common at the time. The memory was made up of an x-y matrix of very small ferrite cores with x-axis and y-axis write lines and a sense and an inhibit line passing through each core. The ferrite cores had discernable hysteresis such that if a current pulse of sufficient amplitude passed through a core, it would cause the rotation of the magnetic field in the core to flip to one direction or the other. The flipping field would induce a current in a sense line passing through the core. The direction of the flux around the core would remain after the current was removed. This property allowed a bit to be written and then read out at a later time. Rotation of the magnetic field in one direction represented a data "1" and rotation of the field in the other direction represented a data "0."

A particular core (a particular bit of memory) was selected by applying current pulses of magnitude of half that required to flip the magnetic field to the x and y wires in the memory matrix. Only in the cores where there was a coincidence of the x and y current pulse was there a sufficient current to flip the magnetic sense of the core.

To read out a bit, pulses were applied to the lines in such a polarity to set the flux rotation to the "0" bit direction. If the bit were a "1," the flux direction would flip, and this would cause a current pulse to appear in the sense line. If the bit were a "0," the field would not flip, and there would not be a pulse in the sense line. The bit was always left in the "0" state after readout. To write a data "1" to the memory, the direction of the current pulses was opposite of that used for readout. To write a data "0," current was sent through an inhibit line, and that prevented the flux direction from flipping.

Fixed (read-only) memory

The fixed memory was a core rope type selected because it required less space and power consumption for a given amount of memory than other types available at the time. Whereas the erasable memory described above could store one data bit in each core, four 16-bit words could be processed in each core of the core rope memory.

A major disadvantage of the core rope memory was that assembling it was tedious and time-consuming, and once assembled it could not be changed except by rewiring. Each small toroidal core of the memory had to be threaded by dozens of wires that also passed through a series of other cores. Some of the wires bypassed certain cores and passed through others. A fine insulated wire was threaded through the eye of a needle similar to an embroidery needle and the needle was used to thread the wire through the cores. Wiring of the memory was done by women with very dexterous fingers.

A good description of the core rope memory is given in MIT Instrumentation Laboratory report R-393 *Logical Description for the Apollo Guidance Computer (AGC 4)*. That report, written in 1963, described a 12,288 word memory.

The words of the fixed memory were divided into separate groups called "ropes." Each rope involved 1,536 cores and processed four 16-bit words per core or a total of 6,144 words. The 36,864 words of memory required six ropes. Each rope was packaged in a separate module. The ropes were further broken down into banks with 1,024 words per bank.

The nickel-iron toroidal cores used in the memory had distinct hysteresis such that once switched by a current pulse in a wire passing through the core, the magnetic flux in the core would remain in that orientation until switched back. There was a "set line" and a "reset line" that threaded through each core of a rope. In addition, in each bank of memory, there were ten pairs of wires called "inhibit lines" that selectively passed through a core or bypassed it. The signal pattern of each pair of inhibit lines was complementary and controlled by registers. If an energized inhibit line passes through a core, the core could not be switched by the set pulse. The combination of signal pattern on the inhibit lines and whether a inhibit line passes through a core allowed selection of which cores in the rope would be switched by the set pulse.

During readout of a rope of the memory, the reset line was energized and all cores that had been set to a "1" flux direction switched to a "0" direction. The readout operation left all cores in the "0" state. During switching of the core, a current pulse would be induced in a sense line if the sense line passed through the core. If a sense line passed through a core, a current pulse would be induced in it when the core switched and this would indicate a data "1." If the sense line did not pass through the core, there would not be a current pulse on the line when the core switched, and the absence of a current pulse indicated a data "0." The outputs of the sense lines were used to generate logic patterns in registers.

Computer program

The computer unit used in the Command Module was called the Apollo Guidance Computer and the computer unit used in the Lunar Module was called the LM Guidance Computer. The computer units were the same. The computer program in the fixed memory was of course different for the Command Module and Lunar Module because of their different missions. The computer program for the Command Module of Apollo 11 was named COLOSSUS 2A and the program for the Lunar Module was called LUMINARY, revision 99.

Most of the software was written in an assembly language named Yul. Assembly language, a step up from machine language, assigns lines of code to basic central processor functions. Examples are the arithmetic functions of add, subtract, multiply, and divide. A line of code representing "add" would read "AD M," specifying that the content of memory location M is added to the accumulator. The accumulator was registered A in the central processor. Likewise, the operation of multiply would read as "MP M" specifying that the content of memory location M be multiplied the contents of the accumulator. The contents of M and the accumulator were single precision numbers so the product was a double precision number. The most significant word of the product was stored in the accumulator register A and the less significant word of the product was stored in register L of the central processor. Another instruction commonly used was the "read" operation that was indicated by "READ XY." It moved the content of Input/Output register XY into the accumulator.

Other portions of the software involving operations that could be performed slower and often with double precision, such as calculations to determine spacecraft position, were written in a higher-order "Interpreter" language. Interpreter used a collection of assembly language functions. Execution of Interpreter code was slower than if the operations were written in assembly language, but it reduced the amount of memory required, and it simplified the coding process.

A printout of a listing of the COLOSSUS 2A computer program on an 11 by 15-inch fanfold paper was a stack about 6 inches high. Fanfold paper, fed through a line printer, was common at the time.

Display and Keyboard (DSKY)

The display and keyboard (DSKY) provided the man-machine interface between the astronauts and the computer. It allowed the astronauts to select various computer programs and command certain functions of the spacecraft and enter data necessary for computer computations, and it presented readouts of pertinent data during the various phases of the mission.

The DSKY was a rugged piece of equipment with substantial structure to contain the electronics, display, and keyboard. The unit was 8 inches high, 8 inches wide, and 7 inches deep. It weighed 17.5 pounds. A drawing of the face of the DSKY is shown on the next page (Fig. 6.15).

A caution panel occupied the upper left region of the face of the DSKY and displays occupied the upper right region. A keyboard occupied the lower portion.

Two DSKY units were used in the Command Module. One was mounted on the instrument panel for access by the commander and the Command Module plot. The other was mounted at the navigation station near the optical telescope and sextant. That DSKY was used in conjunction with celestial sightings by the sextant to perform alignment of the inertial platform.

Communication between the crew and the computer was conducted through a simple but effective system of "Verbs" and "Nouns" that were described earlier. Each Verb and each Noun was identified by a two-digit number. Verbs generally requested an action be taken and Nouns indicated to what entity the operation is to be taken. In addition, a particular computer program was selected by keying in a two-digit number.

Referring to the drawing of the DSKY, the display region on the upper right side of the DSKY displayed computer activity, program number, and three lines of five-digit data plus a sign for each line. In the top row, a caption CMPTR ACTY would light up when the computer was occupied running an internal sequence.

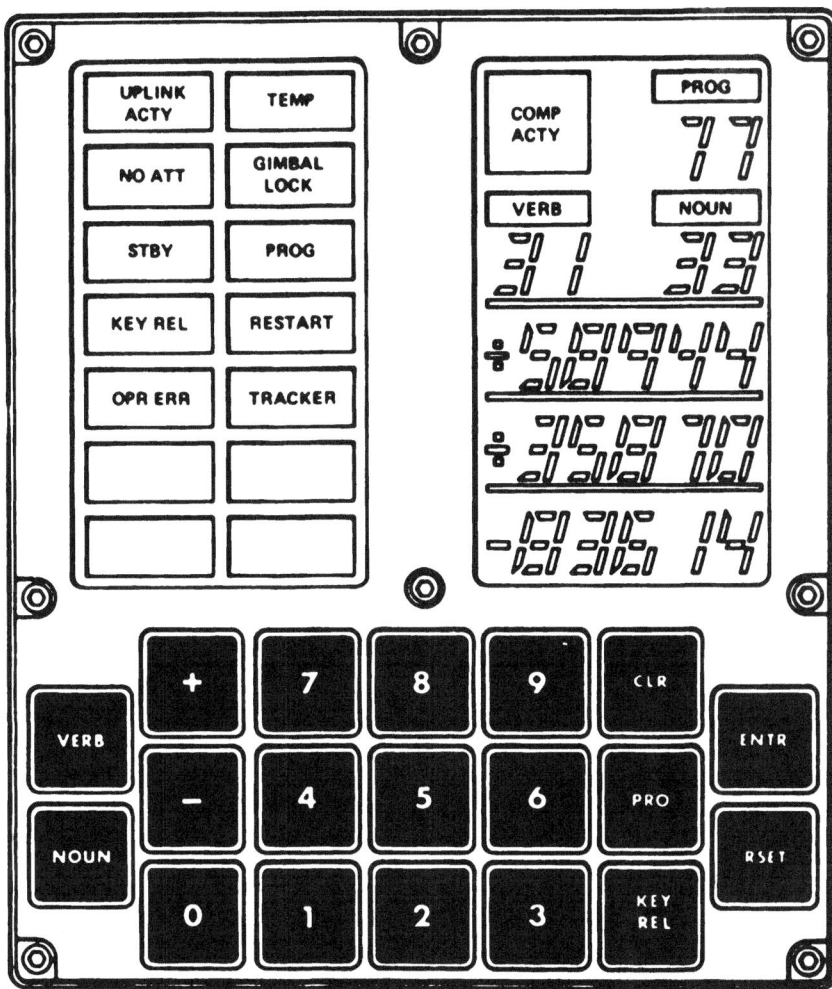

Fig. 6.15 Drawing of face of display and keyboard (NASA graphic)

The program number being run and numbers representing specific Verbs and Nouns were displayed under the captions indicated. The digits were composed of seven-segment green electroluminescent display elements. The display elements required high excitation voltage that was fed through an individual relay to each element.

The upper left side of the Command Module DSKY contained a Warning/Caution panel for ten different events. A particular caption would light up when the computer sensed an error or that a warning was needed.

The astronauts communicated with the computer by first requesting a particular computer program by entering a two-digit number. The action desired from that program was then selected by pressing VERB followed by two numbers and then pressing NOUN followed by two numbers. There were 44 different programs that could be requested from the Command Module computer along with 99 different verbs and 99 different nouns.

The use of a keyboard with VERB and NOUN functions identified by two numbers rather than use an alphanumeric keyboard simplified the DSKY and its interface with the computer. NASA documents indicate that the arrangement was relatively easy to use and it was liked by the astronauts.

The computer programs in the fixed memory were labeled P00 through P79. Several numbers were not used and there were a total of 44 active programs. A few of these programs used early in the mission and their names are listed below as examples.

P00 – AGC Idling

- P00 was used to maintain the computer in a condition of readiness to accept other programs. It was also used when data was being uploaded from the ground.

P01 – Prelaunch or Service Initializing

- P01 was used on the launch pad to perform coarse alignment of the inertial platform. When finished, program P02 was automatically selected by the computer.

P02 – Prelaunch or Service Gyrocompassing

- P02 was used on the launch pad to perform gyrocompassing to seek true north and to fine align the inertial platform.

P11 – Earth Orbit Insertion Monitor

- P11 was automatically selected by the computer when a discrete signal indicating liftoff of the launch vehicle was received. The program could also be selected by manually keying in P11. P11 performed several functions:

    - It generated attitude error signals for the flight Director Attitude Indicator needles.
    - It allows the computer to takeover steering of the Saturn first stage in case of failure of the Saturn guidance system.
    - It computed and displayed trajectory parameters. Verb 06 and Noun 62 were selected to read out three parameters on the three five-digit display fields.

        - Register 1: Inertial velocity (in feet per second)
        - Register 2: Altitude rate (in feet per second)

- • Register 3: Altitude above launch pad (in nautical miles with the last digit representing 0.1 miles)
- – After burnout of the third stage, Verb 16 and Noun 44 were selected in P11 to monitor orbital parameters in the three display registers.
  - • Register 1: Apogee altitude (in nautical miles with the last digit resenting 0.1 miles)
  - • Register 2: Perigee altitude (in nautical miles with the last digit presenting 0.1 miles)
  - • Register 3: Time for freefall (TFF) (in minutes and seconds). TFF is time to freefall to 300,000 feet where atmospheric heating would begin if orbital speed has not been obtained and the engines shut down.

Other keys on the DSKY and their function are listed below.

ENTR – The Enter key was pressed after keying in a program number, verb number, or noun number to enterinformation into the computer.
CLR – The Clear key was pressed to clear the data just entered.
RSET – The Reset key was used to extinguish caution lights on the DSKY that had been set by the computer.
KEY REL – The Key Release key was used to release the displays from keyboard control so the computer could display internal program information.
PRO – The Proceed key was used to inform the computer that the crew accepts the data presented and to proceed.

The Warning/Caution panel in the Command Module DSKY had ten captions. A particular caption would light up when the computer sensed an error or that a warning was needed. The meaning of the captions is given in the table below (Table 6.1).

Table 6.1 Captions on DSKY caution panel

Caption	Meaning
UPLINK ACTY	Data being received from ground
TEMP	Temperature of stable platform out of tolerance
NO ATT	Attitude reference not available from IMU
GIMBAL LOCK	Middle gimbal angle greater than 70 degrees
STBY	Computer is on standby
PROG	Computer waiting for information to be entered by crew
KEY REL	Computer needs control of DSKY to complete a program
RESTART	Computer is in restart program
OPR ERR	Computer detected a keyboard error
TRACKER	One of the optical coupling units failed

Inertial Subsystem

The inertial subsystem was the primary navigation element in the Command Module. The state vector was continuously updated by the computer from data from the inertial subsystem. Periodic platform realignment was performed by the astronauts using star sights by the sextant. The main elements of the inertial sub-system were the navigation base and the inertial measurement unit (IMU).

Navigation Base

The navigation base provided a stable base upon which the inertial measurement unit and the optical sextant and scanning telescope were mounted in precise align-ment. The navigation base was a rigid structure 27 inches long, 22 inches wide, and 4.5 inches thick and weighed 17.4 pounds. It was mounted to the side of the Command Module so that the optical elements had clear lines of sight to the out-side. Shock mounts were used to isolate the optics and IMU from shock and vibra-tion from thrusting. A NASA drawing showing the location of the navigation base, inertial measurement unit, and optical elements on the Command Module is shown on the next page (Fig. 6.16).

Inertial Measurement Unit

The inertial measurement unit (IMU) established a stable platform fixed in inertial space from which measurements of spacecraft attitude and acceleration were made. The stable platform was maintained in its set alignment in inertial space by using output signals from three inertial rate integrating gyros. Spacecraft attitude was determined by measuring the angular differences between the coordinate axes of the spacecraft and that of the stable platform. Velocity in the stable platform coordinate system was determined by integrating the outputs of three accelerom-eters mounted on the stable platform. The platform was generally aligned with its X-axis parallel to the direction of thrust, which was the X-axis of the spacecraft.

The inertial measurement unit that flew to the moon and back on Apollo 17 is presently on display at the Udvar-Hazy Center of the Smithsonian National Air and Space Museum (NASM) near Washington, DC. A photograph of that IMU package is shown on the next page. The IMU package was 13 inches wide and 10.5 inches high. It weighed 42.5 pounds (Fig. 6.17).

The main element of the IMU was a three-gimbal stable platform. The platform held three orthogonally mounted Inertial Rate Integrating Gyros (IRIGs) and three orthogonally mounted pulsed-integrating pendulous accelerometers (PIPAs). The three gimbals gave the platform 3 degrees of freedom. The stable platform was

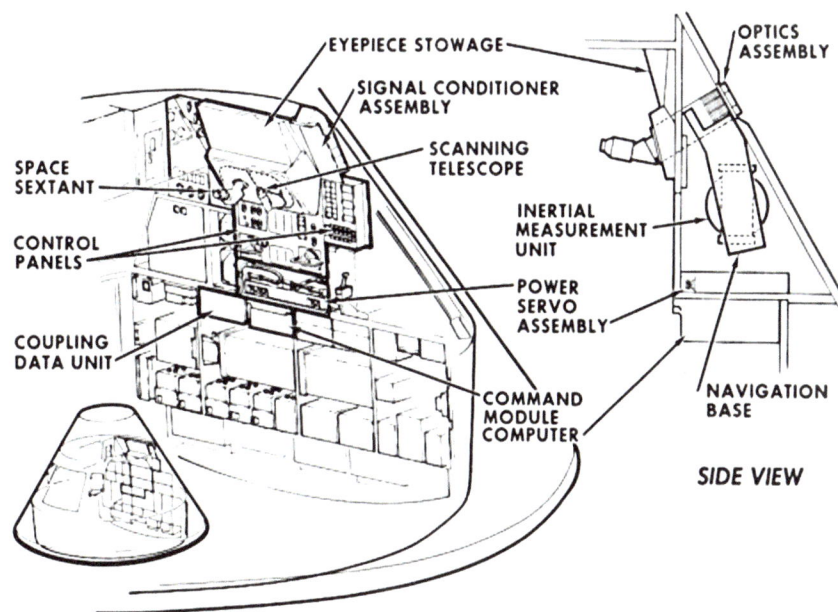

Fig. 6.16 Mounting of navigation base, inertial measurement unit, and optics assembly (NASA graphic)

Fig. 6.17 Inertial measurement unit that flew on Apollo 17 (NASM photograph)

maintained in alignment in inertial space by using error signals developed from its three gyros. Those signals fed three stabilization loops that drove three torque motors that in turn rotated individual gimbals to maintain the platform fixed in inertial space.

A schematic drawing of the IMU from MIT Report R-500 is shown below. The stable platform was mounted to the inner gimbal of the three-gimbal assembly (Fig. 6.18).

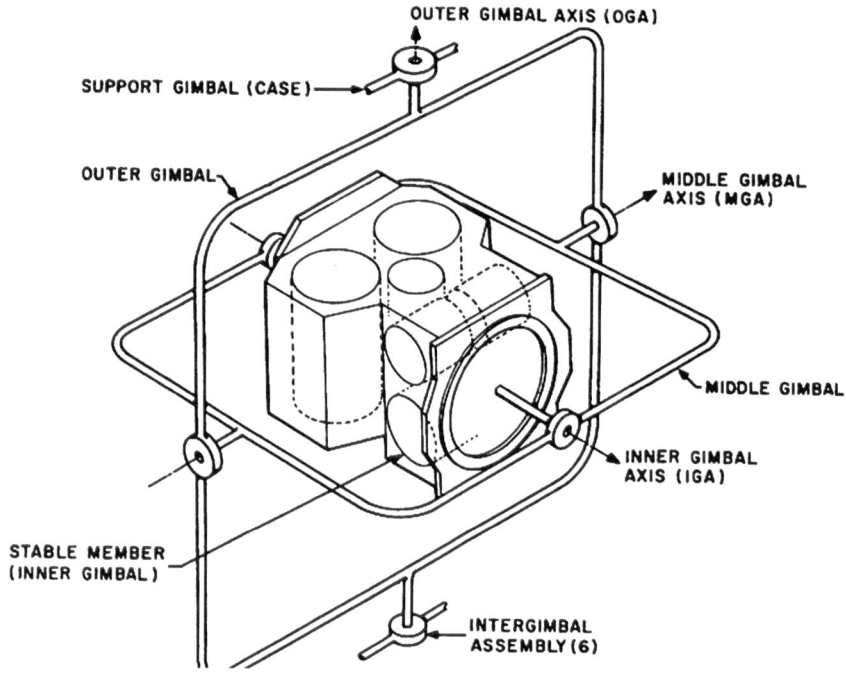

Fig. 6.18 Schematic of inertial measurement unit (MIT graphic)

The stable platform was machined from a block of Beryllium with holes bored to mount the gyros and accelerometers. Beryllium was chosen because of its relatively light weight, good temperature stability, and high strength. Compared with aluminum, for example, it is about a factor of 1.5 lighter, has a factor of 1.9 less linear expansion with temperature, and has 1.9 times higher tensile strength. The platform was irregular in shape with an extent of about 6 inches.

The gyros were cylindrical in shape with a diameter of 2.5 inches and length of about 3.3 inches. The accelerometers were cylindrical with diameter of 1.6 inches and length of about 2 inches. The three gyros were mounted mutually perpendicular to each other on the platform structure. The three accelerometers were mounted mutually perpendicular on the platform and in alignment with the gyros.

A torque motor was attached to each of the gimbal shafts to maintain the platform in its fixed orientation in space as the spacecraft maneuvered. A resolver, which converted angular motion to an electrical signal, was attached to each gimbal shaft to allow measurement of spacecraft attitude relative to that of the stable platform.

The accelerometers on the platform sensed change in velocity along their sensitive axis and outputted a pulse train proportional to the change in velocity. These pulse trains were integrated by the computer to maintain a running update of velocity along each axis. The computer then integrated velocity to obtain constantly updated position of the spacecraft.

The gyros had a slight drift and that required periodic realignment of the platform. The mission checklist called for the first realignment to be made after the spacecraft reached earth orbit and began a coast period. This early realignment was made since the considerable acceleration and vibration associated with the rocket thrusting could disturb the alignment.

The verbal transcript of the early portion of the Apollo 13 flight contains Jack Swigert's transmission to Houston of the angle differences found during the first realignment. Those differences in the X, Y, and Z planes were -00.067, 00.000, and 00.162 degrees, respectively. Spica and Antares were the primary stars used for the realignment. The difference in the measured angle between stars and the actual angle between stars from the computer tables was reported by Swigert to be 00.000 degrees.

Upon receiving the data, Joe Kerwin, CAPCOM in Houston, chided Swigert for the excellent readings by saying "Well, that sounds marginally acceptable." To which, Swigert replied "For a new CMP (Command Module Pilot), it ain't too bad." The gyros were torqued to take out the small drift errors. The alignment procedure is described in the Optics Subsystem section of this chapter.

Analysis by MIT given in MIT report R-500 "Space Navigation Guidance and Control" indicates that the probable accuracy of aligning the Apollo stable platform using star sights was within 0.1 milliradians (0.0057 degrees). They concluded that this accuracy exceeded requirements by a comfortable margin.

There was a good opportunity to measure the drift rate of the gyros used in the IMU, while the Lunar Module of Apollo 11 was sitting on the lunar surface. The IMUs used in the Command Module and in the Lunar Module were the same. Data analyzed during a 17.5 h period between alignments while on the lunar surface indicated gyro drifts in the X, Y, and Z axes of 0.707, −0.73, and 0.623 degrees, respectively. The largest drift, −0.73 degrees, corresponds to a respectably low drift rate of 0.0417 degrees per hour.

Optical Subsystem

The main elements of the Optical Subsystem were a sextant and a scanning telescope. Both instruments were mounted to an optical base with their optical axes aligned. The optical base in turn was mounted to the navigation base. Pointing the optical axis of the optical instruments toward a star or to a landmark required maneuvering the spacecraft.

The scanning telescope had unity magnification and a relatively wide field of view of 60 degrees. It was used for general celestial viewing and to perform coarse alignment of the inertial platform. The telescope line of sight had 2 degrees of freedom. One was rotation of the instrument about the optical axis and that was called the shaft drive. The other degree of freedom was achieved by rotating a prism at the input to the instrument that shifted the line of sight to the side. The prism was rotated by the trunnion drive. The objective lens of the telescope was 32 mm in diameter with a 27.4 mm focal length. An illuminated reticle in the optical path assisted the operator in aligning the telescope to the target.

The optical path of the sextant had a magnification factor of 28 and a field of view of 1.8 degrees. The objective lens was 40 mm in diameter. An illuminated reticule in the optical path assisted in aligning the instrument to stars.

The sextant had a fixed and a moveable line-of-sight reticle. The fixed line of sight was coincident with the axis of the instrument. The fixed line of sight was pointed at a star or landmark by positioning the spacecraft. A second line of sight in the sextant was achieved by rotating the shaft of the instrument and by rotating an indexing mirror with a trunnion drive to point to a second star. The indexing mirror directed the image of a second star to a reflecting mirror. The mirrors allowed sighting a target up to 57 degrees displaced from the shaft axis. The image from the reflecting mirror was passed through a beam splitter along with the fixed line-of-sight image and through the optics of the sextant.

The image of a second star could be brought into coincidence with that of a star in the direct line of sight by rotating the shaft and trunnion drive of the instrument. The angle between the two stars was computed from the angular displacement of the shaft and trunnion drives required to obtain coincidence. The accuracy of measuring angle between two stars by the sextant was reported to be within 10 arc seconds (0.0028 degrees).

The optical assembly was mounted in the Command Module so that the eyepieces for the telescope and sextant were about eye level for an astronaut standing on the aft bulkhead. The aft bulkhead was just above the main heatshield at the bottom of the spacecraft. The operator would be standing with his body approximately parallel to the spacecraft X-axis. The center seat would be folded up to give him room to stand.

A control panel for the optical assembly was located just below the panel that held the telescope and sextant. A display and keyboard (DSKY) was mounted just to the right of the optical subassembly to allow the operator to communicate with the computer during star sights and alignment of the inertial platform. Mounting provisions for a hand rotation controller was located just below the DSKY. The rotation controller was used to maneuver the spacecraft to align the telescope and sextant to a star or landmark. The portion of the spacecraft containing the optics and controls was known as the navigation station.

A photograph of astronaut Jim Lovell looking through the telescope during flight of Apollo 8 is shown below. Lovell is standing at the navigators station with his right hand on the joystick controller for fine attitude change of the spacecraft (Fig. 6.19).

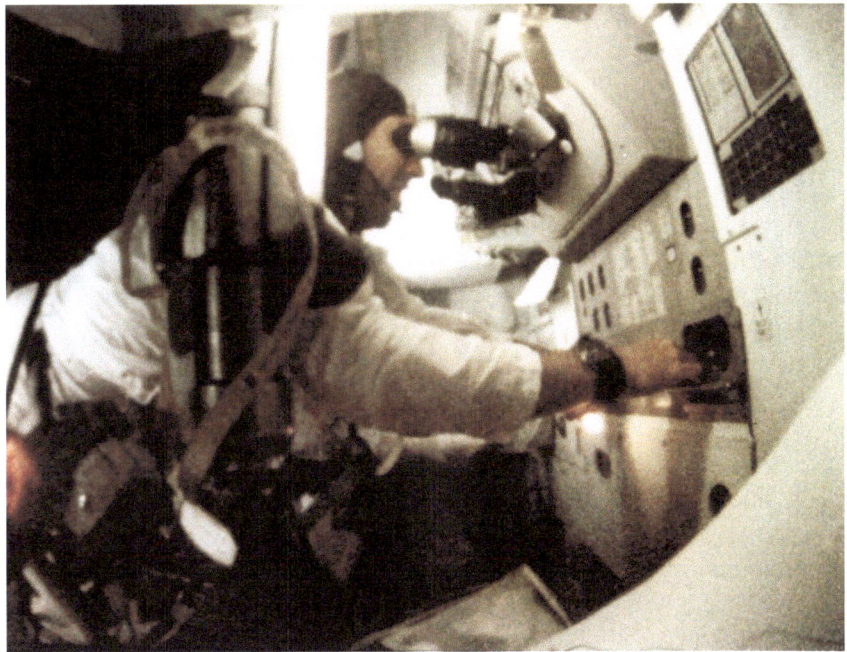

Fig. 6.19 Astronaut Jim Lovell looking through telescope during Apollo 8 mission (NASA photograph)

STABILIZATION AND CONTROL SYSTEM

The stabilization and control system (SCS) controlled spacecraft attitude, controlled thrusting for orbit control, and provided the man-machine interface with controls and displays for the astronauts. The SCS was comprised of the attitude reference subsystem, attitude control subsystem, and thrust vector control subsystem.

Attitude Reference Subsystem

The attitude reference subsystem provided a backup attitude reference in case of failure of the inertial measurement unit. It also provided attitude and attitude rate data to the Flight Director Attitude Indicators and to the thrust vector control subsystem.

The attitude reference system included two body-mounted attitude gyro (BMAG) assemblies identified by "1" and "2." Each assembly contained three gyros aligned with the three spacecraft axes of roll, pitch, and yaw. The gyro assemblies sensed either attitude displacement or to attitude rate. Rate was sensed by caging the gyro.

In one common mode of flight, "attitude hold," attitude signals from the three gyros in gyro assembly "1" were processed and used to operate the reaction control jets to return the spacecraft to the desired attitude. The error signal could also be displayed on the needles of one or both FDAI units to allow manual control to the desired attitude if that mode is selected. Outputs from gyro assembly "2" were used to measure attitude rates, and those signals could be read on the flight directors and also used by the computer to compute attitude changes.

The output signals from the two gyro assemblies were applied to a Gyro Display Coupler (GDC) which conditioned the signals for application to the computer and to the FDAI. The outputs of the GDC could be aligned to the IMU or other references set by the Attitude Set Control Panel by pressing the GDC ALIGN pushbutton on the main control panel.

Attitude Control Subsystem

Normally, most of the control functions of the CSM, including attitude control, were orchestrated by the guidance computer. However, manual control of key functions could also be selected. The hand controllers managed firing of reaction control system engines when the spacecraft was in a manual control mode. There were four groups of a cluster of four reaction engines around the periphery of the service module. These clusters, which were 90 degrees apart on the periphery, were used to maneuver the Command and Service Module. There were also 12 reaction control engines mounted on the Command Module, but those engines were only used after separation of the Command Module from the Service Module just prior to returning to earth. The reaction control system is described later in this chapter.

A control panel, referred to as the Attitude Set Control Panel allowed the astronauts to set in the desired attitude of the spacecraft. The panel was located on the lower left corner of the portion of the main instrument panel in front of the Commander. The panel contained three thumbwheels and three associated displays of angle. The displays were labeled ROLL, PITCH, and YAW. Roll and pitch displays were marked from 0 to 359 degrees and yaw display was marked from 0 to 90 degrees and 270 to 359 degrees. In addition, a narrow rotating drum with marks each 0.2 degrees was present to the right of the numerical displays.

The attitude set control panel developed signals that were differences between the set-in attitudes of roll, pitch, and yaw and attitude information from the IMU or from the GDC. The source of input information, IMU or GDC, was selected by the ATT SET switch located on the main instrument panel in front of the Commander. The angle differences were sent to the FDAI needles and that indication allowed the astronauts to maneuver the spacecraft to reduce the errors to zero. The difference could also be used to align the GDC to the set-in angles.

Thrust from the reaction control engines was either on or off. As a result, the attitude of the spacecraft would slowly oscillate between set limits when in steady flight. This control mode was referred to as a limit-cycle mode. The digital autopilot (DAP) implemented in software in the computer accommodated limit cycling as a stable control operation. The angle between limits, referred to as the deadband, could be set from switches on the main instrument panel. The angular rate of motion between limits could also be selected. In the automatic mode with the computer controlling the attitude and the ATT DEADBAND switch set to MIN and the RATE set to LOW, the deadband was ±0.5 degrees and the rate was ±0.2 deg/sec. In the DEADBAND switch position of MAX and RATE set to LOW the deadband was ±5.0 degrees.

Hand Controllers

Two types of hand controllers were used to maneuver the CSM by the astronauts. The first was a rotational controller used to rotate the spacecraft about its pitch, yaw, and roll axes. The second was a translational controller used to maneuver the spacecraft in a rectilinear fashion along the X, Y, and Z axes. A close-up picture of the rotation controller on the right side of the commander's seat and the translation controller on the left side of the seat is shown on the next page (Fig. 6.20). The close-up view was extracted from the photograph of the interior of the Command Module given previously.

The rotational controller at the right of the Commanders seat had an electrical cable nine feet long that let it be mounted in various places in the crew compartment, including the navigation station. A dovetail arrangement at the back of the mounting box allowed quick mounting to a complementary dovetail fixture.

Similar to a control stick in an aircraft, fore-aft motion commanded attitude change in the pitch plane and side-to-side motion commanded roll attitude change. Yaw attitude change was commanded by twisting the grip. A push-to-talk switch was located on the upper front side of the grip.

The translation controller was mounted to the left armrest of the Commander's couch by a dovetail type fitting. It had a "T"-shaped handle that can be seen in the photograph. Pushing the handle in or pulling it out caused translation of the

Fig. 6.20 Close-up view of rotational and translational controllers (NASM photograph, cropped by author)

spacecraft along the X-axis. Moving the handle up and down caused Z-axis translation. Moving the handle from side to side caused translation along the Y-axis of the spacecraft.

The translation controller served other important mission functions as well. Rotating the handle clockwise transferred thrust-vector control from the Primary Guidance, Navigation, and Control System (PGNCS) to the stabilization and control subsystem (SCS). The SCS was a backup system to the PGNS. Rotating the handle counterclockwise and holding for 3 seconds issued a CSM/Saturn IVB abort command to the mission sequencer. A serious command indeed.

Switches used to select control modes of the rotational controller were located on the instrument panel down and to the left of the Commander's flight director attitude indicator. Three switches labeled Roll, Pitch, and Yaw under the heading Manual Attitude could be set to one of three settings: acceleration command, rate command, or minimum impulse. In the acceleration position moving the rotational controller commanded a rotation rate along that axis that continued to accelerate until the control was brought back to center position.

In the rate position, moving the rotational controller caused an angular rate proportional to the amount of deflection in a given axis. This was the usual mode of operation. The maximum rate at full deflection was set by a switch with settings for low or high. In the low position, the maximum rates were 0.7 degrees per second along all axes. In the high position, the maximum rates were 7.0 degrees per second in the pitch and roll axis and 20 degrees per second in the roll axis. In the minimum impulse position, only one thruster was fired and for only 15 milliseconds when the rotational controller was moved. This mode was used for fine adjustment of attitude.

REACTION CONTROL SYSTEM IN COMMAND MODULE

The reaction control system in the Command Module enabled attitude control after the Command Module separated from the Service Module just before reentering the earth's atmosphere. Attitude of the spacecraft relative to the velocity vector was critical for successful entry into the atmosphere.

The Command Module contained two independent reaction control subsystems for redundancy referred to as subsystem 1 and subsystem 2. Each subsystem had a separate set of six rocket engines directed to provide clockwise roll, counter-clockwise roll, +pitch, −pitch, +yaw, and -yaw.

The reaction engines were 11.7 inches long, and the nozzle exit diameter was 2.13 inches. Each engine generated 93 pounds of thrust. The engines were mounted internally with exits of the engines flush with the outer surface of the Command Module. Since the engines were mounted internally, they were ablatively cooled and that limited their operating life. The engines, which were rated for service life of 200 seconds and 3,000 operational cycles, were developed and built by the Rocketdyne Division of North American Aviation.

The engines used monomethyl hydrazine (MMH) for fuel and nitrogen tetroxide (N_2O_4) as the oxidizer. This combination ignited upon contact with one another. Each of the two redundant reaction systems had a fuel tank, an oxidizer tank, and a helium tank. Helium was used to pressurize the fuel and oxidizer. The helium tank was about 9 inches in diameter and contained helium at a pressure of 4,150 psi. A pressure regulator reduced the pressure to about 180 psi to pressurize the fuel and oxidizer tanks.

Fuel was contained in cylindrical tanks 17 inches long and 12.5 inches in diameter with hemispheric domes. The tanks were built out of titanium alloy. Each tank held 45 pounds of fuel. The oxidizer tank was similar to the fuel tank but 3 inches longer at 20 inches long. Each oxidizer tank held 89 pounds of oxidizer. The tanks had Teflon bladders that held the fuel and oxidizer. When helium pressure was applied to the tanks, it squeezed the bladder and forced the propellants into manifolds that fed the engines.

The fuel and oxidizer lines connected to feed injector valves near each engine. The injector valves were spring loaded closed, and they were opened to admit propellants by energizing solenoid coils. There were two sets of coils in each injector, one for automatic control and one for direct control from the hand rotational controller. Control signals for the automatic solenoid coils were generated by the Reaction Jet Engine ON-OFF Control Assembly (RJ/EC).

The engines could be pulse fired for short impulses lasting less than 70 milliseconds or fired continuously for several seconds. Input signals from the guidance computer instructed operation of the RJ/EC when the spacecraft was under computer control, and signals from the Electronic Control Assembly instructed operation of the RJ/EC when the spacecraft was being controlled by the stabilization and control system.

A backup manual control mode could be selected by the astronauts by turning the "Direct RC" switch on the instrument panel to the ON position. The rotational hand controller was then positioned against the stop in the axis to be controlled to activate the "Direct" switch. Closing the Direct switch powered the direct solenoid coils in the injector valves of the appropriate thrusters so that they could be operated by the rotational controller.

The reaction control system used 41 pounds of propellants during maneuvering to establish the proper attitude for reentry of the Apollo 11 Command Module into the earth's atmosphere. A problem was encountered with one of the negative yaw engines while in the automatic mode, but the engine responded normally to commands when in the manual mode. NASA Technical Note TN D-7151 indicates that 65 seconds after separation from the Service Module, one of the two redundant reaction control systems was deactivated, presumably the one with the engine problem.

The hypergolic propellants posed a risk during a hard landing so the remaining RCS propellants were burned off while the spacecraft was descending on the main parachute. The tanks and lines were then purged by flowing helium through them.

ELECTRICAL POWER

Three storage batteries referred to as entry and post-landing batteries powered the Command Module after separation from the Service Module. They were also used to augment power from the fuel cells during peak load demands such as when positioning the gimbals of the SPS rocket engine. The three batteries were referred to as battery A, battery B, and battery C.

The batteries, which were produced by the Eagle-Picher Company, were silver oxide/zinc type. Each battery consisted of 20 cells packaged into a battery 11.75 × 5.75 × 6.88 inches in size and weighing 28.5 pounds. The batteries were rated at

40-ampere hours each. The open circuit voltage of the battery was 37.2 volts and the nominal voltage at a heavy load of 25 amperes was about 29 volts.

Battery A was connected to battery bus A and battery B was connected to battery bus B. The battery buses in turn could be connected to the main DC power buses. Normally, only batteries A and B were connected to the main power buses. Battery C was available as backup.

The batteries were charged by a battery charger that was turned on and off manually by the astronauts. The battery charger used inputs from the 28 volt bus as well as from the 115 volt, 400 Hz three-phase ac bus to develop a charging voltage of about 40 volts. The battery to be charged was disconnected from the battery power bus, and the charging voltage was applied through a battery charge selector switch.

Monitoring of the voltages of the batteries, fuel cells, and the two main DC busses was selected by a DC INDICATOR switch on the instrument panel and displayed on a voltmeter on the panel. The current flowing in the selected circuit was displayed on an ammeter on the panel.

There were also two relatively small silver oxide/zinc batteries in the Command Module that was used to power a series of pyrotechnic devices. Those devices were used to jettison the escape tower at after launch, separate the S-IVB booster from the CSM, separate the Command Module from the Service Module, and deploy and separate the parachutes during the landing. The batteries were rated at 0.75 ampere hours each and the open circuit voltage was 37.2 volts. They were $6.25 \times 2.64 \times 2.87$ inches in size and weighed 3.5 pounds. The pyrotechnic batteries were not charged during the flight. Pyro battery A was connected to Pyro bus A, and Pyro battery B was connected to Pyro bus B.

Inverter

Some components of the spacecraft required alternating current (AC) power at 400 Hz, and that was generated by solid-state inverters. The inverters converted 28 volt dc power from the dc power busses to ac power. Devices using ac power included the various pump motors, fans within the cryogenic tanks, compressors for the pressure suits, and the battery charger.

There were three inverters, each capable of producing 1,250 volt-amperes of three-phase, 115 volt, 400 Hz power. Two of the inverters were used at a time to feed two ac power busses. The third inverter was a backup in case of failure. The inverters were controlled by switches on the instrument panel and performance was monitored by an AC voltmeter through a selector switch on the instrument panel.

The inverters were quite substantial at $15.0 \times 14.75 \times 5.0$ inches in size and weighing 53 pounds each. They were mounted to cold plates and cooled by the water/glycol cooling loop that transferred heat to radiators on the Service Module.

TELECOMMUNICATIONS SYSTEM

An effective telecommunications system was essential to the Apollo mission. The telecommunications system in the Command Module (CM) included two major subsystems to communicate outside of the CM: the Unified S-Band communications subsystem and the VHF communications subsystem. In addition, it included an intercom for communications among the astronauts and the capability to record audio communications and spacecraft data by a tape recorder.

The term "S-band" refers to a frequency band extending from 2.0 to 4.0 GHz. The term VHF is known as the very high frequency band extending from 30 to 300 MHz. The telecommunications system used discrete frequencies in these two frequency bands.

The unified S-band carrier system was used for long-distance voice and data communications between the Apollo spacecraft and the Manned Spaceflight Center (MSC) in Houston. That communications link was conducted through the Manned Spaceflight Network (MSFN). The MSFN was a network of ground stations around the earth with large steerable antennas capable of communicating with distant spacecraft. The S-band system was employed in both the Command Module and the Lunar Module to communicate with MSC in Houston via the MSFN.

The VHF link was used to communicate between the CM and the LM and between the CM and the MSFN when near the earth. VHF was also used for communication during the recovery process at the end of the mission as the Command Module descended under the parachutes. VHF communications is well suited for applications requiring wide angular coverage using relatively simple small antennas on the spacecraft. VHF has long been used for civil aviation communications for these very reasons.

A functional block diagram of the telecommunications system in the Command Module is shown on the next page (Fig. 6.21). The diagram is a composite of diagrams from different sources assembled by the author to show the various functions in a simplified form.

Telecommunications Antennas

The Unified S-band subsystem included a high-gain antenna and four flush-mounted omnidirectional antennas. The omnidirectional antennas were spaced 90

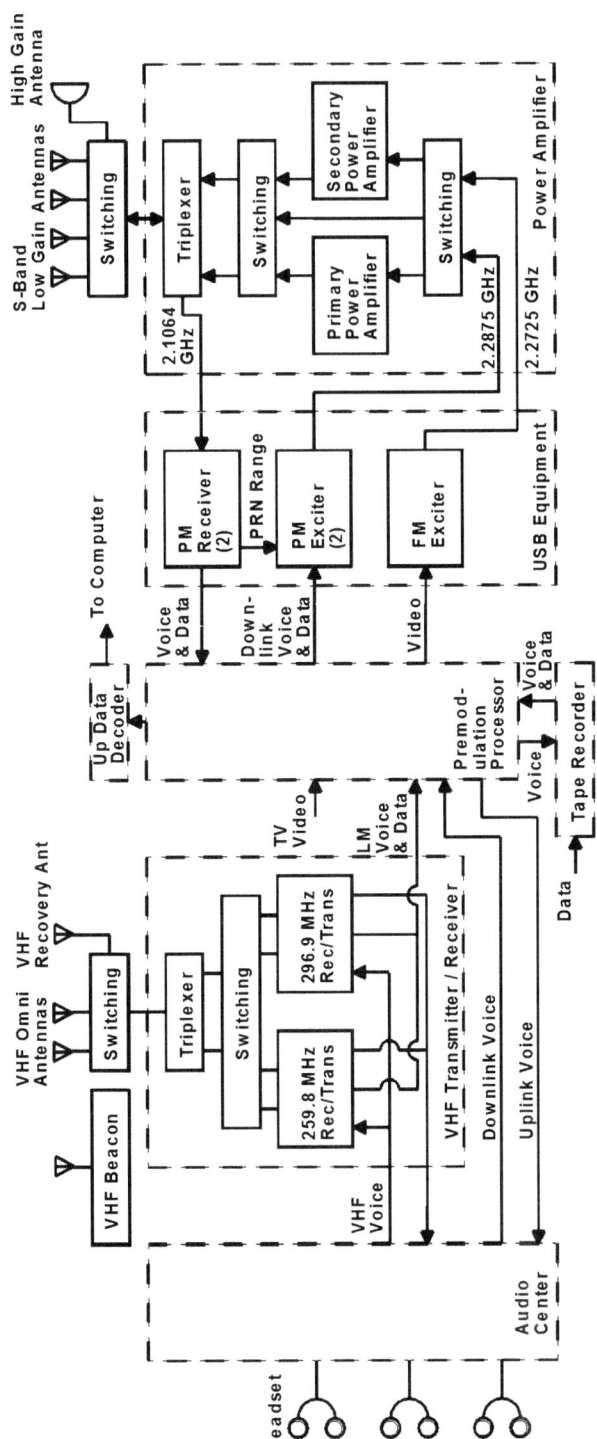

Fig. 6.21 Functional Block Diagram of Telecommunications System in Command Module

degrees apart on the periphery of the Command Module. Those wide beamwidth, low-gain antennas were used during earth orbit and up to the time of deployment of the high-gain antenna after insertion into lunar transfer orbit. Switches on the instrument panel in front of the Lunar Module pilot allowed the astronauts to select the omni antenna most favorably oriented toward a MSFM site or to select the high gain antenna.

The high-gain antenna was steerable and it had selectable beamwidths. It was made up of four truncated parabolic antennas surrounding a square horn antenna with a diagonal dimension of 11 inches. It was mounted near the engine end of the Service Module. Switches on the instrument panel in front of the LM pilot allowed selecting the horn, one of the parabolic dishes, or the combined four parabolic dishes. The 3-dB transmitted beamwidths for those three conditions were 40 degrees, 11.3 degrees, and 4.4 degrees, respectively. The corresponding transmit antenna gains were 8.0 dB, 18.0 dB, and 25.7 dB, respectively.

A control subpanel for the high-gain antenna was located on the center right portion of the instrument panel in the Command Module. The subpanel contained a switch to select MAN (manual) or AUTO (automatic) positioning of the antenna or to select REACQ (reacquire) to cause the antenna to search in angle for the signal. There was also a beamwidth selection switch with positions of WIDE, MED, and NARROW. Once the antenna was positioned within few degrees of a line to the MSFM site, it could be switched to automatic angle tracking.

The subpanel contained two knobs to position the antenna in pitch and yaw. There were two corresponding displays that gave pitch angle and yaw angle of the antenna. A signal strength meter on the panel assisted in positioning the antenna for best signal strength and it gave readout of signal strength during automatic tracking.

Continuing with antennas on the Command Module, the VHF subsystem communicated through two scimitar-type antennas mounted on opposite sides of the Service Module. Each antenna generated a nearly hemispheric radiation pattern. One or the other could be selected by the VHF Antenna switch located on the instrument panel to provide nearly full spherical coverage. The antennas were known as a scimitar type because their shape resembled a scimitar sword blade. There was a third antenna switch position that allowed switching VHF communications to the recovery whip antenna.

Two whip-type antennas, 11 inches long, deployed automatically from the top of the Command Module shortly after the main parachutes opened during the recovery process at the end of the mission. One of those antennas was connected through the VHF antenna switch to the VHF transmitter and receiver and used for voice communications during recovery. The other whip antenna was connected to a VHF recovery beacon.

S-Band Uplink Signal from the Earth

One of the MSFN stations on the rotating earth with the best angle to the spacecraft transmitted an uplink carrier at a frequency of 2.1064 GHz. The carrier was phase modulated by a 30 kHz subcarrier and a 70 kHz subcarrier. The 30 kHz subcarrier was frequency modulated by voice communication from MSC in Houston. The 70 kHz subcarrier was frequency modulated by uplink data for the spacecraft that was developed at MSC.

The uplink signal was picked up by the selected S-band antenna on the CSM and applied to a triplexer and onto the receiver within the unified S-band equipment as shown in the block diagram. NASA documentation refers to the collection of S-band receiving and transmitting equipment as the unified S-band equipment (USB). A phase-locked loop within the receiver developed a coherent reference signal from the uplink carrier that was used to extract the voice and data subcarriers. Those subcarriers were applied to the premodulation processor where the voice and data signals were demodulated. The voice signal was applied to the audio center which conditioned it for the astronaut's headsets.

The data signal was extracted by the premodulation processor and applied to an up-data decoder. The decoded data signal was made available to the computer. One of the key pieces of data uplinked was the state vector of the spacecraft as developed by very accurate tracking equipment back on earth. The uplink carrier could also be phase shift keyed modulated with a pseudorandom code as part of a range measurement arrangement between the ground station and the spacecraft. The code was demodulated in the spacecraft and applied to a phase shift keyed modulator on the downlink signal. A range tracker on the ground determined the time delay for peak correlation of the transmitted pseudo random bit pattern with that returned by the spacecraft. The time delay was equated to range.

The bit rate of the code was 992 kilobits per second and the code was about 5.443 seconds long giving an unambiguous range measurement of about 440,600 nautical miles. The unambiguous range was just beyond the maximum two-way range to the moon. The accuracy of the range measurement was reported in NASA Technical Note D-6723 to be about 30 meters (94 feet). This was very respectable accuracy considering that the range to the spacecraft could be over 219,000 nautical miles.

S-Band Downlink Communications

The CSM transmitted two downlink S-band carriers, one at 2.2725 GHz and one at 2.2875 GHz. The 2.2875 GHz carrier was derived from the uplink carrier and coherent with it. It was developed from the uplink carrier by down converting to an intermediate frequency, extracting the downshifted carrier by a phase-locked loop, and then frequency multiplying back up to S-band to develop a downlink carrier that was higher in frequency than the uplink carrier by a ratio of 240/221.

There were two redundant receivers for the uplink signal and two redundant exciters that developed the 2.2875 GHz downlink signal. The downlink 2.2875 GHz carrier from the two receivers was applied to the two exciter units where it was phase modulated by a 1.024 MHz subcarrier and by a 1.25 MHz subcarrier. The 1.25 MHZ subcarrier was frequency modulated by voice communication from the astronauts. That subcarrier was also modulated by biomedical telemetry data when voice was not being used. Telemetry data from various spacecraft systems was biphase modulated onto the 1.024 MHz subcarrier for transmission back to earth. The output power of the two exciter units was about 220 milliwatts.

The 2.2725 GHz downlink carrier was generated by a separate exciter and used to transmit television signals or scientific data. Television video signals were modulated onto the carrier by wideband frequency modulation.

The power amplifiers developed about 20 watts of S-band power when the astronauts selected HIGH power. The amplifier was turned off and bypassed when the LOW power mode was selected. Switches on the instrument panel in front of the LM pilot allowed him to select any of the three exciters to be connected to either of the two power amplifiers. He could select the PRIM (primary) power amplifier or the SEC (secondary) power amplifier. Another set of switches labeled S-BAND ANTENNA allowed selection of omni antenna a, b, c, or d or the high-gain antenna.

The 2.2875 GHz downlink carrier, which was coherent with the uplink carrier, was extracted by the MSFM equipment back on earth, and the two-way Doppler shift was measured. The Doppler shift was proportional to spacecraft velocity. The spacecraft velocity, which could be as high as 34,000 feet per second, was determined to within a fraction of a foot per second at the MSFN.

After a particular station in the MSFN network had acquired the signal from the spacecraft and determined range by means of the pseudorandom code, further change in range was determined by integrating velocity. The biphase pseudorandom code modulation on the uplink carrier was then turned off to lessen the chance of interference with other uplinked data.

VHF Communications

The VHF link was used to communicate between the CM and the Lunar Module and between the CM and the MSFN when near the earth. It was also used to communicate between the two astronauts as they explored the surface of the moon and between the astronauts on the surface and the LM.

VHF communications equipment in the Command Module included two separate receivers and two separate transmitters along with switching to select various combinations in case of failure of a particular unit. Normally, voice communication was carried on in a simplex mode where the receiver and transmitter were on the same frequency. One set of receiver/transmitters operated at a frequency of 296.8 MHz and the other set operated at 259.7 MHz. A particular receiver/transmitter could be selected by means of a switch on the instrument panel. Alternatively, selection switches allowed transmitting on 298.8 MHz and receiving on 259.7 MHz or transmitting on 259.7 MHz and receiving on 296.8 MHz. Receive only could also be selected on either frequency.

Voice communications from the astronauts was picked up by the microphone that was part of their headsets, conditioned by the audio center, and applied to the modulator of the VHF transmitters. Likewise, voice signals from the VHF receivers were applied to the audio center where they were conditioned to drive the earphones of the headsets worn by the crew.

The VHF communications equipment used on-off amplitude modulation to convey information. Voice signals were amplified and hard limited before being applied to the modulator. The modulator turned the VHF carrier on and off in response to the hard-limited audio signal. This type of modulation, although subject to some distortion, had a benefit of increasing the rms modulation level. This increased intelligibility for weaker signals.

The transmitters generated about five watts of VHF power. NASA report M-932-68-08 indicates that the communication range between the CSM and earth using VHF was about 2,600 nautical miles. This range was with very large antennas and capable transmitting and receiving equipment on earth.

The VHF communications subsystem also contained a ranging function to measure range between the Lunar Module and the Command Module. This range information was used during rendezvous of the LM with the CM as the ascent stage of the LM returned from the moon. During ranging, the 259.7 MHz transmitter in the Command Module was amplitude modulated with three range tones: 247 Hz, 3.95 kHz and 31.6 kHz. These range tones were received by the VHF receiver in the LM, demodulated, and retransmitted back to the CM on 296.8 MHz. The range tones were demodulated by the receiver and processed to determine time delay. The three tones gave an unambiguous range of 327 nautical miles and an accuracy of about 100 feet.

A tape recorder capable of recording both digital data and voice was used to record voice and spacecraft data when the spacecraft was out of communication with the earth as when it was behind the moon. The tape was played back into the S-band downlink when convenient. The mylar tape used was one inch wide and the recorder could record 14 tracks. Tape speeds of 3.75, 15, and 120 inches per second could be selected. The tape speed was automatically selected based on data rate during playback. The recording time available was two hours at 3.75 inches per second and 30 minutes at 15 inches per second.

A small black and white television camera was carried aboard early Command Modules. A long cable allowed it to be handheld or mounted in three different locations within the capsule. The camera had a frame rate of 10 frames per second with 320 lines per frame. The video signal was conditioned by the premodulation processor and then applied to the S-band downlink equipment where it was frequency modulated onto the 2.2725 GHz downlink carrier. Later Apollo missions carried a color camera that provided a frame rate of 30 frames per second with 525 lines per frame.

ENVIRONMENTAL CONTROL SYSTEM

The environmental control system (ECS) had the supremely important job of keeping the astronauts alive and indeed comfortable during the long journey to and from the moon. The system was designed to support three astronauts for 14 days as well as to maintain temperature of electronics and other critical equipment in a normal operating range. Three important functions were performed by the ECS:

- Atmosphere control in the cabin
- Water management
- Thermal control

Atmosphere Control

One of the important functions of the ECS was to regulate the constituency, pressure, and temperature of the atmosphere in the Command Module. That function included removing CO_2 from the atmosphere, supplying oxygen, and maintaining humidity at a desired level. During periods where pressure suits were worn by astronauts in the Command Module the atmosphere within the suits was likewise controlled. The cabin was maintained in a "shirtsleeve" environment during most of the flight.

The volume of the atmosphere within the cabin was about 320 cubic feet. A pure oxygen atmosphere at a pressure of five pounds per square inch (psi) was maintained in the cabin. The source of oxygen was two cryogenic tanks in the Service Module that also supplied oxygen to the fuel cells. The tanks stored a total of 640 pounds of oxygen, of which 172 pounds was allocated to the environmental control system. A cabin pressure regulator maintained a pressure of five psi in the cabin and accounted for oxygen usage by the astronauts and leakage. An indicator on the instrument panel displayed oxygen flow rate. The normal rate was about 0.43 pounds per hour.

The air in the cabin was circulated by centrifugal fans. Cabin air entered a return air valve and ducting took it through a debris filter and to the input of the fans. Two fans were available, although normally only one was energized at a time. Each fan could move 35 cubic feet of air per minute. Carbon dioxide and odors were removed from the cabin by passing the air at the output of the fans through two canisters containing lithium hydroxide and activated charcoal. The cleansed air at the output of the lithium hydroxide and charcoal canisters was ducted to a heat exchanger where the air was cooled and moisture was condensed and removed by a water-glycol cooling loop. The cleansed, cool air was released to the cabin.

Alternate lithium hydroxide and carbon canisters were replaced every 12 hours during the mission. A sensor in the cabin illuminated a light on the main caution panel "CO_2 PP HI" should the partial pressure of CO_2 in the cabin exceeded 7.6 mm of mercury (0.147 psi).

Pressure suits were worn by astronauts in the Command Module during launch and reentry. The pressure suit had two oxygen inlet ports and two exhaust ports. One of the oxygen inlet ports and one of the exhaust ports was connected by a two-passageway hose to a suit hose connector on a panel in the left-hand equipment bay. The port on the panel was divided into an oxygen supply segment and a return segment. There were three such ports on the panel, one for each astronaut. The exhaust path from the suits flowed through the suit hose connector and joined the ducting from the cabin return air vent. The path then flowed through the fans, CO_2 removal canisters, and heat exchanger and back to the suit hose connector ports on the panel.

Water Management

Water management was another important life support function of the ECS. The fuel cells within the service module generated the considerable amount of electrical power required by the spacecraft. A byproduct of power generation by the fuel cells was production of potable water at a rate of about 0.77 pounds per kilowatt-hour (0.092 gallons/kWh). Typical power generation of 2,200 watts produced about 0.28 gallons of water each hour.

The amount of water needed for drinking and food preparation was stored in a 4.3 gallon potable water tank, and the rest was fed to a 6.7 gallon wastewater tank or dumped overboard. The water resulting from controlling humidity in the capsule was also collected and transferred to the wastewater tank. A portion of the water in the wastewater tank was used for evaporative cooling as part of temperature control.

Temperature Control

A quite sophisticated system was used for temperature control of the electronics, air in the cabin, and in the suits worn by the astronauts. Two water-glycol cooling loops, a primary loop and a secondary loop, circulated coolant in pipes to carry heat away from the electronics and from heat exchangers that conditioned air for the cabin and for the pressure suits. Heat was generated in the cabin by lighting, electronic equipment not mounted to cold plates, and by the astronauts themselves. The electronic subassemblies were cooled by mounting the packages to cold plates that had coolant running through them. There were a total of 22 cold plates for this purpose.

The primary loop was the main cooling loop during the mission. The secondary loop could be activated as well if needed and it was the backup to the primary loop. Two centrifugal pumps were used to circulate about 200 pounds of water-glycol coolant per hour through the cooling loop.

There were two ways of dissipating heat transferred by the cooling loops. One involved radiating heat to outer space and the other involved evaporative cooling. The coolant temperature was maintained at about 45° F by the cooling process.

The space radiators consisted of two curved panels, one on each side of the Service Module near the engine end. Each panel had an area of 50 square feet and extended nearly halfway around the periphery. The panels, with coolant tubes embedded, served as the skin of the Service Module in the areas covered. Depending on the attitude of the spacecraft, one panel could be exposed to the sun while the other panel viewed cold space. To make up for less cooling on the sunny side, more coolant was automatically apportioned to the deep space side. If the radiators cooled the fluid to less than 45° F, a portion of the coolant at the inlet to the radiator bypassed the radiator and mixed with the output at the appropriate ratio to maintain the coolant temperature at 45 degrees.

If the coolant temperature at the output of the radiators reached about 48° F, the evaporator, which followed the radiator in the coolant flow path, was activated to bring the temperature down. The radiators alone may not be able to provide sufficient cooling when the spacecraft was orbiting the earth or the moon, for example.

The evaporator consisted of a series of finned plate sandwiches with passages for coolant flow. Fins were attached to the exterior of the plates as well. Felt-metal wicks were placed between the sandwiches as they were stacked together and manifolded to form the radiator assembly. A distribution system supplied waste-water to the one edge of the wicks. Water evaporating from the wet wicks in a low pressure environment had a strong cooling effect. Steam from evaporation passed through the finned areas and was collected in exhaust ducting that was piped to the outside of the spacecraft. NASA Technical Note TN D-6718 indicates that at an ambient pressure of 0.1 pounds per square inch, a coolant flow of 200 pound per hour was cooled to a temperature of 41.5° F by the evaporator. The temperature could be adjusted by controlling the ambient pressure in the evaporator.

Controls for the environmental control system and indicators to monitor conditions were located on the lower portion of the main instrument panel just to the right of center. A switch allowed selecting the secondary cooling loop rather than the primary loop. Another switch allowed selecting whether primary or secondary loop temperatures and pressures were displayed on dual dial indicators. The temperatures displayed were inlet to and outlet from space radiators and outlet of evaporator. Pressures displayed were steam pressure from evaporator and pressure at outlet of coolant pumps. Oxygen flow rate was also displayed on one of the dual dial indicators.

A dual vertical scale indicator displayed suit circuit temperature and cabin temperature. A second dual vertical scale indicator displayed suit circuit pressure and cabin pressure. A third vertical scale indicator displayed partial pressure of CO_2 in the suite circuit. This was the same as in the cabin when the suits were not connected.

ENTRY MONITOR SYSTEM

The tremendous amount of kinetic energy of the Command Module, traveling at about 36,000 feet per second, had to be dissipated as heat and ablation of the aeroshell as the Command Module descended into the upper atmosphere and its drag slowed the spacecraft. The center of gravity of the Command Module was offset from the X-axis and that caused the aerodynamic trim of the vehicle to settle at an angle of attack with respect to the velocity vector. This angle of attack resulted in lift as well as drag. The direction of the lift vector, e.g., up or down, could be changed by rolling the spacecraft. The orientation of the Command Module at the time of entering the atmosphere was critical.

The entry angle, which was the angle that the velocity vector of the Command Module made with the local horizontal, was closely controlled. There was a narrow allowable entry angle corridor, graphically shown in the Apollo Operations Handbook, that was only 1.6 degrees wide. The corridor was bounded by entry angles of −5.7 degrees and −7.3 degrees. The targeted entry angle was −6.4 degrees. Entry angle less than −5.7 degrees would result in the spacecraft skipping off the atmosphere into an elliptical orbit to reenter somewhere downrange at a later time. An entry angle greater than −7.3 degrees would subject the crew to deceleration greater than the acceptable limit of 10 G. One G is normal gravitational acceleration on the earth's surface.

The Entry Monitor System (EMS) provided a visual monitor of critical parameters during entry back into the earth's atmosphere. Normally, the Primary Guidance Navigation and Control System oriented the Command Module for entry and transit through the upper atmosphere. In case of loss of the PGNCS, a manual control procedure using information from the Entry Monitor System allowed a successful return to earth. A drawing of the Entry Monitor System panel is shown on the next page (Fig. 6.22).

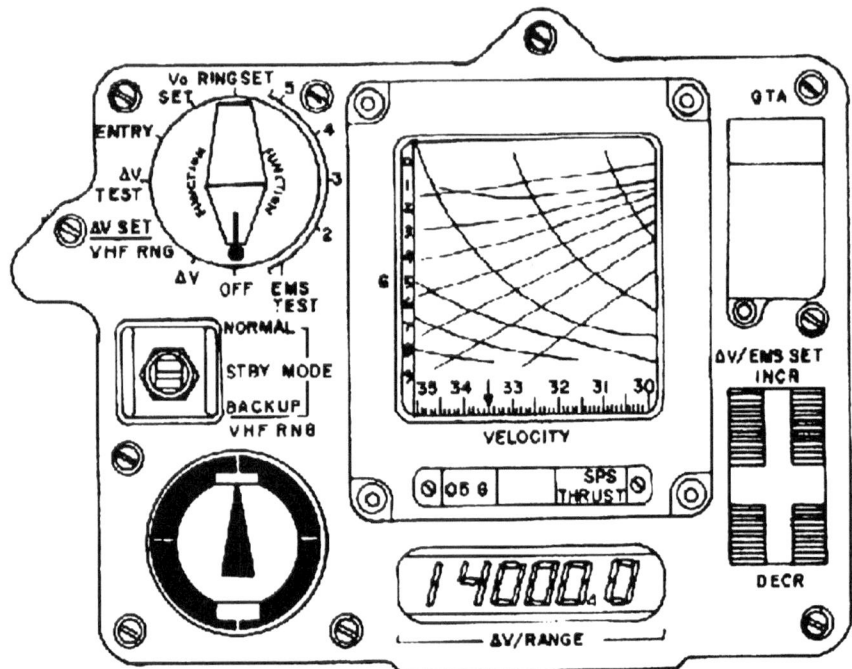

Fig. 6.22 Entry Monitor System control panel (NASA graphic)

The function switch on the upper left of the panel was used to select various functions of the EMS. There were five test positions on the right side of the switch. The V_o switch position was used in conjunction with the ΔV/EMS SET rocker switch on the right side of the panel to set the velocity scroll to the expected inertial velocity at entry. The ΔV SET switch position allowed setting the ΔV display to initial conditions. The RNG SET position allowed setting the RANGE display to initial conditions. The ENTRY position was used during entry to monitor parameters critical to entering and passing through the upper atmosphere.

The MODE switch just below the function switch had a NORMAL position that activated functions selected by the function switch. The BACKUP/VHF RNG position activated displays for manual control and it also enabled VHF ranging information to be displayed. The latter information was used during LM rendezvous with the CSM.

The display at the lower left was the Roll Attitude Indicator. The pointer showed the position of the lift vector. There were two lights on the display referred to as lift vector lights. A corridor determination test was performed 10 seconds after 0.05 G deceleration was sensed by noting which light was lit. The upper light illuminated if the deceleration was greater than 0.26G. The bottom light illuminated if the deceleration was less than 0.26G. Normal control operation resulted in the Roll Attitude Indicator pointing toward the lamp that was lit lit.

For example, if the lower light was on and the roll attitude indicator was pointed toward the light, it would indicate that the PGNCS had rolled the vehicle to a lift-down condition to increase the entry angle. If under manual control, an astronaut would roll the vehicle until the roll attitude indicator was pointed at the light.

The ΔV/RANGE display showed either range to go in miles or ΔV (velocity difference) remaining in feet per second depending on the position of the function switch.

There were two lights at the bottom of the VELOCITY Indicator. The 0.05G light lit up when the deceleration was greater than 0.05G. This indicated that the Command Module was just entering the atmosphere. The SPS THRUST light lit up when the Thrust On command to the SPS engine was enabled.

An important display on the panel was the VELOCITY indicator that presented a trace of deceleration in Gs as a function of velocity in feet per second. The display consisted of a mylar film 90 mm (3.5 inches) wide that was scrolled in the horizontal direction by spacecraft velocity change. The film had an emulsion coating on the back and a series of guide lines printed on the front. The emulsion side of the film was scribed by a stylus driven in the vertical plane by measured deceleration. Curves printed on the film along with the scribed marking allowed the astronauts to monitor or control their entry and passage through the upper atmosphere. A photograph of a small portion of the film from Apollo 8's return to earth is shown below. The white curved line is the scribed trace from the acceleration stylus (Fig. 6.23).

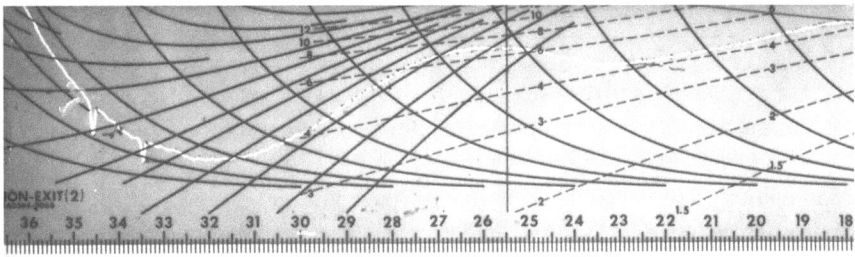

Fig. 6.23 Entry Monitor System scroll from Apollo 8 (graphic from the Apollo 8 Flight Journal)

The full scroll extends from a velocity of 37,000 feet per second to 4,000 feet per second. Deceleration on the vertical scale runs from 0 G at the top to 10 Gs at the bottom. The initial velocity expected at entry into the upper atmosphere was set in before entry. An accelerometer within the EMS unit provided deceleration measurements which were integrated to give velocity difference and that was used to scroll the ribbon past the stylus.

The lines on the scroll that swoop down from the upper left were called excessive G lines, the slope of which represent a peak load of 10 Gs. The slope of the scribed trace should not exceed the slope of the excessive G line for long. The scribed trace from Apollo 8 reflects increasing deceleration until the velocity had decreased to about 32,000 feet per second and then a decrease in deceleration as the velocity further decreased.

The solid lines that swoop up from the lower left were known as excessive range (skip out) lines. A lift-down orientation of the spacecraft was called for if the slope of the scribed trace exceeded the slope of the excessive range lines for long.

The dashed lines starting at a velocity of 30,000 feet per second and moving toward the upper right were called range potential lines. Intersection of the scribed line with those lines indicated the range capability of the spacecraft in hundreds of miles for constant deceleration.

EARTH LANDING SYSTEM

An important system in the Command Module, and last to be used during the Apollo mission, was the Earth Landing System (ELS). The ELS took over the task of further slowing the Command Module after the aeroshell had slowed the velocity of the capsule to about 440 feet per second (300 miles per hour) and the spacecraft had descended to an altitude of about 24,000 feet. The task of the ELS was to further slow the Command Module and deposit it relatively gently into the ocean for recovery.

The ELS contained four different types of parachutes and deployment mechanisms. The parachute subsystems were the following: forward heatshield jettison parachute, drogue parachutes, pilot parachutes, and main parachutes. Deployment of the parachutes was normally automatic and keyed from barometric switches, but manual deployment of the parachutes was available as backup.

A photograph of the Apollo 14 Command Module descending on the main parachutes just before splashdown is shown on the next page (Fig. 6.24).

Chronology of an Earth Landing

In preparation for landing, and as the spacecraft descended through an altitude of about 30,000 feet, a guarded switch, ELS LOGIC, was turned to the ON position by the astronauts to power logic circuitry for the landing events. That guarded switch was located on the lower portion of the instrument panel in front of the

Fig. 6.24 Apollo 14 Command Module descending on main parachutes

Commander. A switch adjacent to the logic switch with AUTO or MAN positions was switched to the AUTO position. If the expected events did not happen automatically, the latter switch was set to MAN (manual) and guarded switches labeled APEX COVER JETT, DROUGE DEPLOY, and MAIN DEPLOY were pushed to manually initiate the events.

The ELS equipment was tightly packed in the forward compartment of the Command Module. The compartment was covered by the forward heat shield and that heat shield had to be jettisoned before the equipment could be deployed. Major events during the landing on earth are summarized below.

1. *Jettison the forward heat shield* – A barometric switch sensed when the spacecraft had descended to 24,000 feet and closed a latching relay to start the landing sequence. After a brief delay of 0.4 seconds, the forward heat shield was jettisoned and a drag parachute was deployed to pull the heat shield away from the Command Module.

2. *Deploy drogue parachutes* – Two seconds after jettisoning the heat shield, two drogue parachutes were deployed to orient and slow the spacecraft. Those parachutes were ribbon types that had openings that let some air flow through to reduce the shock at opening and to avoid bursting the canopy had it been airtight. The velocity was about 440 feet per second when the drogue parachutes were deployed. The parachutes were deployed in a reefed condition to further reduce the shock of opening. Reefing was accomplished by a line running through rings sewed to the inside of the parachute skirt that prevented full opening of the parachute. A pyrotechnic time delay was set off by deployment of the parachute that fired a charge that cut the reefing line after 10 seconds and allowed the canopies to fully open. The canopy diameter was 13.0 feet when fully open and inflated. The drogue parachutes had slowed the spacecraft to about 257 feet per second by the time the spacecraft descended to 10,000 feet.

3. *Deploy main parachutes* – A barometric switch closed when the spacecraft descended to 10,000 feet and that initiated deployment of three pilot parachutes. The pilot parachutes pulled out and deployed the three main parachutes. The main parachutes were initially deployed in a reefed condition. There were two unreefing operations. The first occurred 6 seconds after deployment, and it allowed the canopy to partially open to lessen the shock at opening. The second operation occurred 10 seconds after deployment and allowed the canopy to fully open. The diameter of each of the parachute canopies was 77 feet when fully open and inflated. The parachutes slowed the Command Module to about 31 feet per second by the time it hit the water.

4. *Splashdown* – The main parachute lines were attached to the Command Module such that the module hung at an angle of 27.5 degrees from the vertical. This greatly reduced the shock of impact at splashdown compared to landing broadside on the relatively flat bottom. The main parachute lines were released from the Command Module just after splashdown. Surface winds acting on the parachutes sometimes tipped over the Command Module after splashdown before the lines could be released. The spacecraft would then assume a stable apex-down attitude. If this should happen, the astronauts initiated inflation of three righting bags. These bags, visible in the photograph on the next page of Apollo 11 in the water after landing, caused the spacecraft to rotate to a stable apex-up orientation (Fig. 6.25).

5. *Recovery operations* – To counter the slimmest chance that there may be virulent organisms on the moon, precautions were taken on the first few Apollo missions to minimize the chance of spreading the organisms. On early missions the astronauts donned biological isolation suits while still inside the Command Module after splashdown.

In the case of recovery of Apollo 11, swimmers from the aircraft carrier USS Hornet attached a floatation collar to the Command Module and then the hatch

Fig. 6.25 Apollo 11 Command Module on the ocean after landing

was opened. It must have been wonderful for the astronauts to emerge from the capsule and breathe the sea air after breathing the capsule atmosphere for several days.

Splashdown of Apollo 11 occurred in the Pacific Ocean on 24 July 1969 about 1.7 miles from the target point and about 13 miles from the USS Hornet. Impressive accuracy, given that splashdown was about 1,400 nautical miles from first reentry into the atmosphere and the velocity at reentry, was 36,194 feet per second (24,677 miles per hour).

The astronauts entered a raft manned by one of the swimmers from which they were hoisted one by one up to a hovering helicopter. The helicopter ferried them to the USS Hornet. The astronauts entered a Mobile Quarantine Facility on the Hornet. The quarantine facility with the astronauts inside was taken to Hawaii by the Hornet and then flown to Houston.

NASA had determined that there were no harmful organisms on the surface of the moon by the time of the Apollo 15 mission. The cumbersome biological isolation procedures were dispensed with for Apollo 15 and subsequent missions.

Bibliography

Apollo 15 CM Software, Delco Electronics, Apollo 15 Lunar Surface Journal, undated

Apollo Operations Handbook Block II Spacecraft, NASA document SM2A-03-BLOCK III, October 1969

Apollo Spacecraft & Systems Familiarization, Apollo Training, August 1967

Chilton, Robert G., *Apollo Spacecraft Control System*, Presented at The Symposium on Automatic Control in Peaceful Uses of Space, Stavanger, Norway, June 1965

Gibson, Cecil R. and Wood, James A., *Apollo Experience Report – Service Propulsion Subsystem*, NASA Technical Note NASA TN D-7375

Hoag, David D., *Apollo Navigation, Guidance, and Control Systems*, Presented at the National Space Meeting of the Institute of Navigation, April 1969

Martin, Frederick H. and Battin, Richard H., *Computer-Controlled Steering of the Apollo Spacecraft*, Presented at AIAA Guidance, Control and Flight Conference, August 1967

Munford, Robert E. and Hendrix, Bob, *Apollo Experience Report – Command and Service Module Electrical Power Distribution Subsystem*, NASA Technical Note NASA TN D-7609, March 1974

NASA Apollo Command Module News Reference, Prepared by North American Rockwell Corp., undated.

O'Brien, Frank, *The Apollo Guidance Computer Architecture and Operation*, Springer-Praxis books, Chichester, UK, 2010

Painter, John, and Hondros, George, *Unified S-Band Telecommunications Techniques for Apollo*, NASA Technical Note NASA TN D-2208

Parker, Phil, *The Apollo On-board Computer*, Apollo Flight Journal, Original article published in October 1974

Tomayko, James E., *Computers in Spaceflight The NASA Experience*, NASA Contractor Report 182505, March 1988

7

The Apollo Service Module

A photograph of the Apollo 15 Command and Service Module taken from the Lunar Module while in orbit around the moon is shown on the next page. The Service Module is the dominant cylindrical structure with the rocket motor expansion nozzle attached. The Command Module is the dark looking conical structure attached to the forward end of the Service Module. One of the panels of the Service Module had been jettisoned by pyrotechnics to expose the instruments to the environment (Fig. 7.1).

The Service Module contained the service propulsion engine, reaction control system, and fuel cells that generated electrical power and water for the spacecraft. It also stored most of the consumables for the mission including oxygen, hydrogen, helium, and fuel and oxidizer for the engine.

MECHANICAL CONFIGURATION OF SERVICE MODULE

A drawing of the Service Module is shown on the following page (Fig. 7.2). A drawing of a cross section of the module is shown on the second following page (Fig. 7.3).

Four sets of a group of four reaction control engines used to maneuver the spacecraft were mounted on the periphery of the Service Module. A high-gain S-Band telecommunication antenna and scimitar-type VHF antennas were also mounted on the periphery. The S-Band antenna can be seen in the photograph of the CSM but it is not shown on the drawing.

The outer skin of the Service Module consisted of aluminum honeycomb panels 1-inch thick. Space radiator panels served as the skin near the engine end of the module. Interior support structure consisted of radial aluminum beams that divided

© Springer Nature Switzerland AG 2018
T. Lund, *Early Exploration of the Moon*, Springer Praxis Books,
https://doi.org/10.1007/978-3-030-02071-2_7

Fig. 7.1 Apollo 15 Command and Service Module in orbit around the moon (NASA photograph)

the interior into six sections around a cylindrical core as shown in the cross-section drawing. The service propulsion system (SPS) engine with its nozzle extension was mounted in the center at the bottom of the module. The helium tanks used to pressurize the fuel and oxidizer were also mounted in the center section.

SERVICE PROPULSION SYSTEM (SPS)

The service propulsion system engine in the Service Module generated 20,500 pounds of thrust. The engine was not throttleable but it could be started and stopped quickly. Normally, the computer in the Command Module controlled the firing of the engine, but it could also be fired manually. Fuel for the engine was a 50/50 mixture of hydrazine and unsymmetrical dimethylhydrazine. The oxidizer was nitrogen tetroxide. The fuel and oxidizer ignited spontaneously when brought into contact.

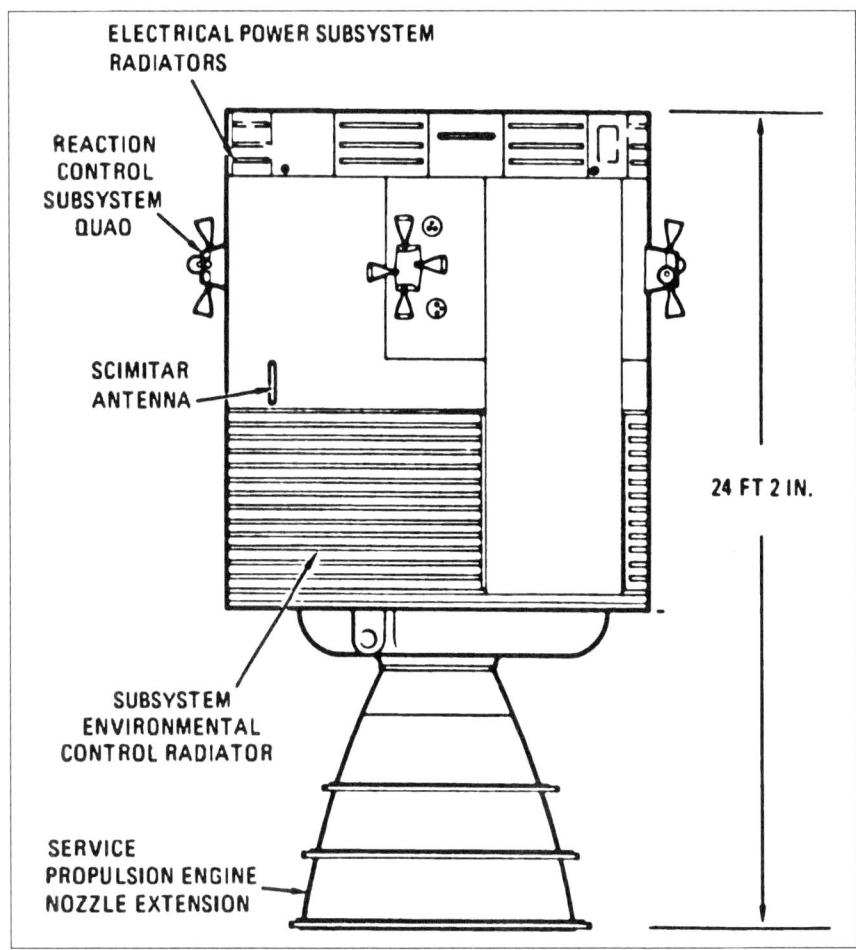

ELECTRICAL POWER SUBSYSTEM
RADIATORS

REACTION
CONTROL
SUBSYSTEM
QUAO

SCIMITAR
ANTENNA

24 FT 2 IN.

SUBSYSTEM
ENVIRONMENTAL
CONTROL RADIATOR

SERVICE
PROPULSION ENGINE
NOZZLE EXTENSION

Fig. 7.2 Apollo Service Module (NASA drawing cropped by author)

The thrust chamber of the engine contained the injectors for fuel and oxidizer, the combustion chamber, and the exhaust nozzle. A long flared extension was bolted to the exhaust nozzle. The combustion chamber was ablatively cooled. Normal combustion chamber pressure was between 95 and 105 pounds per square inch when the engine was firing. This pressure was displayed on a gauge on the instrument panel, and it was monitored by the astronauts during a burn.

The engine was designed for a 750-second service life with a minimum of 36 restarts available. The engine was 160 inches long and the nozzle extension exit diameter was 98 inches. It weighed about 650 pounds. The engine was held in a gimballed mount that allowed angular control of the thrust vector.

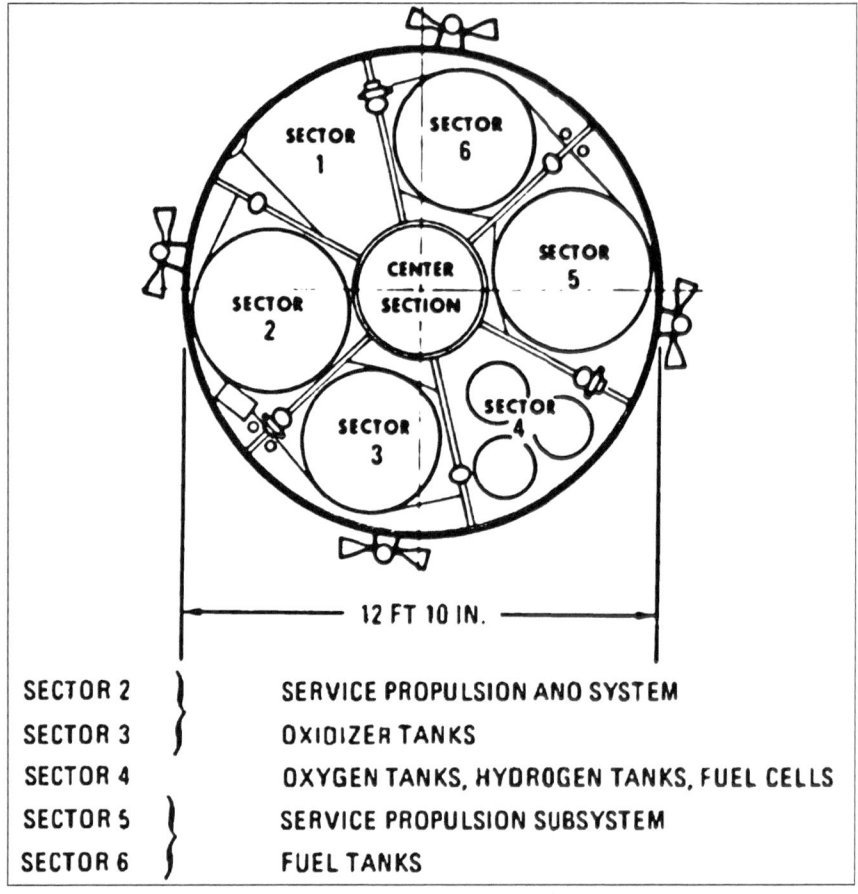

SECTOR 2   }
SECTOR 3   }   SERVICE PROPULSION ANO SYSTEM
              OXIDIZER TANKS
SECTOR 4   OXYGEN TANKS, HYDROGEN TANKS, FUEL CELLS
SECTOR 5   }
SECTOR 6   }   SERVICE PROPULSION SUBSYSTEM
              FUEL TANKS

Fig. 7.3 Cross section of the Service Module through propellant tanks (NASA graphic)

Fuel for the engine was stored in two tanks that NASA documentation referred to as the storage tank and the sump tank. The two tanks were cylindrical in form with hemispheric endcaps. The fuel storage tank was 155 inches long and 45 inches in diameter and held 7,058 pounds of fuel. The fuel sump tank was 154 inches long and 51 inches in diameter and held 8,708 pounds of fuel. Total fuel carried was 15,766 pounds.

Oxidizer for the engine was also carried in a storage tank and a sump tank. The dimensions of those two tanks were the same as for the fuel tanks. The storage tank contained 11,285 pounds of oxidizer and the sump tank contained 13,924 pounds. Total oxidizer carried was 25,208 pounds. The nominal ratio of oxidizer to fuel flow by weight during engine firing was 1.6:1.

The storage tanks and sump tanks were connected in series by pipes at the bottom of each tank. Propellants to the engine were fed out of the sump tanks. The tanks were pressurized by helium from two helium storage tanks connected in parallel. The helium tanks were spherical with inside diameter of 40 inches. Each held 19.4 cubic feet of helium, and they were initially pressurized to 3,600 pounds per square inch. A primary regulator reduced the pressure to about 186 pounds per square inch before the helium gas was applied to the fuel and oxidizer storage tanks for pressurization. A secondary regulator was available in case of failure of the primary regulator.

Controls for the SPS engine were located on the control panel in front of the Commander. Monitoring gauges for SPS parameters were located on the panel in front of the Lunar Module pilot.

The SPS engine was normally controlled by the Apollo Guidance Computer. It could also be controlled by the Stabilization and Control System (SCS). The SCS included provisions for manual control.

The engine could be fired manually by the Commander by first switching the SPS THRUST DIRECT switch to ON. A pushbutton labeled THRUST ON located just to the right of the switch was then pushed to fire the engine.

The engine was supported in the spacecraft by gimbals that allowed steering the engine in the pitch and yaw planes to maintain thrust through the center of gravity of the spacecraft. Normally, the gimbals were controlled automatically by the computer, but they could also be controlled from the Stabilization and Control System using inputs from one of the rotational controllers.

The gimbals were displaced in pitch and roll by using separate motor-driven, clutched actuators. The motors were first started by switches on the panel and then clutches were activated to drive the actuators in either the plus or minus directions. While thrusting, the gimbals had an angular range of ±4.5 degrees about a +1 degree offset in yaw and an angular range of ±4.5 degrees about a +2 degrees offset in pitch. The reason for the angular offsets was that the center of mass of the spacecraft was offset from the centerline of the vehicle.

Before firing the engine in the SCS ΔV mode, the gimbals were adjusted by the astronauts so that the thrust vector would pass through the center of gravity of the vehicle. The offset angles were set in by pitch and yaw thumbwheels on the instrument panel. The proper values for pitch and roll offsets were brought up from the computer and displayed on the DSKY.

During the Apollo 11 mission, the engine was fired five times for a total of about 532 seconds. All firings were performed under control of the Command Module computer. The firing durations in seconds and resulting velocity change in feet per second were:

Evasive maneuver – 2.9 seconds, ΔV = -19.7 fps
Midcourse correction – 3.1 seconds, ΔV = -20.9 fps

Insertion into lunar orbit – 357.5 seconds, ΔV = -2,918 fps
Circularize lunar orbit – 16.9 seconds, ΔV = -159 fps
Trans-Earth injection – 151.4 seconds, ΔV = 3,279 fps

REACTION CONTROL SYSTEM

The reaction control system in the Service Module fired small rocket engines to control the attitude and position of the Command and Service Module. Attitude could be controlled in the pitch and yaw planes and around the roll axis. Translation maneuvers could be conducted along the X, Y, and Z axes.

The Service Module contained four separate and largely independent reaction control subassemblies. There were four such subassemblies mounted 90 degrees apart around the periphery of the Service Module. Each subassembly contained a grouping of four rocket engines on the outside of the panel and propellant tanks, associated controls, and feed lines for the engines on the inside of the panels.

A photograph of the Apollo 15 CSM in orbit over the moon with two sets of reaction control engines visible is shown below (Fig. 7.4). The engine clusters are

Fig. 7.4 Apollo 15 CSM showing reaction control engines (NASA photograph cropped by author)

the appendages with four nozzles mounted toward the Command Module side of the Service Module.

The reaction engines were 13.4 inches long, and the nozzle exit diameter was 5.6 inches. Each engine generated about 100 pounds of thrust. The engines, which were rated for service life of 1,000 seconds and 10,000 operational cycles, were developed and built by the Marquardt Corporation.

The engines used monomethyl hydrazine (MMH) for fuel and nitrogen tetroxide (N_2O_4) as the oxidizer. This combination ignited upon contact with one another. Each of the four reaction control subassemblies included two fuel tanks, two oxidizer tanks, and a helium tank. Helium was used to pressurize the fuel and oxidizer. The helium tank was about 12.4 inches in diameter and contained 1.35 pounds of helium at a pressure of 4,150 psi. A pressure regulator reduced the pressure to about 180 psi to pressurize the fuel and oxidizer tanks.

Fuel for each of the reaction control subassemblies was contained in two titanium alloy cylindrical tanks with hemispherical caps. The primary fuel tank was 23.7 inches long and 12.6 inches in diameter. It held 69.1 pounds of fuel. The secondary fuel tank was 17.3 inches long and 12.6 inches in diameter. It held 42.5 pounds of fuel. The total amount of fuel was 111.6 pounds. The primary oxidizer tank was 28.6 inches long and 12.6 inches in diameter. It held 137 pounds of oxidizer. The secondary oxidizer tank was 19.9 inches long and 12.6 inches in diameter. It held 89.2 pounds of oxidizer. Total amount of oxidizer was 226.2 pounds.

The fuel and oxidizer tanks had Teflon bladders that held the fuel and oxidizer. Helium pressure was applied to the tanks to squeeze the bladder and force the propellants into feed lines for the engines.

The fuel and oxidizer lines connected to feed injector valves near each engine. The injector valves were spring loaded closed, and they were opened to admit propellants by energizing solenoid coils. There were two sets of coils in each injector, one for automatic control and one for direct control from the hand rotational controller. Control signals for the automatic solenoid coils were generated by the Reaction Jet and Engine ON-OFF Control Assembly (RJ/EC).

The engines could be pulse fired for short impulses lasting less than 70 milliseconds or fired continuously for several seconds. Input signals from the guidance computer instructed operation of the RJ/EC when the spacecraft was under computer control, and signals from the Electronic Control Assembly instructed operation of the RJ/EC when the spacecraft was being controlled by the Stabilization and Control System.

A backup manual control mode could be selected by the astronauts by turning the "Direct RC" switch on the instrument panel to the ON position. The rotational hand controller was then positioned against the stop in the axis to be controlled to activate the "Direct" switch. Closing the Direct switch powered the direct solenoid coils in the injector valves of the appropriate thrusters so that they could be operated by the rotational controller.

A total of 560 pounds of propellant was used for attitude and translation control of the CSM during the Apollo 11 mission. This was comfortably less than the 1,151 pounds of propellant available.

ELECTRICAL POWER SYSTEM

The primary source of electrical power for the Command Service Module was fuel cells. There were three fuel cell assemblies in the Service Module. Each generated up to 1,420 watts of power at a nominal voltage of 28 volts DC. The fuel cells used oxygen and hydrogen carried in cryogenic liquid form as reactants.

In addition to fuel cells, the spacecraft carried three 37-volt silver oxide-zinc storage batteries with a capacity of 40 ampere-hours and two 0.75 ampere-hour batteries. The fuel cells and the associated hydrogen and oxygen tanks were contained in the Service Module. The five batteries were mounted in the Command Module.

The three 40 ampere-hour batteries were maintained in a charged condition by a battery charger controlled by the astronauts. Those batteries supplemented fuel cell power during peak loads and they provided power for the Command Module after it was separated from the Service Module near the end of the mission. The two smaller batteries were reserved to supply surge power for pyrotechnics during the mission and they were not recharged during flight.

Switches for control of the power system and indicators to display important parameters of the power system were located on the right side of the main instrument panel of the Command Module.

Fuel Cells

The three fuel cell units were alkaline type built by Pratt & Whitney Aircraft Company. A photograph of the fuel cells mounted to a shelf in the Service Module is shown on the next page (Fig. 7.5). The photograph appears in the Cortright Commission's report: *Report of Apollo 13 Review Board*. The fuel cells were quite substantial units: 22 inches in diameter, 44 inches high, and each weighed 245 pounds. The lower cylindrical portion of the units contained 31 individual fuel cells in a pressurized container and the upper portion contained accessories and controls.

The three fuel cell units were referred to as Fuel Cell 1, Fuel Cell 2, and Fuel Cell 3. Fuel Cells 1 and 3 are in the front of the photograph, and Fuel Cell 2 was mounted behind them. The mounting shelf was located in a truncated pie-shaped

Fig. 7.5 Fuel cells for Apollo spacecraft (NASA photograph)

compartment referred to as Sector 4 in the Service Module. Oxygen and hydrogen tanks supplying reactants for the fuel cells were mounted lower in the compartment. A sketch of the Sector 4 compartment is given on the next page (Fig. 7.6).

Each fuel cell unit consisted of 31 individual fuel cells connected in series. The voltage output of each cell was one volt when lightly loaded, resulting in 31 volts for the overall unit. The individual fuel cells consisted of an electrolyte solution of potassium hydroxide and water between a porous nickel anode and a porous nickel-nickel oxide cathode. Nickel was chosen because it resisted the corrosive effects of potassium hydroxide and it acted as a catalyst for hydrogen oxidation at the anode.

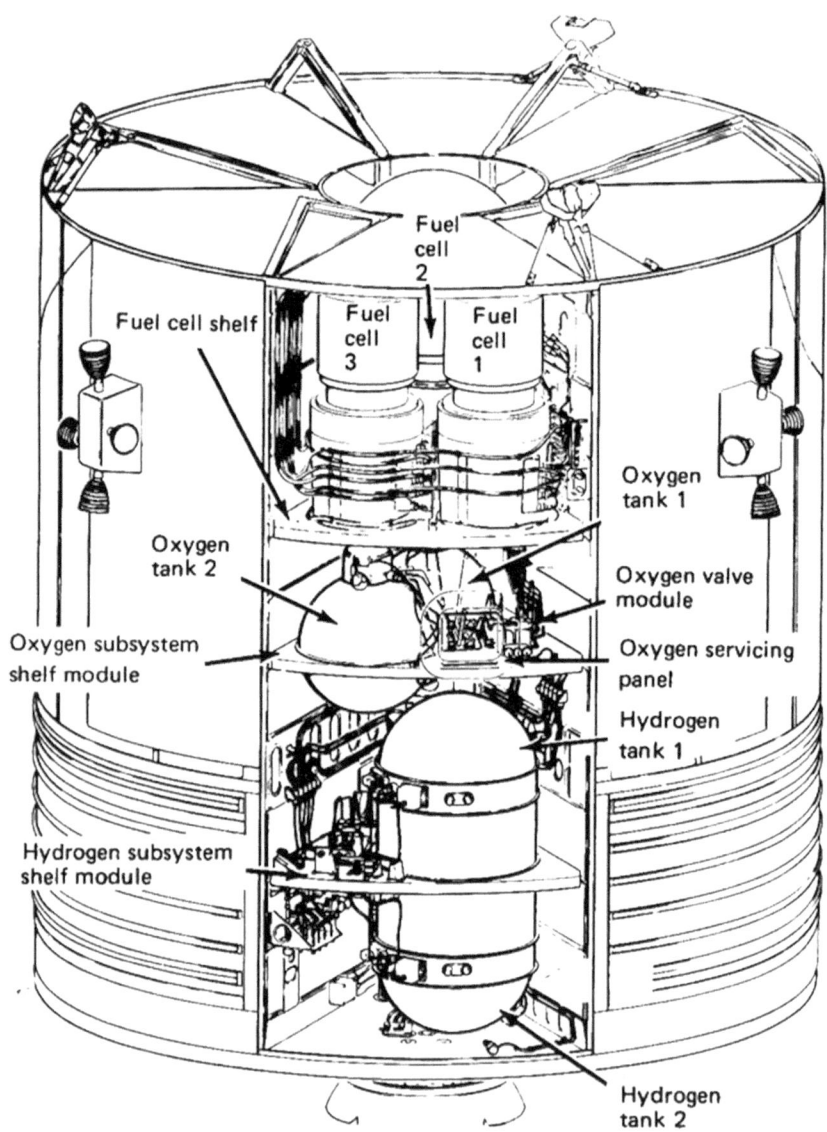

Fig. 7.6 Sector 4 of the Service Module holding fuel cells and reactants (NASA graphic)

At a normal ratio of potassium hydroxide to water in the electrolyte of about 83% to 17% by weight, the electrolyte was solid at room temperature. Consequently, the temperature had to be elevated to keep it in liquid form. The temperature of the fuel cell stack was regulated to about 200°C (392°F). The fuel cell was pressurized to keep the electrolyte from boiling.

The tank was initially pressurized to 1,500 psi and a pressure regulator reduced the pressure applied to the diaphragm and hence to the electrolyte to 53 psi. Oxygen was applied to the fuel cell through a differential regulator that maintained oxygen pressure at 9.5 psi above the nitrogen pressure, or at 62.5 psi. Hydrogen was applied to the fuel cell through a differential regulator that maintained hydrogen pressure at 8.5 psi above the nitrogen pressure, or at 61.5 psi.

Hydrogen flowed through a hydrogen regenerator to input of the fuel cells. Water vapor (steam) was carried away from the fuel cells by the hydrogen flow. Hot hydrogen and water vapor at the exit of the fuel cells was directed to a bypass valve that diverted a portion of hot gas to the regenerator where it warmed the incoming hydrogen gas. The operating temperature of the fuel cell was controlled by controlling the amount of hot gas bypassed through the regenerator which in turn controlled the temperature of the hydrogen entering the fuel cell.

The unbypassed hot hydrogen gas was applied to a condenser where heat was transferred to the glycol-water cooling loop and thence to the space radiators on the periphery of the Service Module. Much of the water vapor was condensed by the cooler temperature in the condenser. The output of the condenser was applied to a centrifugal water separator where the water was extracted and ported off to a potable water storage tank. The hydrogen gas at the output of the water separator was applied to a pump that raised its pressure to the original pressure. The repressurized gas was combined with new hydrogen gas at the input to the fuel cell.

Typical electrical power consumption of the Command Service Module was about 2,200 watts. At a voltage of 28 volts, the typical total current supplied by the three fuel cells would have been 78.6 amperes. The Apollo Operations Handbook gives hydrogen consumption of 0.00257 pounds per ampere per hour, oxygen consumption of 0.0204 pounds per ampere per hour, and water generation of 0.0297 pounds per ampere per hour for the fuel cells. At a total load of 78.6 amperes, hydrogen consumption would have been 0.202 pounds per hour, oxygen consumption would have been 1.603 pounds per hour, and water production would have been 2.33 pounds (0.28 gallons) per hour.

Controls for the fuel cells and display of pertinent parameters were located on the right side of the Command Module instrument panel 2 and in the lower portion of instrument panel 3. FUEL CELL REACTANTS switches 1, 2, and 3 at the left bottom of panel 3 allowed oxygen and hydrogen to flow to the particular fuel cell when in the ON position. Indicators above the switches showed striped lines if the reactants were cut off. Other switches controlled heaters in the fuel cells and amount of coolant applied to the radiators. Instrumentation concerning the fuel cells included vertical indicators displaying flow rates of oxygen and hydrogen and indicators showing temperature of critical items. A selector switch allowed monitoring those parameters for Fuel Cells 1, 2, or 3. The pressure and quantity remaining in the cryogenic hydrogen and oxygen tanks were displayed on vertical displays in panel 2.

The electrical power outputs of Fuel Cells 1 and 2 were connected through switches to DC MAIN BUS A and the electrical power outputs of Fuel Cells 3 and 2 were connected through switches to DC MAIN BUS B. The DC buses furnished DC power to the spacecraft.

Cryogenic Hydrogen and Oxygen Storage Tanks

The Service Module contained two cryogenic hydrogen tanks and two cryogenic oxygen tanks. The oxygen tank fed the fuel cells in the Service Module and the environment control system in the Command Module. The spherical oxygen tanks were 26 inches in diameter, and each tank held 320 pounds of useable oxygen in liquid form. The hydrogen tanks were cylindrical at 31.75 inches in diameter and about 19 inches long. One end of the tank had a hemispherical endcap and the other end had a flat cap. The two hydrogen tanks were mounted vertically on a shelf with the flat ends back to back to the shelf. Each hydrogen tank held 28 pounds of usable hydrogen in liquid form.

The properties of the liquid form of these common gases are interesting. Liquid hydrogen has a freezing point of -259.3°C (-434.8°F or 13.9K), boiling point of -252.9 °C (-423.2°F or 20.3K), and a density of 4.42 pounds per cubic foot at atmospheric pressure. Liquid oxygen has a freezing point of -218.8 °C (–61.9°F or 54.8K), a boiling point of -183.0 °C (-297.4°F or 90.2K), and a density of 73.23 pounds per cubic foot at atmospheric pressure. The caption K is Kelvin. The Kelvin scale is equivalent to °C +273.18. Zero Kelvin (K) is absolute zero.

The cryogenic tanks were constructed similar to thermos bottles with an inner shell separated from the outer shell by a vacuum region. The vacuum was maintained by a vacuum-ion pump to hold heat leakage to very low values. The pressure in the oxygen tanks was maintained at 900 ±35 pounds per square inch absolute (psia). Pressure was controlled by heating the contents of the tank. A 114 watt heater was wound around a cylindrical tube in the middle of the tank as shown in the cross-section sketch of the tank given on the next page (Fig. 7.7).

Two fans, one mounted near the bottom and one near the top of the cylinder, circulated the liquid oxygen around the heater and prevented stratification. A pressure switch activated and turned the heater and fans on when the pressure in the line leading from the tank dropped below 865 psia, and it turned the heater and fans off when the pressure increased beyond 935 psia. A capacitance-type quantity gauge and a temperature gauge were also located in the tank as shown in the sketch.

The hydrogen tank had a similar arrangement for the heater, fan, and quantity gauge.

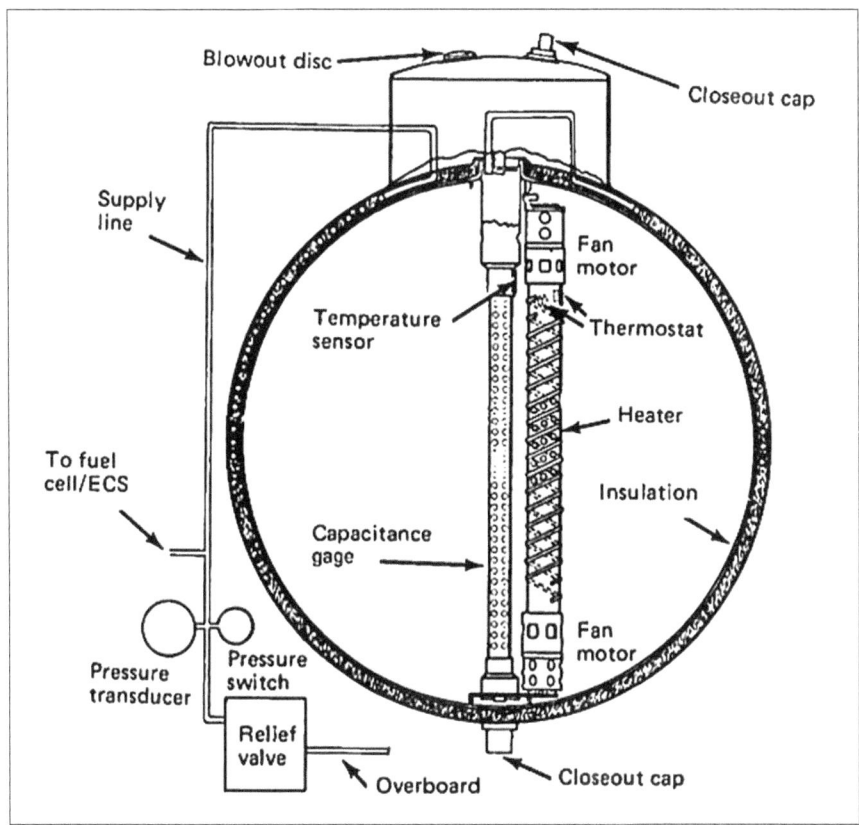

Fig. 7.7 Sketch of cross section of the Apollo oxygen tank (NASA graphic)

Normally, the heaters and fans were controlled automatically by the pressure switches. As backup, the heaters and fans could be controlled manually. Either automatic (AUTO) or manual (ON) control could be selected by the O_2 HEATERS and H_2 HEATERS switches on the instrument panel.

It was important to monitor the pressures in the tanks while the heaters and fans were controlled manually. Pressures in oxygen tanks 1 and 2 and in hydrogen tanks 1 and 2 were displayed on vertical display indicators on the instrument panel 2. The quantity in the two oxygen tanks and the two hydrogen tanks were displayed on vertical indicators next to the pressure indicators.

Oxygen Tank Explosion in Apollo 13

High drama in the Apollo program occurred when oxygen tank 2 exploded in the Service Module of Apollo 13. The explosion, which occurred en route to the moon, put the astronauts at mortal risk.

The explosion was caused by a combination of factors including a work-around for a loose filler tube in the oxygen tank. The loose filler tube problem was uncovered during prelaunch testing. As part of the work-around, liquid oxygen in the tank was boiled off by applying 65 volts to the heater for several hours. Unfortunately, the tank was built to an earlier specification that called for 30 volts maximum for the heater.

Contacts supplying current to the heater welded shut as they tried to open when the temperature limit of 80°F was reached because of the high current resulting from the 65 volt potential. The temperature in the tank became extremely hot and likely damaged the insulation on the wiring to the fan. The overheating was not noticed and the tank was refilled with liquid oxygen prior to launch.

The early portion of Apollo 13 flight to the moon was normal. Procedures called for the fans to be periodically turned on for a short time by the astronauts to prevent stratification of the cryogenic liquids in the tanks. During one of these turn-ons, vibration caused the fan wires with damaged insulation in the oxygen tank to contact and arc. That ignited the Teflon insulation, and the resulting fire in the oxygen environment caused very high temperature and pressure in the tank, and it exploded. The explosion caused extensive damage within the Service Module.

Decisive action by the astronauts and mission control personnel in Houston, aided by technical inputs from contractors, averted looming tragedy. The explosion in oxygen tank 2 caused a leak in oxygen tank 1 as well, and soon power from a remaining operating fuel cell was lost. The astronauts transferred guidance information and the state vector from the computer in the command module to the computer in the Lunar Module. They then closed down power to the Command Module to save power in the batteries for use during reentry into the earth's atmosphere.

The astronauts moved into the Lunar Module where they endured numbing cold with minimal electrical power. Power was minimized to extend LM battery life for the long trip back to earth. Engineers on the ground ran a series of computer simulations to determine the best trajectory to expedite return to earth. The Lunar Module descent engine was fired to cause the spacecraft to swing around the moon and head back toward earth on the selected trajectory. Finally, when near the earth the astronauts reentered the Command Module. The Command Module was powered up on batteries and separated from the Lunar Module and from the Service Module. A successful entry into the atmosphere and splashdown near the recovery ship followed.

The human factor had carried the day! *The New York Times* was apt in declaring: "Only in a formal sense will Apollo 13 go into history as a failure."

SCIENTIFIC INSTRUMENT MODULE (SIM) BAY

The Scientific Instrument Measurement (SIM) Bay in the Service Module of Apollo 15, Apollo 16, and Apollo 17 contained a series of instruments to obtain additional information about the moon while the CSM was in lunar orbit. The SIM Bay in all three missions included a panoramic camera with a 24-inch lens, a mapping camera with a 3-inch lens, and a laser altimeter. The cameras provided panoramic and mapping photographs of the lunar surface. The laser altimeter measured range to the surface while in orbit and developed a profile of the surface. The cameras and laser altimeter were operated in segments of the orbit during several orbits to examine selected portions of the lunar surface.

The mapping camera assembly contained a camera with a 76 mm (3-inch) focal length lens that imaged the lunar surface and a 76 mm focal length camera that imaged the star field 96 degrees to the side to establish the exact attitude of the mapping image. The assembly also contained the laser altimeter instrument. The assembly, which was quite substantial, weighed 275 pounds.

The mapping camera had a field of view of 74 degrees. Images were captured on film with a frame size of 4.5 by 4.5 inches. From an orbital altitude of 110 km, the imaged patch on the surface was 167 km square. Film for the mapping camera was on a roll 1,500 feet long contained in a removable canister. The film from all of the cameras was retrieved by an EVA by the Command Module Pilot during the return trip to earth.

The panoramic camera had a focal length of 610 mm (24 inches) that resulted in a field of view of about 10.8 degrees. The resolution of the surface image was about two meters. During imaging, the lens was rotated in an arc perpendicular to the line of flight. The arc was 108 degrees in extent centered at the nadir. The shutter was opened during the rotation and a slit focused the image on a moving strip of film during the scan.

In addition to the cameras and laser altimeter, the SIM Bay of Apollo 15 and Apollo 16 included a gamma ray spectrometer, X-ray spectrometer, and an alpha particle spectrometer. The SIM Bay of Apollo 17 contained cameras, laser altimeter, UV spectrometer, infrared scanning radiometer, and a radar lunar sounder.

Details of these instruments and of photographs and other data gleaned are rather extensive. That material will not be covered in this book in deference to discussion of hardware for the main mission of Apollo of landing men on the moon.

Bibliography

Apollo Operations Handbook Block II Spacecraft, NASA document SM2A-03-BLOCK III, October 1969

Gibson, Cecil R. and Wood, James A., *Apollo Experience Report – Service Propulsion Subsystem*, NASA Technical Note NASA TN D-7375

Munford, Robert E. and Hendrix, Bob, *Apollo Experience Report – Command and Service Module Electrical Power Distribution Subsystem*, NASA Technical Note NASA TN D-7609, March 1974

Warshay, Marvin and Prokopius, Paul R., *The Fuel Cell in space: Yesterday, Today and Tomorrow*, NASA Technical Memorandum 1023266

8

The Apollo Lunar Module

The Lunar Module (LM) was the vehicle that fulfilled President Kennedy's challenge to land men on the moon. It was a complex yet functionally elegant spacecraft that separated from the Command and Service Module while in orbit around the moon and under computer control augmented by human assistance, descended to a soft landing on the lunar surface. A photograph of Apollo 11 Lunar Module *Eagle* at Tranquility Base (Mare Tranquillitatis) is shown on the next page. Astronaut Buzz Aldrin is removing an experiment package from the LM in the picture (Fig. 8.1).

The Lunar Module was composed of a descent stage and an ascent stage so named for the action of their main rocket engines. The descent stage, which was the lower stage and included the landing legs, used its engine to slow the spacecraft from lunar orbit to a powered descent for landing. The ascent stage, which was the upper stage and included the crew compartment, used its engine for ascent from the moon and rendezvous with the orbiting Command Service Module. The descent stage served as the launch platform for the ascent stage during that daunting departure from the lunar surface.

A drawing of the Lunar Module with captions identifying important components is shown on the following page (Fig. 8.2). The ascent stage begins just above the large rectangular areas of the descent stage.

The Lunar Module was a large vehicle. It was about 23 feet high from the bottom of the landing pads to the top of the S-band steerable antenna, and it was about 14 feet wide and 14 feet deep. The diagonal distance across the landing pads was 31 feet. The descent stage was 10 feet, 7 inches high from the bottom of the landing pads to the top of the upper deck. The main structure of the ascent stage was 9 feet, 3 inches high.

© Springer Nature Switzerland AG 2018

T. Lund, *Early Exploration of the Moon*, Springer Praxis Books,
https://doi.org/10.1007/978-3-030-02071-2_8

Fig. 8.1 Apollo 11 Lunar Module *Eagle* at Tranquility Base

The weight of the Lunar Module with propellants, oxygen, and water was about 33,200 pounds. For comparison, the S-2F twin-engine anti-submarine airplane then in production by Grumman for the US Navy had a maximum takeoff weight of 26,147 pounds. That rugged airplane had a maximum speed of 265 mph and an endurance of 9 hours and carried a crew of four.

The author attended several meetings at the Grumman plant on Long Island during the Apollo program, and often there was an opportunity to examine a full-scale mockup of the Lunar Module. While large and ungainly, the no-nonsense design for utility was impressive.

LUNAR MODULE BACKGROUND

The Lunar Module emerged as a crucial element of the Apollo mission to put a man on the moon after the decision was made to proceed with a lunar orbit rendezvous approach. NASA assembled a set of requirements for the Lunar Module and issued a request for proposals in July 1962. Nine companies submitted proposals. Grumman Aircraft Engineering Corporation in Bethpage, New York, was announced to be the winner of the competition on 7 November 1962. Grumman was an old-line aircraft manufacturing company, renowned for their fighter airplanes for the US Navy.

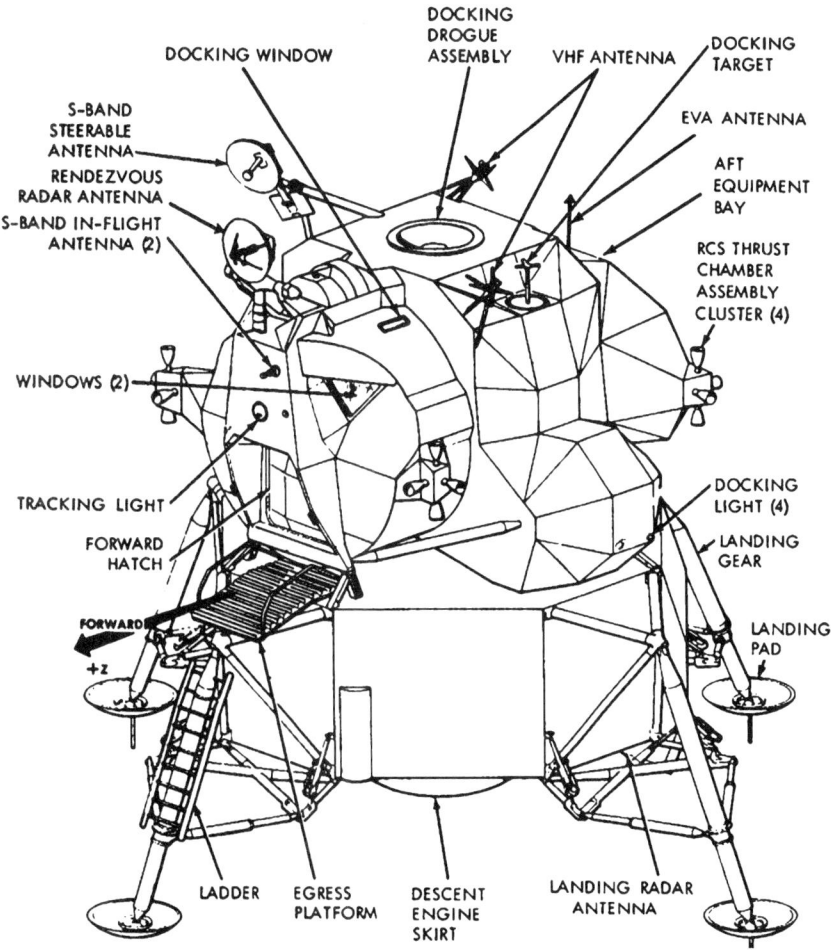

Fig. 8.2 Drawing of Lunar Module (NASA graphic)

The initial contract with Grumman for the Lunar Module was signed in March 1963 for the amount of $387.9 million. The contract called for 10 ground test articles (LTA) and 15 flight articles (LM). Neither Grumman nor NASA fully appreciated the scope and complexity of the job at the time. At the end of the program, the Lunar Module had cost about $2.2 billion. This was about 8.7% of the total cost of the Apollo program.

The Grumman management team was led by Joseph Gavin, the vice president of Grumman who functioned as program director. He was supported by a Lunar Module management team of Robert Mullaney, program manager; John Snedeker, business manager; and William Rathke, engineering manager. The engineering team was led by Tom Kelly, chief engineer for the Lunar Module.

In 1966 Grumman had about 7,500 people working on the program of which about 3,000 were engineers. The Grumman engineering spaces were typical for large aircraft companies of the day. Engineers worked at desks in huge open rooms referred to as bull pens. There were probably about 100 engineers per room. The author met with a section leader at his desk a few times. His desk was distinguished in the huge room because it was by a pillar and he had a filing cabinet.

NASA set up a separate program office for the Lunar Module within the Manned Spaceflight Center (MSC) in August 1962. William Rector headed up the office as project manager. Other senior leaders at MSC, including Joe Shea, paid close attention to Grumman's progress and meted out caustic prodding when schedules were not met.

Several LTA test articles were used in a comprehensive test program. LTA-10R and LTA-2R were flown in earth orbit on flights of Apollo 4 and Apollo 7, respectively. Thirteen flight articles were built. Of these, 11 were flown in space. The first was LM-1 flown unmanned as part of Apollo 5. The last was LM-12 that carried Apollo 17 astronauts down to the surface of the moon. LM-13 was under construction for Apollo 18 when the decision was made to cancel the last three Apollo flights. LM-13 was completed and now resides in the Cradle of Aviation Museum on Long Island.

CONFIGURATION OF DESCENT STAGE

The descent stage contained a gimballed descent rocket engine and fuel and oxidizer tanks for the engine. The stage also contained five batteries for electrical power and a water tank and a gaseous oxygen tank for life support for the crew. A four-legged, folding landing gear was attached to the stage.

There were storage compartments around the bottom of the descent stage for the Apollo Lunar Surface Experiment packages and for the Lunar Roving Vehicle on later Apollo missions.

Life-Support Items

The descent stage contained significant amounts of oxygen and water for the crew. The oxygen supply in the descent stage supplied oxygen needs during descent and during the lunar stay. Oxygen was stored in gaseous form at an initial pressure of 2,800 pounds per square inch in a tank about 27 inches in diameter. Oxygen from the pressurized tank was regulated to 900 psi before being piped to the Oxygen Control Module in the ascent stage.

Water was stored in a large tank in the descent stage and in two smaller tanks in the ascent stage. Water was forced out of the tanks by nitrogen pressure acting against a bladder in the tanks. The water tank in the descent stage held 44 gallons.

Landing Gear

The landing gear, which consisted of four hinged struts with footpads on the bottom, folded to fit within the shroud of the launch vehicle during launch from earth. The astronauts extended the landing gear prior to descent to the lunar surface by operating a switch in the Lunar Module. The struts of the four landing legs contained crushable aluminum honeycomb that acted as shock absorbers to reduce the shock of landing. Footpads were attached to the landing struts by a ball joint that allowed pivoting of the footpads. The pads were 37 inches in diameter, and the upturned edges around the periphery extended 7 inches. Once landed, the legs stabilized the spacecraft.

The landing gear supported the spacecraft so that the bottom of the descent engine bell was about 18 inches above the surface before compression of the crushable honeycomb. After the landing of Apollo 11, Neil Armstrong reported that there was about 12 inches clearance between the engine bell and the lunar surface.

There were thin probes 67 inches long that hung down under three of the landing pads to alert the astronauts of proximity to the lunar surface. Switches on the probes were activated upon contact with the lunar surface and illuminated two lunar contact lights on the instrument panel. A ladder was attached to the landing leg located below the hatch of the ascent stage. The ladder allowed the astronauts travel between the ascent module and the lunar surface. The leg with the ladder did not have a contact probe under the pad.

Descent Rocket Engine

The descent rocket engine was mounted on gimbals in the center of the descent stage. The gimbals allowed vectoring the thrust over a range of ±6 degrees in the pitch and yaw planes. The engine was throttleable with a nominal maximum thrust of 9,900 pounds. There were restrictions on the throttleable range. The operating chamber pressure was about 100 pounds per square inch absolute (psia).

The engine was 90.5 inches long with nozzle extension, and the diameter at the exit of the nozzle extension was 59 inches. The engine weighed 394 pounds. The rocket nozzle extension, which provided an exit area ratio of 47.5 to 1, was radiation cooled. It was designed to collapse by up to 28 inches without upsetting the lander should the nozzle extension come down on a large rock, for example.

The throttleable rocket engine for the Lunar Module posed a technical challenge to keep the firing stable over a range of thrust. The final engine design that evolved operated stably within a thrust range of 10% to 65% of maximum thrust and at maximum thrust. However, operation was unstable in the thrust range of 65% to 92.5% of maximum so operation in that range was avoided. The engine was adequate for the mission despite the restrictions.

The engine used a mixture of equal parts by weight of hydrazine (N_2H_4) and unsymmetrical dimethylhydrazine as fuel. Nitrogen tetroxide (N_2O_4) was used as oxidizer. The fuel and oxidizer ignited upon contact with one another in the engine. The commercial name for the fuel was Aerozine 50.

Fuel was carried in two cylindrical tanks with hemispherical endcaps. Scaling from drawings, the tanks were about 51 inches in diameter and 70 inches long. The two tanks held a total of 7,492 pounds of usable fuel. The tanks were pressurized by helium to force the fuel into feed lines. The feed lines from the two tanks were connected together to supply fuel to the engine.

The oxidizer was nitrogen tetroxide (N_2O_4) carried in two tanks the same size as the fuel tanks. The oxidizer tanks were pressurized and also interconnected in the same way as the fuel tanks. The two tanks held a total of 11,957 pounds of usable oxidizer. The ratio by weight of oxidizer to fuel injected into the engine was 1.6 to 1.

Helium for pressurization of the propellants was contained in two helium tanks. One tank stored gaseous helium that was used for initial pressurization of the fuel and oxidizer. The second tank was a cryogenic storage vessel containing helium in a supercritical state. That tank held a large quantity of supercritical helium, and it was used to pressurize the propellant tanks once fuel started to flow. The gaseous helium tank had a volume of 1 cubic foot, and it held 1.12 pounds of helium. The cryogenic tank had a volume of 5.9 cubic feet, and it held 49.4 pounds of usable helium. The diameter of the cryogenic tank was about 27 inches.

Helium flowed through a fuel/helium heat exchanger where flowing fuel was used to warm the very cold gas boiling off of the supercritical helium. Gaseous helium was needed for initial pressurization because the very cold helium from the cryogenic tank could freeze the fuel in the heat exchanger until a good flow rate was achieved. The warmed helium was regulated to a pressure of 245 psia before being applied to the propellant tanks.

The allowable operating time for the engine was 1,100 seconds (18 minutes, 20 seconds). Operating longer than that risked burn-through of the charred combustion chamber. The total operating time of the engine during the Apollo 11 mission was 13 minutes, 05 seconds.

It is instructive to look at descent engine operation during the landing of the Apollo 11 Lunar Module. Firing of the engine was controlled by the computer until just before landing when Neil Armstrong changed the flight path to avoid landing near a large crater.

The Apollo 11 LM descent engine was first fired after separation of the LM from the Command Module while in lunar orbit to lower the perilune and establish an orbit of 58.5 by 7.8 nautical miles. This maneuver was accomplished by firing the engine at reduced thrust for 28 seconds. The firing decreased the spacecraft velocity by 75 feet per second.

The braking phase involved firing the engine at various thrust levels for 8 minutes, 26 seconds and reduced the velocity from 5,560 feet per second to 506 feet per second.

The approach phase began at the end of the braking phase at an altitude of about 7,500 feet and velocity of 506 feet per second. The engine thrust was gradually decreased throughout the 100 seconds of the approach phase. At the end of the approach phase, the altitude was 512 feet and the velocity was 55 feet per second.

The landing phase began at the end of the approach phase. The duration of the landing phase, which lasted until touchdown, was 149 seconds. The landing phase burn had been extended by Armstrong to avoid a large crater surrounded by rocks. Touchdown occurred at 102:45:40 mission elapsed time, and the engine was shut off.

During the Apollo 11 landing, the Lunar Module pilot, Buzz Aldrin, called out altitudes and velocity from the computer display to the commander, Neil Armstrong, who was flying the vehicle. At an altitude of 100 feet and about 1 minute from landing, Aldrin called out fuel quantity remaining of 5%, which he read from a fuel quantity gauge display.

In total, the engine fired for 13 minutes, 05 seconds. NASA Technical Note TN D-7143 indicates that the hover time remaining was 63.5 seconds at the time of landing.

Electrical Power in Descent Stage

Electrical power for the Lunar Module was obtained from large silver-zinc batteries. The ascent stage contained two batteries, each with a capacity of 296 ampere-hours at 28 volts. The descent stage for early missions contained four batteries, each with a capacity of 400 ampere-hours at 28 volts. A fifth battery was added to the descent stage for later flights after the Lunar Module was forced into a lifeboat role during the Apollo 13 mission. The batteries used silver-zinc chemistry with a potassium hydroxide electrolyte.

Each battery in the descent stage was about 17 by 9 by 10 inches in size and weighed 133 pounds. Each battery could supply about 25 amperes at 28 volts for 16 hours, or 11.2 kilowatt-hours. The batteries were not recharged after leaving the earth. Two batteries were paired onto a power bus. There were two DC power busses labeled CDR (commander) bus and LMP (Lunar Module pilot) bus. The names reflected the location of switches and circuit breakers on the instrument panel that controlled the power busses. The fifth battery when present could be connected to either bus.

The descent stage batteries were used to power the Lunar Module when in lunar orbit and during the stay on the moon. Both the descent and ascent batteries were used to power the LM during descent to the lunar surface. The ascent stage batteries were used to power the ascent stage during ascent from the lunar surface and rendezvous with the Command Module.

CONFIGURATION OF ASCENT STAGE

The ascent stage was composed of three sections: crew compartment, midsection, and aft equipment bay. The crew compartment included living space for the crew as well as displays, switches, and controls necessary to fly the spacecraft. It was the command and control center for the Lunar Module. The compartment was cylindrical in form, 92 inches in diameter and 42 inches deep. The midsection was joined to the crew compartment to make up the cabin. The pressurized cabin had a volume of about 235 cubic feet.

The midsection was located directly behind the crew compartment and open to it. It was 54 inches deep, 60 inches high, and it had an elliptical cross section about 56 inches wide. The floor of the midsection was 18 inches above the floor of the crew compartment. The midsection contained the overhead hatch and docking tunnel, environmental control equipment, and stowage areas for the astronauts.

The aft equipment bay, located behind the midsection, contained equipment racks with cold plates for electronics equipment. It also contained two oxygen tanks, two helium tanks to pressurize the propellant tanks, two batteries, two inverters for electrical power, and life-support equipment.

A photograph looking forward into the crew compartment of Lunar Module 2 is shown on the next page (Fig. 8.3). A close-up photograph of the instrument panel of LM-10 that flew the Apollo 15 mission is shown on the following page (Fig. 8.4). LM-2 was scheduled to fly an unmanned test flight in earth orbit, but test results from a flight involving LM-1 were successful, and the flight of LM-2 was canceled. The spacecraft was used for ground testing instead. LM-2 is now proudly displayed in the Smithsonian National Air and Space Museum in Washington, DC.

The two forward-facing windows are prominent in the photograph looking forward into the crew compartment. The window in the overhead at the left was used during rendezvous and docking with the Command Module. The display and keyboard (DSKY) for the LM computer would normally mount in the open gap in the panel in the center of the picture. The DSKY was the same as described for the Command Module.

Sketches of space inside the crew compartment are shown on the following two pages (Figs. 8.5 and 8.6). The two astronauts faced a multi-paneled control/display panel, and each astronaut had a forward-facing window. The commander stood at the left flight station, and the Lunar Module pilot stood at the right flight station. An optical telescope used for star sighting was located at eye level between the flight stations. The sketch of an astronaut standing at the optical telescope shows the midsection of the stage behind him.

The deck of the crew compartment was about 56 inches wide and 36 inches deep. The deck was covered with Velcro pile, and the bottom of the astronaut's boots had Velcro loops to keep the astronauts from floating off the floor in zero gravity.

Fig. 8.3 Interior of LM-2 (NASM photograph)

Fig. 8.4 Instrument panel of LM-10 (NASA photograph)

Originally, seats were planned for the crew but by standing they were closer to the windows, and that resulted in a much wider view of the lunar surface. Tethers attached at the waist of the astronauts were kept taunted to stabilize them in a standing position perpendicular to the floor. The descent and ascent durations were relatively short at 2 hours 33 minutes and 3 hours 41 minutes, respectively, so standing for those periods was not a problem.

The overhead hatch and docking tunnel were located at the top of the midsection. The overhead hatch was 33 inches in diameter and opened inward. The docking tunnel was 32 inches in diameter and 16 inches long. The midsection also contained the environmental control system (ECS) equipment; guidance, navigation, and control subsystem (GN&CS) equipment; the suit liquid cooling assembly; and life support and communication umbilicals. It also contained stowage areas and food containers for the astronauts.

Control and Display Panels

There were a total of 12 control and display panels in the LM. The two main panels, referred to as panels 1 and 2, were side by side at about eye level in the center. Panel 1 on the left contained flight instruments and controls, propellant quantity

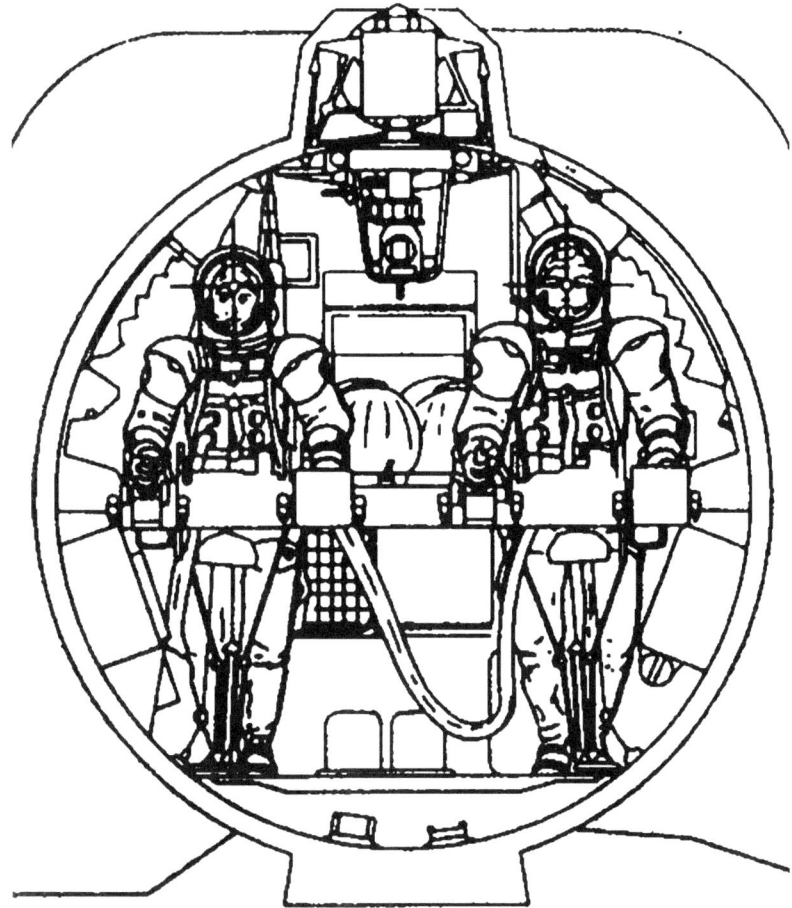

Fig. 8.5 Astronaut's flight stations (NASA graphic)

indicators, and warning lights. It was located in front of the commander. Panel 2 contained flight instruments and controls, reaction control system and environmental control system indicators and controls, and caution lights. It was located in front of the Lunar Module pilot. Drawings of control panels 1 and 2 are shown on the the following two pages (Figs. 8.7 and 8.8). Panels 1 and 2 were split out by the author from the original NASA drawing of several contiguous panels to make the captions on the panels larger and easier to read.

Panel 3, which was located just below panels 1 and 2, contained indicators and controls for the radars, engines, spacecraft stability, event timer, and lighting. Panel 4, located in the center just below Panel 3, contained the display keyboard (DSKY) for the LM guidance computer and indicators for the inertial subsystem. Panel 5, located in front of the commander at waist level, contained engine start and stop push buttons, X-axis translation push button, and mission timer controls.

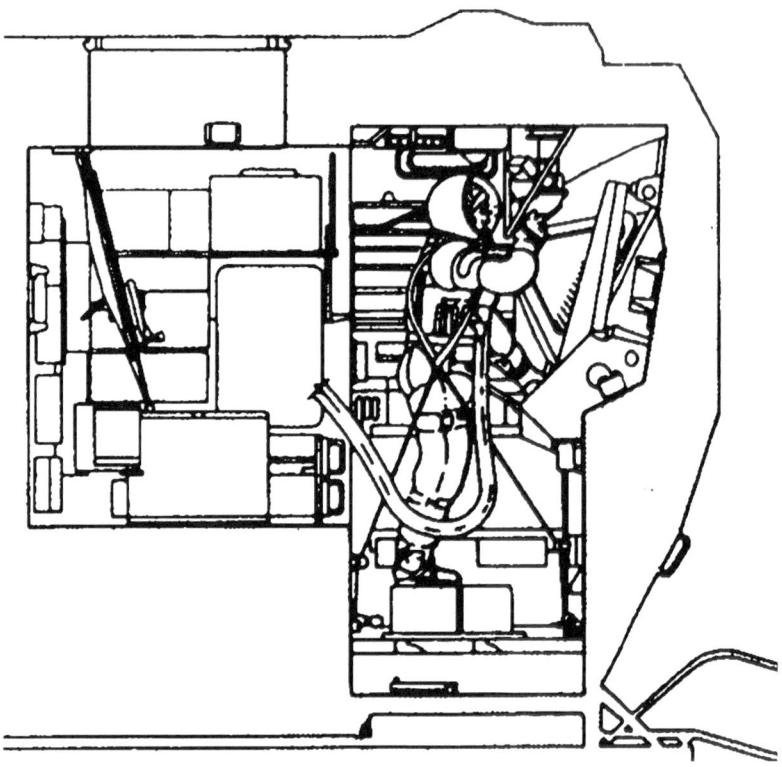

Fig. 8.6 Astronaut standing at optical telescope (NASA graphic)

Panel 6, located in front of the Lunar Module pilot, contained abort guidance controls.

A hand-operated attitude control assembly for the commander was located to the right of panel 5, and a hand-operated translation controller assembly was located to the left. Likewise, a hand-operated attitude control assembly for the Lunar Module pilot was located to the right of panel 6, and a hand-operated translation controller assembly was located to the left. The attitude control assembly and the translation controller assembly were basically the same as described previously for use in the Command Module.

A brief description of major controls and indicators on panels 1 and 2 is given in the following paragraphs along with their use during the mission.

Instrument Panel 1

Panel 1, in front of the commander, contained a caution/warning panel at the top with 14 active caution/warning indicators. Warning indicators lit up red to alert the crew of a problem that could affect crew safety and required prompt attention.

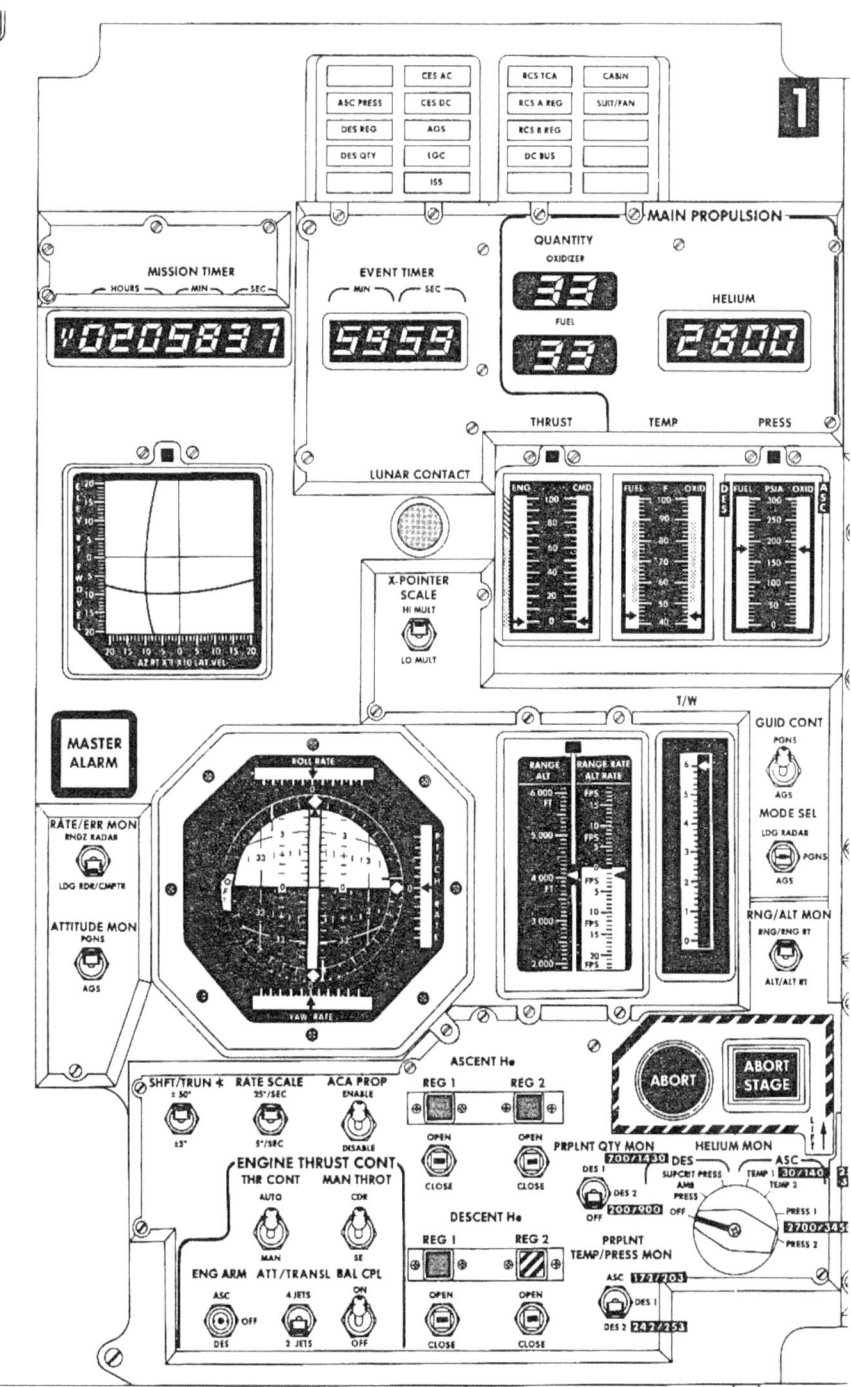

Fig. 8.7 Lunar Module control panel 1 (NASA drawing cropped by author)

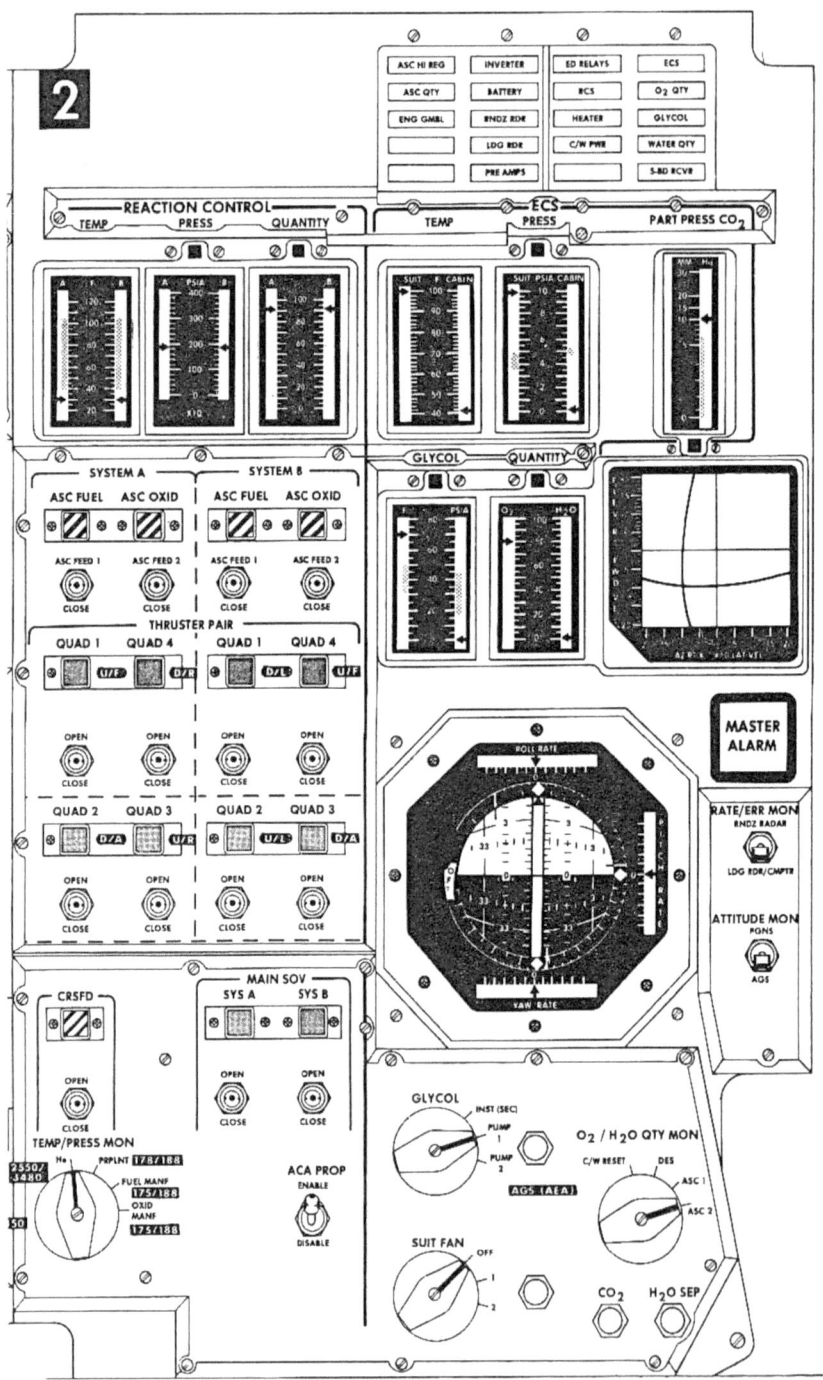

Fig. 8.8   Lunar Module control panel 2 (NASA drawing cropped by author)

Caution indicators lit up yellow to indicate a situation that was not critical to crew safety but to a situation that the crew should be aware of.

Below the caution panel on the left were digital displays for the mission timer and event timer. Controls for the event timer were located on Panel 3 that was mounted just below panels 1 and 2. Controls for the mission timer were located on panel 5. Below the caution panel to the right were digital displays of MAIN PROPULSION showing oxidizer quantity, fuel quantity, and helium tank pressure.

An instrument referred to as X-pointer indicator was located on the left side of the panel. That instrument had a spherical display surface with cross-pointers. It had horizontal and vertical scales with zero at the center and ± 20 at the ends. Various parameters including forward and lateral velocity of the LM and elevation and azimuth trunnion angles for the rendezvous radar antenna could be selected for display by the cross-pointers. A scale factor switch located to the right of the display allowed selecting a high multiplier or a low multiplier.

The X-pointer indicator was likely set to display forward and lateral velocity during the landing phase. It would have been an aid to the astronauts for nulling forward and lateral velocities of the LM just prior to touchdown.

A blue lunar contact light located above the scale switch illuminated when the long probes below the landing pads contacted the lunar surface.

At the same level on the right side of the panel, there were three vertical displays that read out thrust, temperature, and pressure associated with the descent engine. The thrust displays had separate scales for commanded thrust and engine thrust generated. The temperature indicators and the pressure indicators had separate scales for oxidizer and fuel.

A red MASTER ALARM light located on the left side of the panel illuminated whenever a warning was displayed on the caution/warning panel.

A predominant instrument on the left side of the panel was the flight director attitude indicator (FDAI). That instrument had the same functions as described previously in Chapter 6 for the FDAI in the Command Module. A photograph of the FDAI was also given in Chapter 6.

Attitude of the LM was displayed on a rotating ball in the center of the FDAI. Rotation of the ball in the vertical direction represented pitch angle of the spacecraft, and the pitch angle could be read from markings on the ball. Likewise, rotation in the horizontal direction represented yaw angle with respect to a reference attitude, and the yaw angle could be read off of the ball. Roll angle of the LM was indicated by roll of the ball and roll angle could be read off of an angular scale on the instrument just outside of the ball.

A switch to the left of the FDAI selected the source of attitude information either from the primary guidance and navigation system (PGNS) or the abort guidance system (AGS). The PGNS and the AGS are discussed later in this chapter.

Indicators near the periphery of the FDAI displayed roll rate at the top, pitch rate at the side, and yaw rate at the bottom. The scales were unnumbered with zero at the center. The scale factors could be set by a RATE SCALE switch located below the FDAI to either 25 degrees per second or 5 degrees per second for full-scale deflection.

The flight director function of the FDAI was provided by three short yellow needles that appear in the photograph of the FDAI. The needles indicated magnitude and direction of attitude errors in each axis. The vertical needle at top of the instrument gave roll angle error, the horizontal needle gave pitch error, and the vertical needle at the bottom of the display gave yaw error. The astronauts could use these needle deflections to return the attitude of the spacecraft to planned values when under manual control.

There were three vertical displays to the right of the FDAI. The first one read either range or altitude. The next display read either range rate or altitude rate. A RNG/ALT MON switch located on the right side of the panel was used to select either range/range rate or altitude/altitude rate. The range/range rate mode was used when obtaining data from the rendezvous radar, and the altitude/altitude rate mode was used during landing when the landing radar was active. The next vertical indicator over was labeled T/W. This was a thrust-to-weight indicator that gave information pertaining to acceleration along the X-axis.

Two push buttons labeled ABORT and ABORT STAGE were located just below the T/W indicator. Pressing the ABORT button armed the descent engine and initiated the abort program that fired the engine. Pressing the ABORT STAGE button resulted in immediate shutdown of the descent engine and enabled circuits for descent stage separation and ascent engine firing. An ascent engine-on command then initiated stage separation and firing of the ascent engine.

There were a group of switches near the bottom of panel 1 under the heading ENGINE THRUST CONT. The switch labeled THR CONT allowed selecting either automatic or manual control of thrust. The MAN THROT switch allowed selecting either commander or Lunar Module pilot control of manual throttle. The ENG ARM switch allowed selecting arming of the ascent engine or the descent engine or OFF with no arming. The engine was fired by pressing the START push button located on panels 5 and 6.

A rotary switch at the bottom right of the panel labeled HELIUM MON allowed selecting temperatures and pressures of the helium tanks in the descent and ascent stages. These measurements were read on the digital display labeled HELIUM near the top of the panel.

Instrument Panel 2

Panel 2, located on the Lunar Module pilot's side, also had a caution/warning panel at the top of the panel. A series of six vertical indicator displays was located below the caution and warning panel. Three vertical displays at the left displayed parameters of the reaction control system under the heading REACTION CONTROL. There were two separate indicators, labeled A and B, to display parameters of the redundant reaction control systems A and B. The three displays were TEMP (temperature), PRESS (pressure), and QUANTITY. A rotary switch located at the bottom of the panel selected which item to be monitored: helium, propellant, fuel, or oxidizer.

Three vertical displays to the right displayed parameters of the environmental control system under the heading ECS. The dual indicator on the left read temperature of the suit and of the cabin from 20 to 100 degrees Fahrenheit. The next dual indicator read pressure in the suit and in the cabin from 0 to 10 psia. The single indicator at the right read partial pressure of CO_2 from 0 to 30 inches of mercury.

Further down on the panel under REACTION CONTROL were a series of switches to control the reaction control engines.

On the right side of the panel below, the ECS displays were two dual vertical displays, one labeled GLYCOL and the other labeled QUANTITY. The dual display under GLYCOL read temperature and pressure on a scale from 0 to 80. The dual display under QUANTITY read O_2 (oxygen) and H_2O (water) from 0 to 100 percent. A rotary switch labeled GLYCOL at the bottom of the panel allowed selecting INST(SEC), pump 1, or pump 2. The switch position INST(SEC) selected the secondary cooling loop. A rotary switch labeled O_2/H_2O QTY MON at the bottom of the panel allowed selecting DES (descent), ASC 1 (ascent 1), or ASC 2 (ascent 2).

An X-pointer indicator, flight director attitude indicator, and master alarm, which were the same as on panel 1, were located on the right side of the panel. A rotary switch labeled SUIT FAN at the bottom of the panel had positions for OFF, 1, and 2.

Electrical Power

Primary electrical power for the ascent stage was obtained from two silver-zinc batteries. Each battery could supply 50 amperes at 28 volts for 5.9 hours or 8.26 kilowatt-hours. Each battery was 35.8 by 5.0 by 7.8 inches in size and weighed 124 pounds. One battery was connected to the CDR bus and the other to the LMP bus. Batteries located in the descent stage were also connected to those two DC buses. Power for the ascent stage was obtained from descent stage batteries while on the lunar surface. Primary power was switched to the ascent stage batteries just before launch from the moon.

The ascent stage contained two inverters that operated off of the 28 volt DC power busses and generated alternating current (AC) power at 115 volts and 400 Hz. Each inverter was connected to a separate 28 volt DC power bus. The outputs of the inverters were connected to separate AC power busses labeled AC bus A and AC bus B. The AC electrical load in the Lunar Module was about 350 volt amperes, and this was accommodated by one of the inverters. The other inverter was available as backup.

Electrical power was routed to circuit breaker panels in the crew compartment where each major electronic assembly and each heater and lighting circuit had a dedicated circuit breaker. A circuit breaker would automatically open if there was excessive current drawn on that line. A circuit breaker could also be opened manually to unpower a particular piece of equipment. There were two main circuit breaker panels. One, located to the left of the commander's station, contained 89 circuit breakers. The other panel, located to the right of the Lunar Module Pilot's station, contained 71 circuit breakers.

GUIDANCE, NAVIGATION, AND CONTROL SUBSYSTEM

The guidance, navigation, and control subsystem (GN&CS) performed the tasks necessary for guidance and control of the Lunar Module and for navigation from lunar orbit to a designated landing area on the moon. When departing the moon, the navigation function guided the ascent stage to a rendezvous with the Command Module.

Major elements of the GN&CS were the Primary Guidance and Navigation Section (PGNS), the Abort Guidance Section (AGS), and the Control Electronics Section (CES). The major interfaces of these elements with other subsystems in the spacecraft are shown on the simplified block diagram of the overall Lunar Module system shown on the next page (Fig. 8.9). The diagram was adapted from a graphic in Grumman document LMA790-2 *Vehicle Familiarization Manual*.

Primary Guidance and Navigation Section

The Primary Guidance and Navigation Section (PGNS) was made up of the Computer Subsection, Inertial Subsection, and Optical Subsection. The PGNS was similar in function to the Primary Guidance Navigation and Control System used in the Command Module. A block diagram of the Primary Guidance and Navigation Section is given on the page following the overall system diagram (Fig. 8.10).

Computer Subsection

The computer unit was the same as that used in the Command Module. The software was different and focused on landing on the moon and ascending from the moon and rendezvousing with the Command Module. The display and keyboard

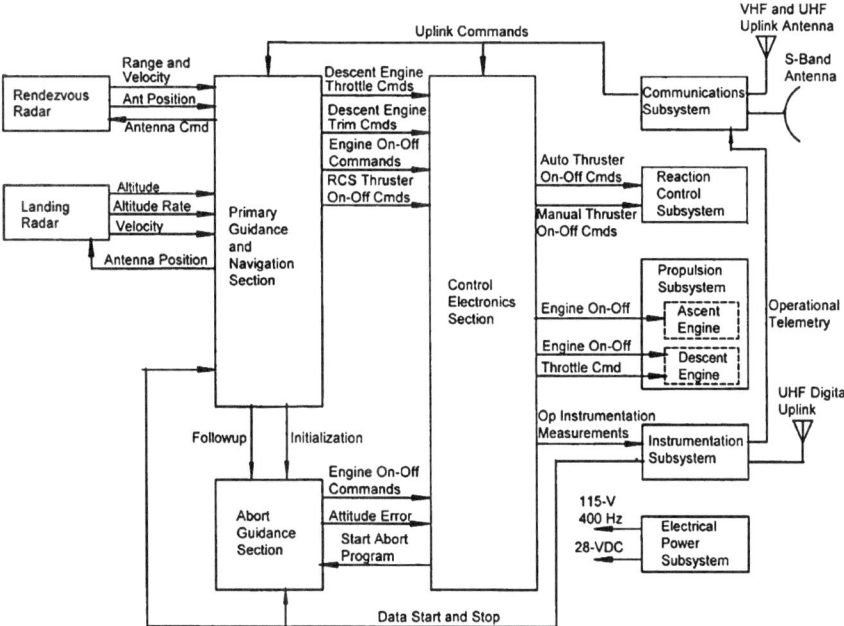

Fig. 8.9 Simplified block diagram of Lunar Module guidance, navigation, and control subsystems (Grumman graphic)

(DSKY) was also the same as used in the Command Module except for two added items to the caution and warning panel.

Both the computer and the DSKY were described in some detail in Chapter 6, so only a few comments will be made here. The computer represented state-of-the-art packaging of computer functions with reliable circuitry in a small size at the time. The computer assembly was 24 inches long, 12.5 inches wide and 6 inches high. It weighed 70.1 pounds and consumed about 55 watts of power at 28 volts DC when operating. It consumed 15 watts during standby A photograph of the computer and DSKY is shown on the following page (Fig. 8.11).

The display and keyboard (DSKY) provided the man-machine interface between the astronauts and the computer. It allowed them to select various computer programs and command certain functions of the LM, enter data necessary for computer computations, and it presented readouts of pertinent data during the various phases of the mission. The DSKY was located in the Lunar Module between the commander and Lunar Module pilot. It was mounted on panel 4 just above the forward hatch.

Communication between the crew and the computer was conducted through a simple but effective system of "programs," "verbs," and "nouns" that were described earlier.

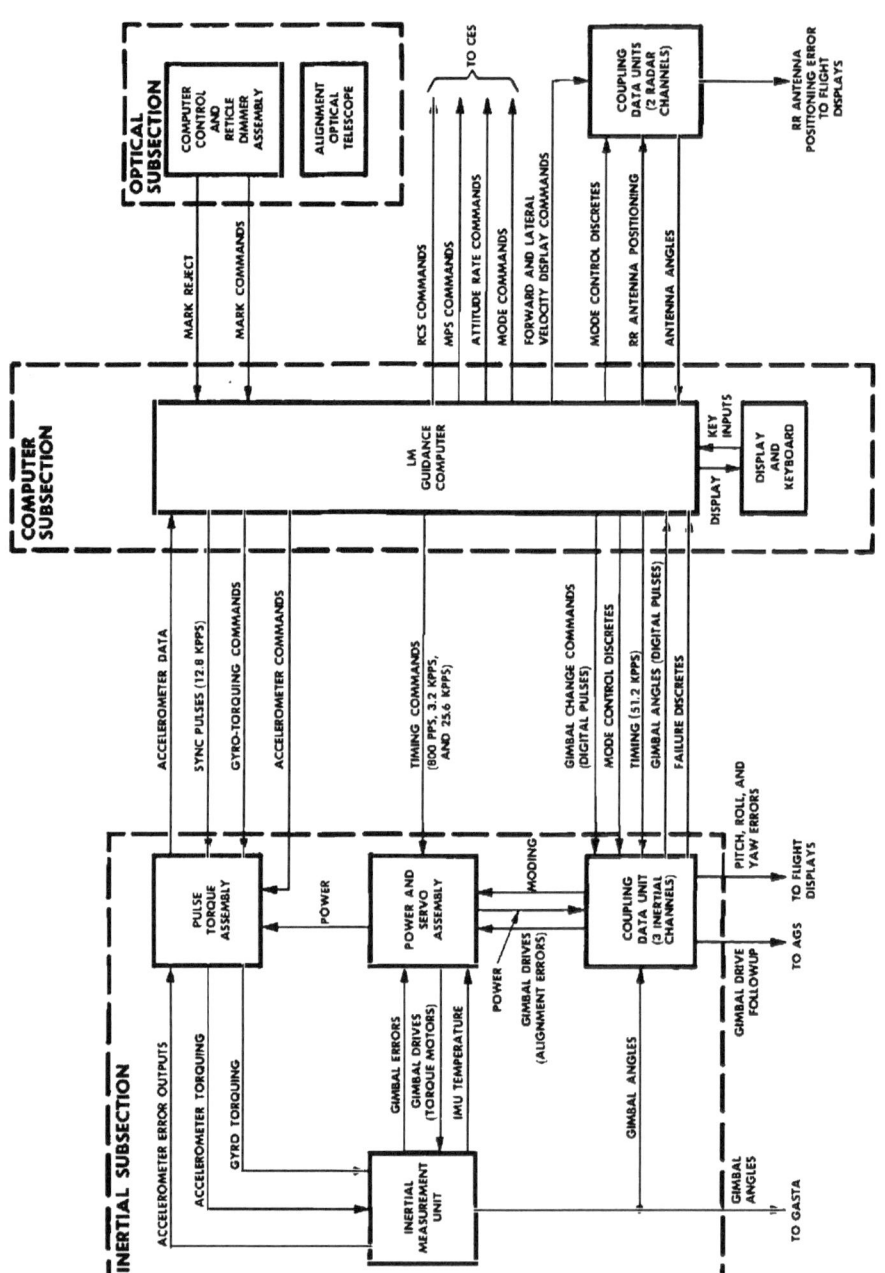

Fig. 8.10 Primary Guidance and Navigation Section (NASA/Grumman graphic)

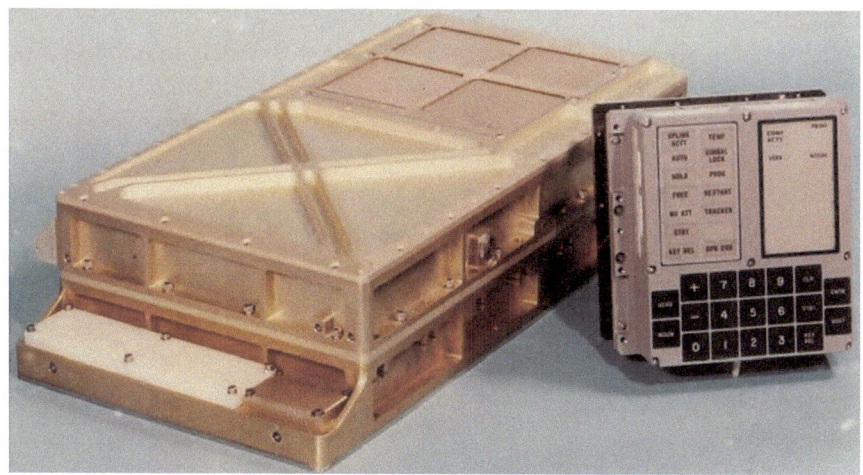

Fig. 8.11 Apollo Computer and DSKY (NASA photograph)

The warning/caution panel for the Lunar Module DSKY had 12 captions. A particular caption would light up when the computer sensed an error or that a warning was needed. The meaning of the captions are shown in the table below. The captions were the same for the CM and the LM DSKY units except that the Lunar Module had two additional items marked ALT and VEL (Table 8.1).

Table 8.1 Captions on DSKY warning/caution panel

Caption	Meaning
UPLINK ACTY	Data being received from ground
TEMP	Temperature of stable platform out of tolerance
NO ATT	Attitude reference not available from IMU
GIMBAL LOCK	Middle gimbal angle greater than 70 degrees
STBY	Computer is on standby
PROG	Computer waiting for information to be entered by crew
KEY REL	Computer needs control of DSKY to complete a program
RESTART	Computer is in restart program
OPR ERR	Computer detected a keyboard error
TRACKER	One of the optical coupling units failed
ALT	Landing radar altitude data no good
VEL	Landing radar velocity data no good

The ALT light would be on and steady when the landing radar was powered but the altitude data good signal was not present. The light would flash if the landing radar altitude reasonability test failed. The VEL light would be on and steady when the landing radar is powered but the velocity data good signal was not present. The light would flash if the landing radar velocity reasonability test failed.

The software loaded into the LM guidance computer (LGC) for Apollo 11 was named LUMINARY 1A. The software was updated for each subsequent Apollo mission up to Apollo 15. Apollo 15, 16, and 17 used LUMINARY 1E.

The astronauts communicated with the computer by first requesting a particular computer program by entering a two-digit program number followed by ENTER. The LUMINARY software for the Lunar Module contained 32 computer programs that were listed as P00 through P77. Not all program numbers were used.

A few of the programs and their titles are given below for illustration.

Program	Title
P00	LGC Idling
P52	IMU Realign
P63	Braking Phase
P64	Approach Phase
P66	Landing Phase (ROD) (Rate of Descent)
P12	Powered Ascent
P20	Rendezvous Navigation

The action desired from a particular program or the format desired for data readouts on the DSKY was selected by pressing VERB followed by two numbers and ENTER. Data to be displayed was selected by pressing NOUN followed by two numbers and ENTER. Verbs generally requested an action, and nouns generally specified what data was to be displayed. There were 99 number pairs allocated to verbs with several pairs marked SPARE. There were 99 number pairs allocated to nouns again with several marked SPARE. Keying in Verb 37 readied the computer to accept a new program.

The computer used information from the inertial measurement unit and various descent software programs to calculate thrust commands for the descent engine and guidance commands for the reaction control jets to follow a preprogrammed path to landing.

The use of the programs, verbs, and nouns will be illustrated by recounting some computer controlled events associated with the descent and landing of Apollo 11. We will start at the time in the mission when the LM was in orbit around the moon and approaching the orbit's perilune of about 50,000 feet.

The Lunar Module pilot, Buzz Aldrin, keyed in Verb 37 to ready the computer to select a new program and then keyed in 63 to select Program P63. Program P63, labeled "braking phase," first controlled the preparation for Powered Descent Initiation (PDI). It maneuvered the spacecraft to align the thrust axis with the velocity vector, and it computed the precise time when engine ignition should occur. Verb 06 and Noun 62 were then selected to present pertinent data on the DSKY. Verb 06 called for displaying decimal data in the three data display fields. Noun 62 called for display of "Absolute Value of Velocity," "Time to Ignition," and "Delta V Accumulated."

The program paused just before time for ignition and requested a "Go" from the crew before igniting the engine. This query for "Go?" was in the form of flashing Verb 99 on the DSKY. The crew had received "you're Go for powered descent" from mission control about 5 minutes previously so they could act on the flashing Verb 99 as soon as it came up. Upon seeing flashing Verb 99, Aldrin pressed PRO (proceed) on the DSKY and the engine fired to begin the braking phase. The engine was throttled to about 10 percent of full thrust for a few seconds at the start of the burn to allow time orient the thrust vector along the center of gravity, and then it was throttled up to nearly full thrust.

Program P63 continued into the braking phase where the software controlled the descent engine to slow the velocity from 5,560 feet per second to 506 feet per second. The slowed spacecraft dropped out of lunar orbit, and by the end of the braking phase, it was at an altitude of about 7,129 feet above the lunar surface. The burn of the descent engine during the braking phase lasted 8 minutes, 26 seconds. A throttle reduction from nearly full throttle to partial throttle was made about 6 minutes, 26 minutes after ignition. Verb 06 and Noun 63 were used to monitor progress by displays on the DSKY during most of this time. Verb 06 specified decimal display of data and Noun 63 called for display of "Absolute Value of Velocity," "Altitude Rate," and "Computed Altitude."

About 4 minutes into the burn, Armstrong rolled the spacecraft to a face-up position, and shortly thereafter the radar altimeter portion of the landing radar achieved lock-on to the lunar surface. Aldrin keyed in Noun 68 which displayed "Slant Range to Landing Site," "Time to Go in Braking Phase," and "LR Altitude– Computed Altitude." The last item gave the difference between altitude measured by the landing radar and that computed by the LM guidance computer. The difference of about 2,800 feet was within the reasonableness bounds given uncertainty in the computed altitude above the lunar surface. The computer had computed altitude based on inputs from the PNGS. Upon approval from mission control in Houston, Aldrin keyed in Verb 57 (permit landing radar updates) and ENTER to allow landing radar altitude data to update the computer.

The braking phase continued to a point in the descent profile referred to as "high gate" where the braking phase ended and the approach phase began. The altitude was 7,129 feet at that point, and velocity was 506 feet per second. The computer automatically switched to program P64 at the end of the braking phase. Program P64, labeled "approach phase," contained software designed to guide the spacecraft from high gate to a point directly above the landing site referred to as "low gate." Nominally, the spacecraft would be at an altitude of about 500 feet at low gate. The display on the DSKY was set to Verb 06 (display decimal data) and Noun 64. Noun 64 called for display of "Time Left for Redesignation – LPD Angle," "Altitude Rate," and "Computed Altitude."

The first data field, "Time Left for Redesignation – LPD Angle," contained a two-digit number giving seconds remaining for redesignation of the landing site followed by a blank on the display followed by a two-digit number giving the "LPD Angle" in degrees. The LPD (Landing Point Designator) sighting grid was a scribed graduated line marked in degrees on the window of the Lunar Module. Sighting past the sighting grid at the LPD angle showed the crew the computer predicted landing area on the surface. A sketch of the window with the scribed sighting grid is shown below (Fig. 8.12).

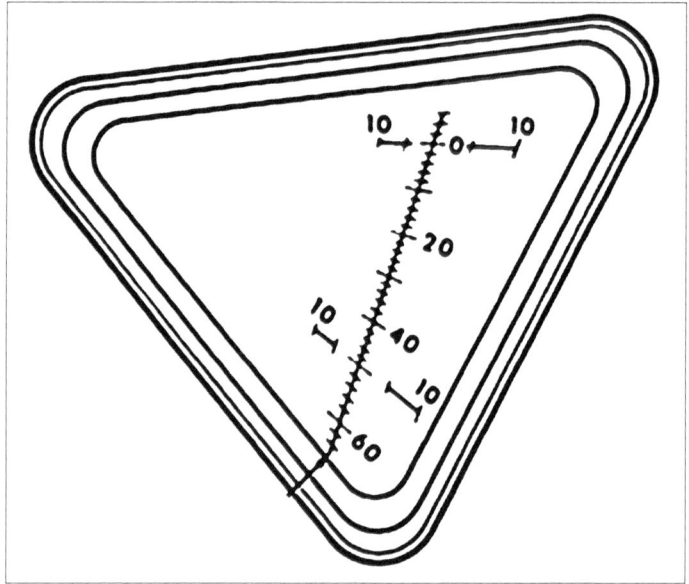

Fig. 8.12   Landing Point Designator sighting grid on window of LM (NASA graphic)

The approach phase program, P64, allowed the crew to resignate the landing site if they did not like the site indicated by the LPD sighting grid on the window at the LPD angle indicated.

At an altitude of about 1,000 feet the DSKY display indicated a LPD angle of 35 degrees. Armstrong did not like the landing area at that LPD since it was dominated by a crater with boulders along its sides. He switched the PGNS MODE switch on panel 3 from AUTO to ATT HOLD (attitude hold) to manually control the spacecraft attitude by the attitude control assembly (ACA). Deflecting the attitude control assembly would cause an attitude rate, and when the ACA was released, the PGNS would maintain (hold) that attitude. The switch to attitude hold was made at about 650 feet altitude.

Armstrong pitched the LM to slow the descent while keeping forward velocity and flew the spacecraft over the crater. Remaining in manual control, he flicked the descent rate switch on panel 5 that, among other things, switched the software to program P66. Program P66 was labeled "rate of descent landing." The spring-loaded descent rate switch, which was shaped like a paddle, was mounted on a pedestal on panel 5 and readily assessable to the commander's left hand. Flicking the switch down commanded the computer to increase the rate of descent by 1 foot per second. Flicking it up decreased the descent rate by 1 foot per second.

Armstrong saw a satisfactory landing area and maneuvered the spacecraft over it. He pitched the spacecraft to null forward velocity and controlled the descent rate to allow the spacecraft to descend to a soft landing. Noun 60 was called up on the DSKY during the landing to show "forward velocity," "altitude rate," and "computed altitude." Aldrin kept Armstrong informed of progress of the landing by calling out those values every few seconds during the landing process.

The descent had moments of concern because of a recurring program alarm. Shortly after the spacecraft had been rolled over to a face-up attitude and 5 minutes, 17 seconds after Powered Descent Initiation, the PROG warning light came on. Aldrin keyed Verb 5, Noun 9 (ALARM CODES) into the DSKY and alarm code 1202 appeared. Armstrong asked mission control for a reading on 1202. Mission control came back with a "Go" on the alarm after about 27 seconds. In other words, continue the landing. There were a total of four 1202 alarms and one 1201 alarm in a period of about 4 minutes. These alarms also showed up in the telemetered data sent down to earth.

Alarm codes 1201 and 1202 were symptomatic of computer overload. Code 1201 was labeled "executive overflow – no core sets," and Code 1202 was labeled "executive overflow – no VAC areas." No core sets indicated that there were no cores unoccupied in the memory for the program then running. Likewise, no VAC areas indicated that there were no vector accumulators available for temporary variables. The software was cleverly written to clear itself if either of these alarms occurred by stopping what it was doing and clearing out low priority data while retaining important information such as the vehicle state vector. It then automatically resumed operation, starting with the highest priority task.

Fortunately, a 1201 alarm had been deliberately triggered during simulations conducted in mission control before the launch of Apollo 11. As a result, the mission control guidance officer, Steve Bales, and software engineer, Jack Garman, had seen these alarms before. Garman wrote down on a piece of paper the alarm codes requiring abort after the simulation session. When the alarms happened during the landing of Apollo 11, Garman knew immediately that Codes 1201 and 1202 were not cause for abort as long as the software cleared them and resumed operation. The "Go" on the alarms was passed from Garman up to Bales and up to the flight director, Gene Krantz, and on to capsule communicator, Charlie Duke, who passed the "Go" on to the crew.

The cause of the computer overload was dithering of signals representing steering angles of the rendezvous radar antenna. This took up computer time to continuously update small apparent position changes even though the antenna was stationary. The rendezvous radar was in "standby" mode, just in case it was needed in a hurry, but the antenna position signals were active. The dithering was a result of the 800 Hz excitation for the position synchros not being phase locked with the 800 Hz reference used by the computer. The alarms began after the landing radar acquired the lunar surface and began providing radar data. This increased the computer workload and the addition of the spurious signals from the rendezvous radar depleted the processing time margin. The alarms stopped after Armstrong switched to attitude hold mode, which reduced the computer processing tasks.

Inertial Subsection

The Inertial Subsection contained the inertial measurement unit and attendant control hardware. The inertial measurement unit (IMU) was the central navigation element in the Lunar Module. Processing of IMU data by the computer kept the state vector of the Lunar Module continuously updated. The state vector contained seven numbers: three numbers representing spacecraft velocity in three axes, three numbers representing spacecraft position in three axes, and one number representing time when the data was gathered.

The inertial measurement unit used in the Lunar Module was identical to that used in the Command Module. The Command Module IMU was described in some detail in Chapter 6 of this book, and only a brief repeat of that information is given here. A photograph of the inertial measurement unit that flew to the moon and back in the Command Module of Apollo 17 was given in Chapter 6.

The inertial measurement unit (IMU) established a stable platform fixed in inertial space from which measurements of spacecraft attitude and acceleration were made. Spacecraft attitude was determined by measuring the angular differences between the coordinate axes of the spacecraft and that of the stable platform. Velocity in the stable platform coordinate system was determined by integrating the outputs of three accelerometers mounted on the stable platform.

The gyros of the stable platform had a slight drift and that required periodic realignment of the platform. An alignment was performed using star sights a few minutes after the Lunar Module separated from the Command Module prior to descending to the lunar surface.

The IMU was aligned five times while on the lunar surface. Different techniques were used to evaluate alignment techniques. Two alignments were made in the conventional manner using sights on two stars. Two alignments were made using the gravity vector of the moon along with one star sight. One alignment was made using the gravity vector of the moon and azimuth stored from a previous alignment. All alignments were satisfactory.

Optical Subsection

Star sights necessary for alignment of the inertial platform in the Lunar Module were taken by the alignment optical telescope (AOT). The AOT was less complicated than the sextant used in the Command Module for the same purpose. A sketch of the AOT is shown below (Fig. 8.13). A sketch of controls for the alignment optical telescope is shown on the next page (Fig. 8.14).

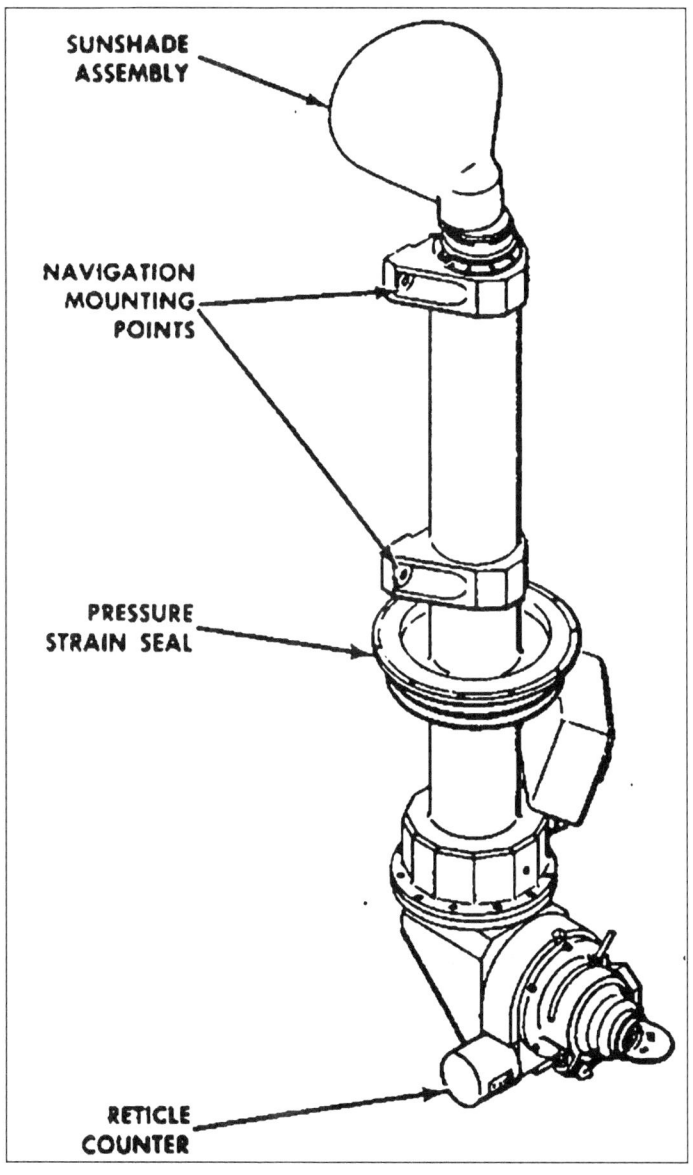

Fig. 8.13 Sketch of alignment optical telescope (NASA graphic)

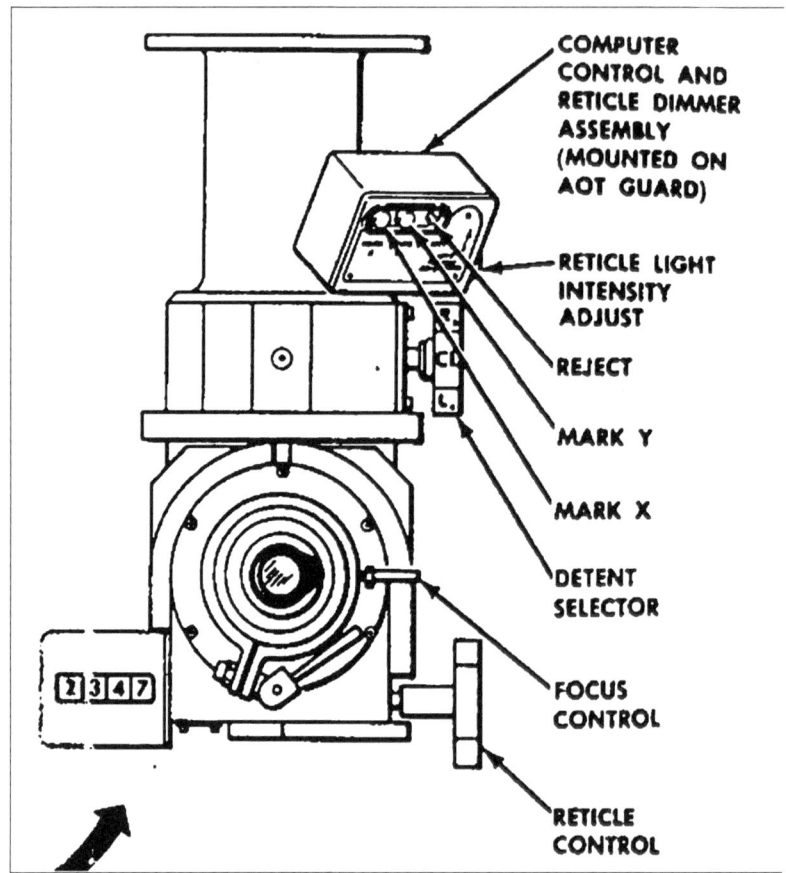

Fig. 8.14 Sketch of controls for alignment optical telescope (NASA graphic)

The AOT was mounted on a Navigation Base along with the IMU to assure a fixed angular relationship between the two. The AOT was mounted with its shaft parallel to the spacecraft X-axis. The telescope functioned as a periscope with unity magnification. It had a lighted reticle to assist in the star sighting process. The field of view of the telescope was 60 degrees centered and 45 degrees above the Y-Z plane. The center of the field of view could be positioned over 360 degrees in azimuth in six fixed detent positions. The positions were selected by rotating a detent selector knob on the instrument.

A sketch of the field of view arrangement of the AOT is shown on the following page (Fig. 8.15). A view of the reticles of the AOT is shown on the next following pages (Figs. 8.16 and 8.17). The graphic of the reticles appears in the Grumman publication *Apollo Lunar Module News Reference*.

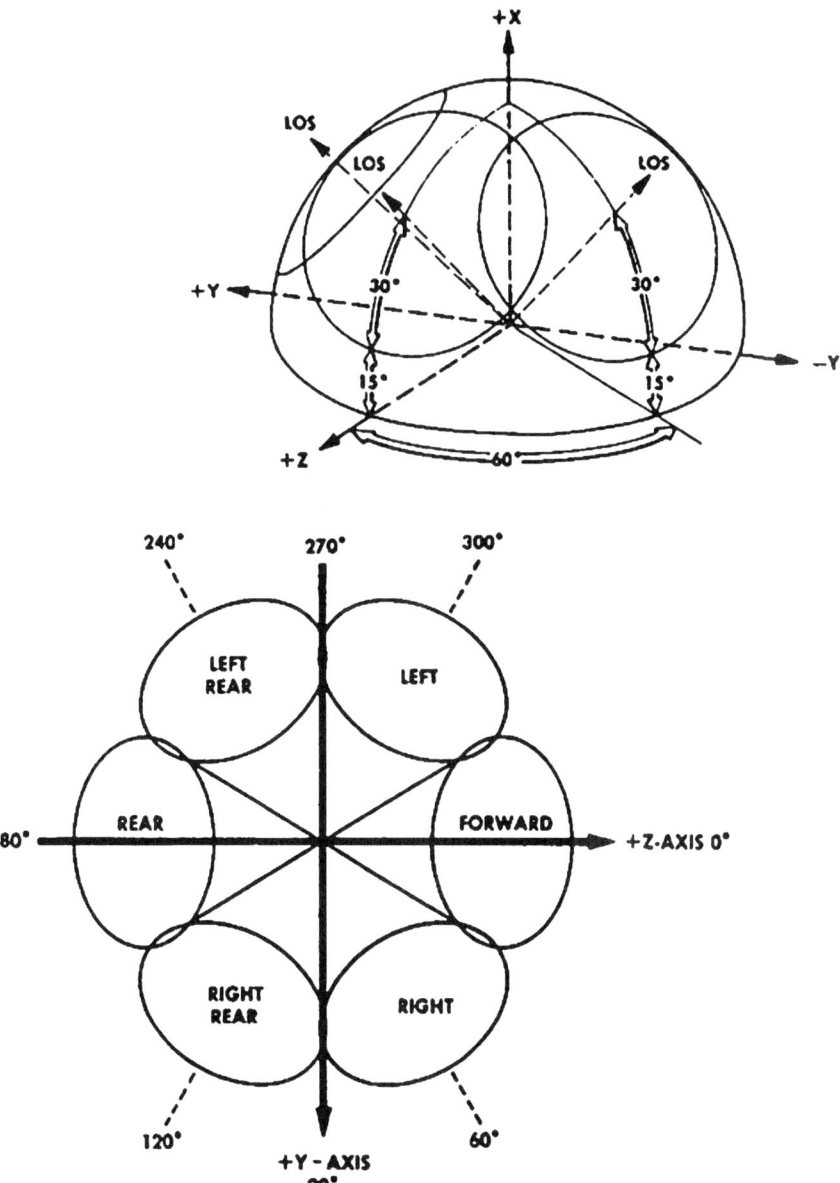

Fig. 8.15 Alignment optical telescope fields of view (NASA graphic)

The reticle had cross hairs oriented along the X and Y axes of the instrument. The portion of the reticle extending from zero towards the + Y axis consisted of dual radial lines. The reticle also included an Archimedian spiral with two parallel lines that spiraled from the center to the periphery.

Star sights were taken while the LM was in free flight by setting the azimuth detent position so that the star was in the field of view, preferably near the center. The spacecraft was then maneuvered so that either the vertical or horizontal segment of the cross hair intersected the star. Assuming the X-axis segment was intersected first, the astronaut pressed the MARK X push button on the control box. He then maneuvered the spacecraft to intersect the star with the Y segment of the reticle and pressed the MARK Y push button. The spacecraft attitude from the IMU was recorded by the computer at each pressing of the mark button. The two angles defined the location of the star relative to the IMU axis. The procedure was repeated for a second star, and the computer used star sight data to calculate the orientation of the IMU with respect to the reference coordinate system.

A different star sighting approach was used when the Lunar Module was sitting on the surface of the moon since the spacecraft could not be maneuvered. The operator again selected an azimuth detent that contained the star in the field of view and entered the detent and the code number of the star into the DSKY. He then used the manual reticle control knob to rotate the reticle until the star appeared in the space between the two radial lines. The shaft angle to the star was displayed on the counter located to the left of the eyepiece. The operator entered that number into the DSKY. He then rotated the reticle until the star appeared between the two parallel spiral lines and entered the value on the counter into the DSKY. The rotation angle to the spiral could be used to compute angle from the center to the star. The procedure was repeated for a second star.

Abort Guidance Section

The Abort Guidance Section (AGS) was the backup to the Primary Guidance and Navigation Section (PGNS). If the PGNS failed, the mission had to be aborted and the AGS took its place. The AGS was required to take over from a failed PGNS any time after separation of the Lunar Module from the Command and Service Module, including while on the lunar surface.

The abort guidance system was developed by TRW who were an important aerospace company at the time and well experienced in building spacecraft and components for spacecraft.

Two push buttons labeled ABORT and ABORT STAGE were located on panel 1 in front of the commander. Pressing the ABORT button armed the descent engine and initiated the abort program that fired the engine.

If an abort was required during descent to the lunar surface, the ABORT STAGE button was pressed, and that resulted in immediate shutdown of the descent engine and enabled circuits for descent stage separation and ascent engine firing. An ascent engine-on command then initiated stage separation and firing of the ascent engine.

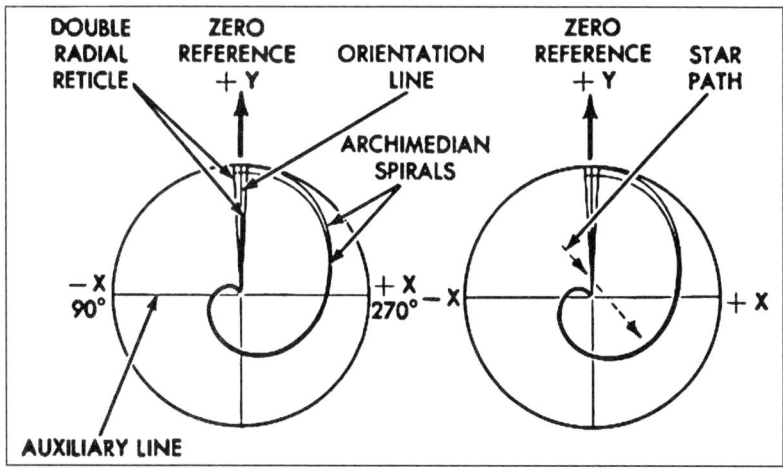

Fig. 8.16 Reticle pattern used for in-flight star sighting (Grumman graphic for NASA)

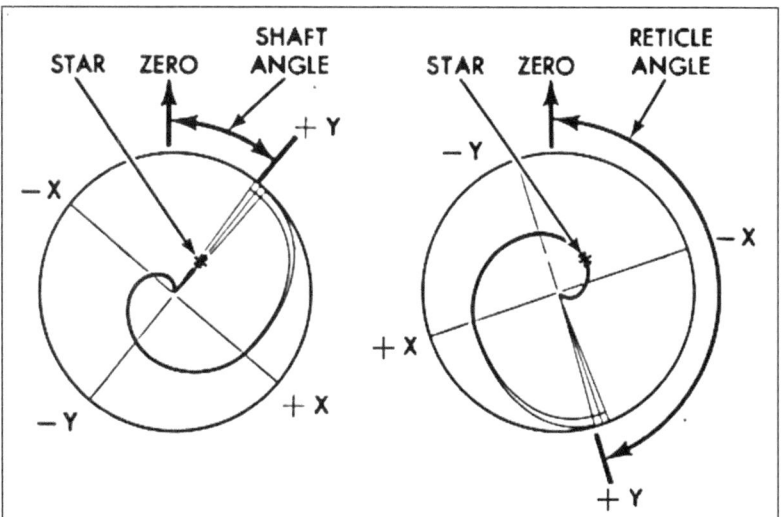

Fig. 8.17 Reticle pattern used for star sighting when the LM was on the lunar surface (Grumman graphic for NASA)

To transfer control of the LM from a failed PGNS to the AGS, the astronauts switched the GUID CONT (guidance control) switch on the right side of panel 1 from PGNS to AGS. The AGS MODE CONTROL switch located on panel 3 was moved from OFF to either ATT HOLD or AUTO. The switch was set to AUTO to allow the AGS to provide guidance steering and engine control.

The Abort Guidance Section contained a strapped-down inertial system for navigation. The strapped-down inertial system was not as accurate as the PGNS, but it was adequate to provide navigation and guidance from any point in the mission to successful rendezvous and docking with the Command and Service Module. The AGS was considerably lighter than the PGNS, and it required less electrical power. The AGS consisted of the Abort Sensor Assembly (ASA), Abort Electronics Assembly (AEA), and the Data Entry and Display Assembly (DEDA).

Abort Sensor Assembly

The Abort Sensor Assembly contained a strapped-down inertial system that included three strapped-down, pulse-rebalanced, pendulous accelerometers and three strapped-down, pulse rebalanced, rate-integrating gyros. One accelerometer and one gyro were provided for each of the X, Y, and Z axes. The accelerometers sensed acceleration along a particular axis and the gyros sensed rotation about that axis. The term "strapped-down" means that the accelerometers and gyros were mounted directly to the structure that was attached to the spacecraft rather than to an inertially stabilized platform as in the case of the inertial measurement unit. The ASA included electronics to operate and read out the sensors.

The accelerometers and the gyros used current pulse torqueing to rebalance the outputs. The pulse torqueing servo amplifiers that supplied the balancing current pulses also generated pulses proportional to incremental velocity and angle sensed by the ASA. Those pulses, which were scaled at 2^{-16} radians per pulse for angle and 0.003125 feet per second per pulse for velocity, were supplied to the Abort Electronics Assembly for processing.

The Abort Sensor Assembly was mounted on the navigation base along with the inertial measurement unit. The ASA was 5.1 by 9.0 by 11.5 inches in size and weighed 20.7 pounds. It consumed 74 watts of power when operating.

Abort Electronics Assembly

The Abort Electronics Assembly was a compact general purpose computer with 4,096 words of ferrite core memory. Half of the memory (2,024 words) was hard-wired (ROM), and half was erasable (RAM) and used for temporary storage. The word size of the memory was 18 bits, and the cycle time of the memory was 5 microseconds. The computer used 18-bit words consisting of 17 bits plus a sign. Parallel arithmetic achieved an "add" time of 10 microseconds and a "multiply" time of 70 microseconds.

The software package for the abort guidance system was called Flight Program 6. That updated program was first used on LM-5 that flew on the Apollo 11 mission. The software included a set of 27 instructions such as ADD, MPY (multiply), and STQ (store Q register). The software was written in LEMAP (LEM Assembly Program). The Lunar Module was called the Lunar Excursion Module (LEM) early in the Apollo program.

Major functions performed by the AGS computer when the Abort Guidance Section was active were:

- Maintain attitude reference
- Perform LM and CSM navigation
- Solve guidance equations
- Provide steering commands
- Provide engine commands
- Provide automatic alignment
- Drive attitude and navigation displays
- Provide telemetry data

The astronauts could call up submodes and guidance routines in the software by keying addresses and numerical words into the Data Entry and Display Assembly. Address 400 allowed selecting AGS submodes. Address 410 allowed selecting guidance routines. A TRW document *LM/AGS Operating Manual Flight Program 6* contains details of the computer program including a listing of submodes and guidance routines. Some examples of submodes are given below.

Address	Numerical value	Description
400	+00000	Attitude hold (maintains inertial attitude)
400	+10000	Orients LM to the desired thrust direction
400	+20000	Orients the Z-axis of the LM in the direction of the CSM
400	+30000	AGS is aligned to the IMU in the PNGS
400	+40000	Lunar align mode (while on lunar surface)

A few of the guidance routines that could be called up by the astronauts to guide them to a rendezvous with the Command Module are given below.

Address	Numerical value	Description
410	+00000	Orbital insertion mode Guides the LM to an orbit around the moon (typically at an altitude of 30,000 feet) that is coplanar with the CSM orbit.
410	+100000	Coelliptic Sequence Initiate mode Computes the magnitude of a horizontal burn to set up the coelliptic maneuver
410	+200000	Constant delta h mode Computes the maneuver to place the LM in a trajectory such that the altitude difference between the LM and CSM is constant.
410	+300000	TPI search mode Used prior to the Terminal Phase Initiate (TPI) maneuver to determine when TPI should be performed
410	+400000	TPI execute mode Performs Terminal Phase Initiate maneuver

Navigation for both the LM and the CSM was performed by the computer continuously from the last state vector information for the LM and the CSM provided by the PNGS or entered manually. Closed form orbital equations were used to develop continuous navigation data for the CSM, and inputs from the strapped-down navigation system were used for LM navigation. As a result, the abort guidance system had up-to-date computed state vectors of the LM and the CSM at all times.

Data Entry and Display Assembly (DEDA)

The Data Entry and Display Assembly (DEDA) was the human/computer interface for the abort guidance system. It allowed the astronauts to select the modes of operation of the AGS by using the keyboard and to enter data manually by the keyboard. It had electroluminescent displays to inform the astronauts of the address called up in computer memory and the value of the data at that address.

A drawing of the DEDA is shown below. The DEDA was mounted in panel 6, which was located in front of the Lunar Module pilot. The drawing shown was copied from a TRW document although essentially the same drawing occurs in NASA documents. The arrow on the upper left side of the drawing points to a STOP switch that could be used by the Lunar Module pilot to shut off either the descent engine or the ascent engine (Fig. 8.18).

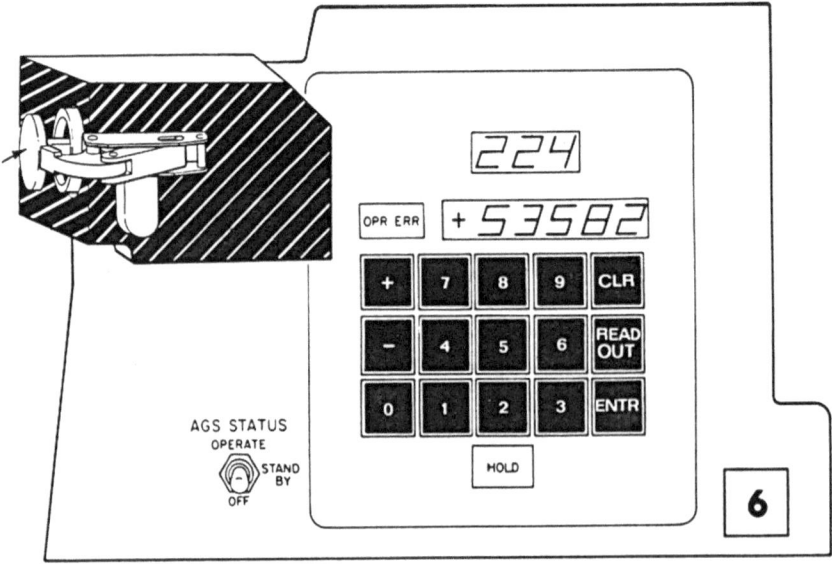

Fig. 8.18 Lunar Module panel 6 showing the Data Entry and Display Assembly (TRW drawing)

The DEDA had two rows of electroluminescent displays. The upper row of three digits displayed the address of the desired operation or data and the lower row of five digits plus sign displayed the value of the item addressed. A 16-button keyboard was located below the displays. The keyboard consists of 10 numerical keys labeled 0 through 9, plus and minus keys, and four dedicated purpose keys. The dedicated purpose keys were:

CLR (clear) – Initialized (cleared) the DEDA and blanked all lighted characters. CLR needed to be pressed prior to entering data into the DEDA.
READOUT – Commanded the three-digit address of the selected data to be displayed along with the sign and five-digit value of the data.
HOLD – Hold the value on the display until either the READOUT or CLR buttons were pressed.
ENTR – Entered the address and data set in by the astronaut.

The procedure used to insert data started by pressing CLR followed by pressing three digits corresponding to the address of the AEA memory where the data was to be inserted. Next the sign of the data was entered followed by the five-digit numerical value of the data. If the address and the numerical value on the display were correct, the operator would press the ENTR button to enter the data into the computer.

To read out data from the computer, the operator would press CLR followed by three digits representing the address of the data. He would then press READOUT to display both the address and the numerical value of the data at that address. For example, the three digit address "317" would bring up range from the LM to the CSM with units of nautical miles displayed to the nearest tenth of a mile. Likewise, the three digit address "440" would bring up range rate between the LM and the CSM with units of feet per second displayed to the nearest tenth of a foot per second and with a sign.

Radar Subsystem

The Lunar Module carried two radar units: the landing radar and the rendezvous radar. The landing radar provided range and velocity of the Lunar Module relative to the lunar surface. The range and velocity data were fed to the PNGS and to the displays to allow an automatic or manually controlled approach and soft landing on the moon. The rendezvous radar provided range, range rate, and angle from the Lunar Module to the Command and Service Module. That data was provided to the PNGS and to displays to enabled guidance of the ascent stage of the Lunar Module to a rendezvous with the Command and Service Module.

Landing Radar

The landing radar was developed by the Ryan Aeronautical Company who had developed a similar radar for the Surveyor spacecraft that landed on the moon before Apollo. The author was technical lead for both at the Ryan Aeronautical Company. Much of material and photographs of the landing radar presented here are from the author's files.

The inertial navigation system was the central element of the PNGS in the Lunar Module. It provided good-quality information on the state vector of the spacecraft in inertial space, but a sensor with direct contact with the lunar surface was necessary to make a soft landing. To illustrate, the Apollo 11 Mission Report states that there was a difference of about 2800 feet between the altitude above the surface computed by the PNGS and the actual altitude as measured by the radar when the radar altimeter first acquired the lunar surface at about 44,000 feet altitude. A mission rule called for the descent to be aborted if an altitude update by the landing radar had not been established by the time the PGNCS estimated altitude had decreased to 10,000 feet.

The landing radar was composed of two assemblies; an antenna assembly and an electronics assembly. A photograph of the antenna assembly resting on a bench and the electronics assembly being carried is shown on the next page (Fig. 8.19). The antenna assembly had not yet had a thermal blanket installed over the honeycomb fiberglass frame when the photograph was taken.

The landing radar consisted of a Doppler velocity sensor and a radar altimeter. The Doppler velocity sensor determined velocity by measuring the Doppler shift along three narrow beams of continuous wave (CW) microwave energy that was transmitted to the surface. The Doppler shift on the signal reflected back to the radar, f_D, was proportional to velocity along the beam as given by the formula $f_D = 2V/\lambda$ where V is velocity and λ is the wavelength of the transmitted signal.

The transmitted frequency was 10.51 GHz, the corresponding wavelength was 0.0936 feet, and the Doppler scale factor was 21.37 Hz per foot per second of velocity along the antenna beam. The Doppler shifts along the three beams were accurately measured by frequency trackers, and the results were resolved into velocity components along the X-, Y-, and Z-axes.

The radar altimeter function of the radar operated by transmitting a narrow beam of frequency modulated microwave energy to the surface and comparing the frequency of the signal reflected back to the radar to that being currently transmitted. The frequency modulation was a linear sawtooth function centered about a transmitted frequency of 9.58 GHz. The frequency difference between reflected and transmitted signals was proportional to the time delay between transmitted and received signal plus the Doppler shift. In equation form, the frequency difference, f_{R+D}, could be written $f_{R+D} = 2SR/c + 2V/\lambda$ where S is the slope of the frequency modulation, R is the range to the surface, and c is the propagation velocity

Fig. 8.19 Antenna assembly of landing radar resting on a bench and electronics assembly being carried (from author's files)

(velocity of light). The slope, S, was equal to 1,141 MHz per second at altitudes of 2,500 feet and above and 5,705 MHz per seconds below 2,500 feet (Fig. 8.19).

The linear sawtooth frequency modulated transmitted signal had a deviation of 8.0 MHz at altitudes above 2,500 feet and 40 MHz at altitudes below 2,500 feet. The higher deviation at lower altitudes resulted in increased accuracy of the range measurement. The duration of the linear sweep was 7.0 milliseconds, and it was followed by a flyback time of 0.7 milliseconds. The range scale factor was equal to 2.32 Hz per foot at ranges above 2,500 feet and 11.60 Hz per foot at ranges below 2,500 feet. The Doppler shift, $2V/\lambda$, was equal to 19.48 Hz per foot per second at the altimeter transmitted frequency. The Doppler component was removed in subsequent signal processing by using data from the Doppler velocity sensor.

The radar altimeter was required to operate over an altitude range of 40,000 feet to 10 feet. The velocity sensor was required to operate over an altitude range of 25,000 feet to 5 feet. The upper altitude range of the velocity sensor was constrained by the attitude of the Lunar Module to keep the antenna beams at a favorable incidence angle with the lunar surface.

The accuracy of the velocity sensor was specified in terms of velocities in the antenna coordinate system (V_{xa}, V_{ya}, and V_{za}). The relationship between the antenna coordinate system and the vehicle coordinate system is illustrated in the figure on page 269. The V_{xa} axis was aligned with the X-axis of the spacecraft in the case of antenna position 2.

The specified 3σ accuracy in V_{xa} was 1.5% of the total velocity or 1.5 feet per second, whichever was greater for altitudes from 25,000 feet to 5 feet. The specified 3σ accuracy in V_{ya} and V_{za} was 2.0% of total velocity or 2.0 feet per second, whichever was greater for altitudes above 2,000 feet. At altitudes below 2,000 feet, the specified 3σ accuracy in V_{ya} and V_{za} was 2.0% of total velocity or 1.5 feet per second, whichever was greater. The stipulation "3σ" for accuracy means that 99.73 percent of all measurements are within the specified accuracy.

The specified 3σ accuracy of the radar altimeter was 1.4% of range ± 15 feet at altitudes above 2,000 feet and 1.4% of range ± 5 feet at altitudes of 2,000 feet and lower.

The physical size of the electronics assembly was 15.75 inches long, 6.75 inches wide, and 7.38 inches high. The antenna assembly was 20.0 inches long, 24.6 inches wide, and 6.5 inches high. The total weight of the landing radar was 42.0 pounds. The maximum power consumption was 132 watts. The antenna assembly was made from magnesium to reduce weight. Ryan Aeronautical Company pioneered the process of dip brazing magnesium to assemble the antenna from dozens of machined pieces.

A simplified block diagram of the landing radar from the author's files is shown on the next page (Fig. 8.20).

Antenna Assembly

The antenna assembly contained the transmit and receive antennas, transmitters, modulator for the altimeter transmitter, microwave receiver, and Doppler preamplifiers.

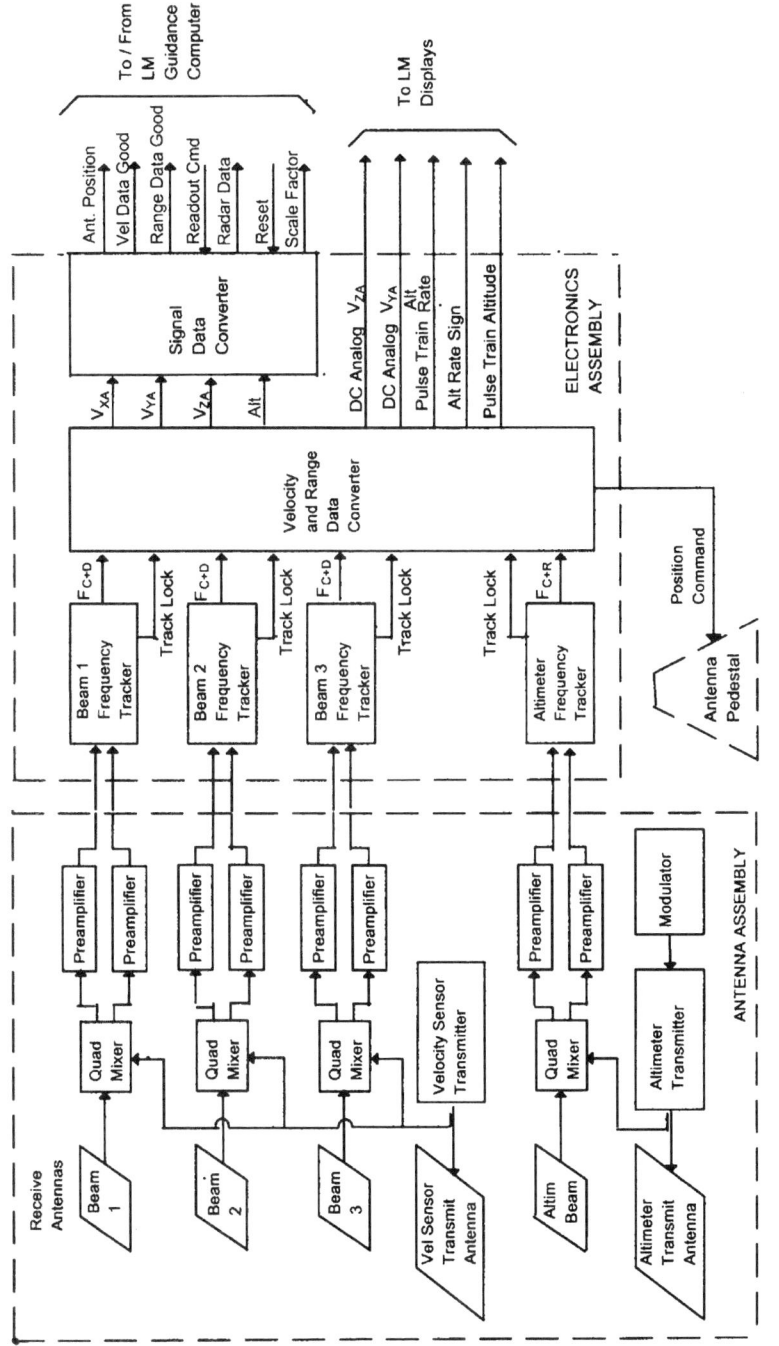

Fig. 8.20 Simplified block diagram of landing radar for the Lunar Module (from author's files)

Antennas

The velocity sensor used a slotted array transmit antenna interlaced with a slotted array transmit antenna for the radar altimeter. The two interlaced transmit antennas are the large structures in the center of the antenna in the photograph above. The velocity sensor transmit antenna array generated four narrow antenna beams deployed as shown in the drawing on the following page. Only three of the four beams were used. The radar altimeter transmit antenna generated two narrow beams with one of the beams oriented in the plane of two of the velocity sensor beams as shown in the drawing. The other antenna beam was not used.

There were four independent receive array antennas directed at the angles of the active transmit beams. These antennas are grouped about the transmit arrays as shown in the photograph shown previously. The radar altimeter receive array was at the upper left in the photograph. The receiver arrays generated single antenna beams broadside to the arrays.

The velocity sensor beams were spayed out 24.55 degrees from the centerline of the antenna beams. The beams were displaced ±20.38 degrees from the centerline in the antenna X-Z plane and displaced 14.88 degrees from the centerline in the antenna X-Y plane. The altimeter beam was located between beams 1 and 2 at an angle of 20.38 degrees from the centerline of the antenna beams (Fig. 8.21).

The two-way gain of each of the three velocity sensor antenna beams was 49.2 dB and the two-way beamwidth was 3.7 by 7.3 degrees. The two-way gain of the radar altimeter beam was 50.4 dB, and the two-way beamwidth was 3.9 by 7.5 degrees. The narrow antenna beamwidth was in the spacecraft Z-axis which was the in the direction of travel during powered descent.

The antenna assembly was mounted on an antenna pedestal that allowed the antenna to be positioned in one of two positions to accommodate the attitude of the Lunar Module relative to the local vertical. The positions of the pedestal were either zero degrees with the centerline of the velocity sensor beam group along the X-axis of the vehicle or with the centerline of the beam group displaced 24 degrees from the X-axis. The centerline of the beams is the dashed line in the figure showing antenna beam orientation. The antenna pedestal was mounted to the descent stage such that the Y-axis of the antenna was six degrees displaced from the Y-axis of the Lunar Module.

A switch labeled LDG ANT on control panel number 3 had switch positions of AUTO, DES, and HOVER. Normally, the switch was set to the AUTO position and the LM guidance computer controlled switching of the antenna pedestal. The pedestal could be manually switched to DES (descent), (antenna position 1), or HOVER (antenna position 2). In the AUTO position, the LM guidance computer switched the pedestal from antenna position 1 to antenna position 2 at the time of LM reached the "high gate" point at about 7,600 feet altitude.

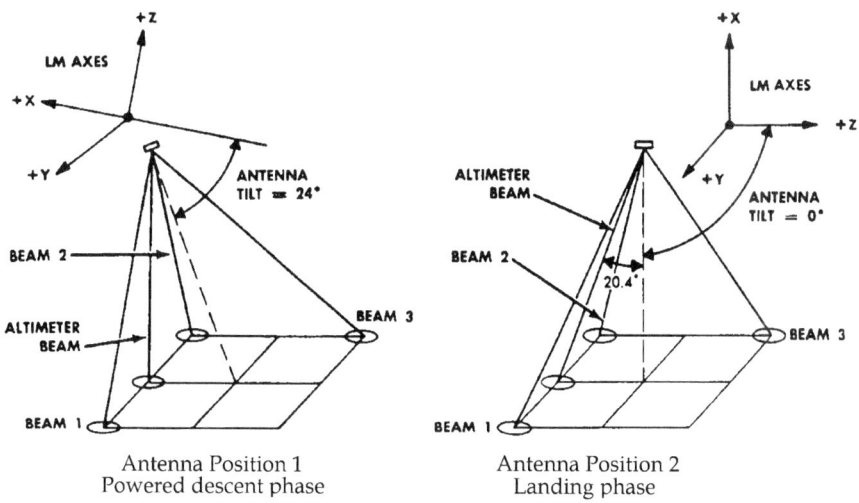

Fig. 8.21 Landing radar antenna beam orientation (NASA graphic)

During the braking phase of the powered descent, the X-axis of the vehicle was tilted forward to allow the engine to thrust against the velocity vector. At the time that the spacecraft was started to roll to a windows-up attitude at about 45,000 feet altitude, the X-axis of the spacecraft was about 73 degrees from the local vertical. The windows-up position allowed the landing radar antenna beams to impinge on the lunar surface. The antenna pedestal was set to the 24 degrees displaced position during this time so that the radar altimeter beam and velocity sensor beams 1 and 2 had contact with the lunar surface at a favorable angle. The radar altimeter beam was oriented about 28.6 degrees forward of vertical. During the landing of Apollo 11, the radar altimeter achieved lock on the surface at 44,000 feet before completion of the roll to windows-up attitude.

The displacement of the X-axis of the spacecraft from the vertical decreased as the spacecraft descended and all three beams of the velocity sensor acquire signals at about 28,000 feet. This was signified by generation of the velocity data good signal.

The antenna was switched to position 2 at the start of the approach phase at about 7,500 feet. The centerline of the antenna was aligned with the spacecraft X-axis in position 2. At the end of the switching time to position 2, the spacecraft X-axis was about 45 degrees forward of the local vertical.

A photograph taken by the author of the antenna assembly mounted on the bottom of the descent stage of Lunar Module LM-2 is shown next page (Fig. 8.22). The antenna pedestal is in position 2. Lunar Module LM-2, which was used for ground testing, is on display in the Smithsonian National Air and Space Museum (NASM).

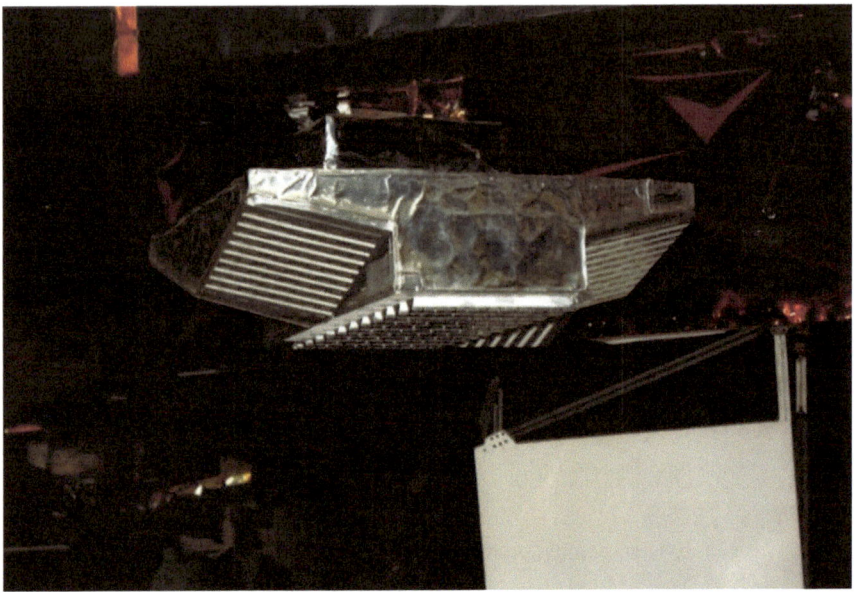

Fig. 8.22 Antenna assembly mounted to descent stage of LM-2

Transmitters

The microwave energy transmitted by the antenna was developed by two transmitters. One, operating in a continuous wave mode, was used for the velocity sensor, and the other, operating in a frequency modulated mode, was used for the radar altimeter. The transmitters were built by RCA. RCA developed the rendezvous radar for the Lunar Module, and it so happened that the power output and operating frequency of the transmitter for that radar was similar to that required by the landing radar. To avoid duplicating effort, RCA modified their basic transmitter design for use by the landing radar.

The velocity sensor transmitter generated a minimum of 200 milliwatts of continuous wave power at a frequency of 10.51 MHz. The transmitter for the radar altimeter was frequency modulated with a center frequency of 9.58 GHz. It accepted a linear frequency modulated input signal with a center frequency of at 99.78 MHz from the modulator. The duration of the linear frequency sweep was 7.0 milliseconds, and it was followed by a flyback time of 0.7 milliseconds. The transmitter multiplied and amplified this signal to a minimum output power level of 175 milliwatts and center frequency of 9.58 GHz. The total frequency deviation at the transmitter output was 8.0 MHz at altitudes above 2500 feet and 40 MHz at 2,500 feet and below.

Receivers

The Doppler frequency was extracted in the velocity sensor receiver by mixing the receive signal with a sample of the transmitted signal. The received signal from each of the receive antennas was applied to two mixers which were excited from a sample of the transmitted signal in phase quadrature. The resulting quadrature pair of Doppler signals was applied to preamplifiers. At higher altitudes when the signal level was low, the preamplifiers had a gain of 88 dB (factor of 25,119). When the signal strength increased to a preset threshold at lower altitudes, the gain was automatically reduced to 55 dB (factor of 562).

There were two reasons for using quadrature channels for the Doppler signals. First, it allowed determining the sign of the signal. That is, whether the velocity along the antenna beam was closing or opening. Second, it allowed discriminating in the frequency trackers between true Doppler signals and spurious signals such as may be reflected from a vibrating structure.

The radar altimeter receiver extracted the range plus Doppler signal by mixing the received signal with a sample of the transmitted signal just as in the case of the velocity sensor. Quadrature mixers were followed by dual preamplifiers.

Electronics Assembly

The electronics assembly processed signals from the antenna assembly and provided velocity and range data to the LM guidance computer and to displays in the cockpit.

Frequency trackers

The quadrature signals at the output of the preamplifiers were processed by frequency trackers. Separate frequency trackers were used for each beam. The quadrature input signal was upshifted in a single sideband modulator in the frequency tracker. Signal acquisition was based on comparing the amount of energy in the two upshifted sidebands. True Doppler or true range signals were single sideband in nature and appeared in one sideband of the mixing process, while receiver noise or spurious signals from vibrating spacecraft structure appeared in both sidebands. Comparison of energy in the two sidebands allowed measurement of signal-to-noise ratio. True Doppler or true range signals were accepted if the signal-to-noise ration exceeded 3 dB in the tracking filter. Spurious signals were automatically rejected.

The finite antenna beamwidth resulted in a spread of Doppler frequencies. The width of the Doppler frequency spectrum was a function of spacecraft velocity and the angle between the velocity vector and the antenna beam. The frequency tracker searched for the Doppler signal over the frequency band of expected signals with a tracking filter that encompassed the expected spectral width of the Doppler signal. Initially at high velocities, the width of the tracking filter was 2,800 Hz. As the spacecraft slowed and the signal spectrum width became narrower, the width of the tracking filter was automatically reduced to 400 Hz. The radar altimeter frequency tracker used a tracking filter width of 3,200 Hz at high altitudes and switched to 400 Hz at lower altitudes.

Once acquired, the center of the frequency spectrum was traced with good accuracy. The velocity sensor frequency trackers indicated that they had acquired and locked onto the Doppler signal by issuing a discrete signal called tracker lock. When all three of the velocity sensor channels had acquired Doppler signals, a velocity data good discrete signal was generated and sent to the LM guidance computer. When the radar altimeter channel had acquired the range plus Doppler signal and Doppler velocity sensor beams 1 and 2 had also acquired signals, a range data good signal was sent to the LM guidance computer. Data from velocity sensor beams 1 and 2 were necessary to remove the Doppler shift of the range plus Doppler signal.

Velocity and range data converter

The outputs of the velocity sensor frequency trackers were applied to the velocity and range data converter where the velocity components along the three antenna beams were resolved into an orthogonal set referenced to the antenna coordinate system. That velocity coordinate system was denoted V_{xa}, V_{ya}, and V_{za}. The V_{xa} axis of the antenna was aligned with the X-axis of the spacecraft when the antenna was rotated to position 2. In antenna position 1, the V_{xa} axis was displaced 24 degrees from the X-axis of the spacecraft.

The output of the altimeter frequency tracker was also applied to the velocity and range data converter. The Doppler component on the altimeter signal was removed by resolving the velocity components along velocity sensor beam 1 and beam 2 into the orientation of the antenna beam and then subtracting the scaled frequency from the altimeter signal.

Velocity and range data were developed by the velocity and range data converter and provided to the Signal Data Converter in pulse train form. The velocity and range data was also provided in forms needed to drive the various displays on the LM control panels. The velocity components V_{ya}, and V_{za} were provided in DC analog form to the X-pointer Indicator that displayed fore-aft and lateral velocities when the antenna was in the landing position (position 2). Altitude and Altitude Rate (V_{xa}) were supplied to the Range/Alt and Range Rate/Alt Rate meters on the control panel in pulse train form. The pulse repetition rate was proportional to range and range rate. A DC discrete signal was provided to indicate the sense of Altitude Rate.

Signal data converter

The velocity components and range information developed in the velocity and range data converter were provided to the signal data converter in pulse train form. The signal data converter interfaced with the LM guidance computer (LGC) by accepting strobe signals from the LGC and using the strobes to assemble and read out the range and velocity data in 15-bit serial binary form to the LGC. Other information provided to the LGC by the signal data converter included antenna position, velocity data good, range data good, and range scale factor.

Rendezvous Radar

The rendezvous radar was used to track the Command Service Module while the Lunar Module was on the lunar surface and during flight of the ascent stage towards rendezvous with the CSM. Radar tracking was aided by a transponder in the CSM. The transponder received the signal sent from the rendezvous radar and retransmitted it at a different frequency. As a result, the signal received by the rendezvous radar was much stronger than it would have been had it relied on the signal reflected off of the CSM (skin tracking). The rendezvous radar determined range, range rate, line-of-sight angle, and angular rate to the CSM. These measurements were provided to the LM guidance computer and to displays on the control panels of the LM.

The rendezvous radar was composed of an antenna assembly and an electronics assembly. The antenna assembly was mounted on top of the LM, and the electronics module was mounted on the inside. The antenna assembly featured a 24-inch parabolic reflector mounted on a pedestal with 2 degrees of freedom. The antenna could be put into an angular search pattern to search for the transponder signal in the CM and once acquired, track the CM with good angular accuracy.

The rendezvous radar transmitted a continuous wave signal at an X-band frequency of 9,832.8 MHz, and the transponder in the CM responded with a continuous wave signal at 9,792.0 MHz. The nominal radiated power by the rendezvous radar and of the transponder was 300 milliwatts. The continuous wave signal transmitted by the rendezvous radar was phase modulated by three tones: 200 Hz, 6.4 kHz, and 204.8 kHz. The transponder extracted the tones and retransmitted them as phase modulation of its transmitted signal. The signal processor in the electronics assembly of the rendezvous radar measured the delay in the received tones to determine range to the CM. The unambiguous range capability of the rendezvous radar was 400 nautical miles (nm). The minimum range capability was 80 feet.

The specified angular tracking accuracy was given in terms of a bias error and a random error. The allowable bias error in milliradians (mr) was 8 mr. The specified allowable random error was 5.3 mr at a range of 400 nm and 4.7 mr at a range of five nm. One milliradian is equal to 0.0573 degrees.

The range accuracy was also given in terms of a bias error and a random error. The allowable bias error was ±500 feet for ranges greater than 50.6 nm and ± 120 feet for ranges less than 50.6 nm. The allowable random error was 300 feet plus 0.25% of range for ranges between 400 nm and 5 nm and 80 feet plus 1.0% of range for ranges from 5 nm to 80 feet. The angular rate accuracy in milliradians per second was 0.4 mr/sec at a range of 400 nm and 0.2 mr/sec at ranges from 100 nm to 80 feet.

A simplified block diagram of the rendezvous radar, adapted from a NASA graphic, is given on the next page (Fig. 8.24).

Antenna Assembly

The antenna assembly was mounted at the top of the Lunar Module above the forward hatch. A photograph of the antenna assembly taken by an Apollo 14 astronaut while standing on the ladder leading to the lunar surface is shown below. The dark circular item with the white ring around it in the foreground is the optical telescope (Fig. 8.23).

Fig. 8.23 Rendezvous radar antenna. Photo taken from ladder of LM 14 after an EVA (NASA photograph cropped by author)

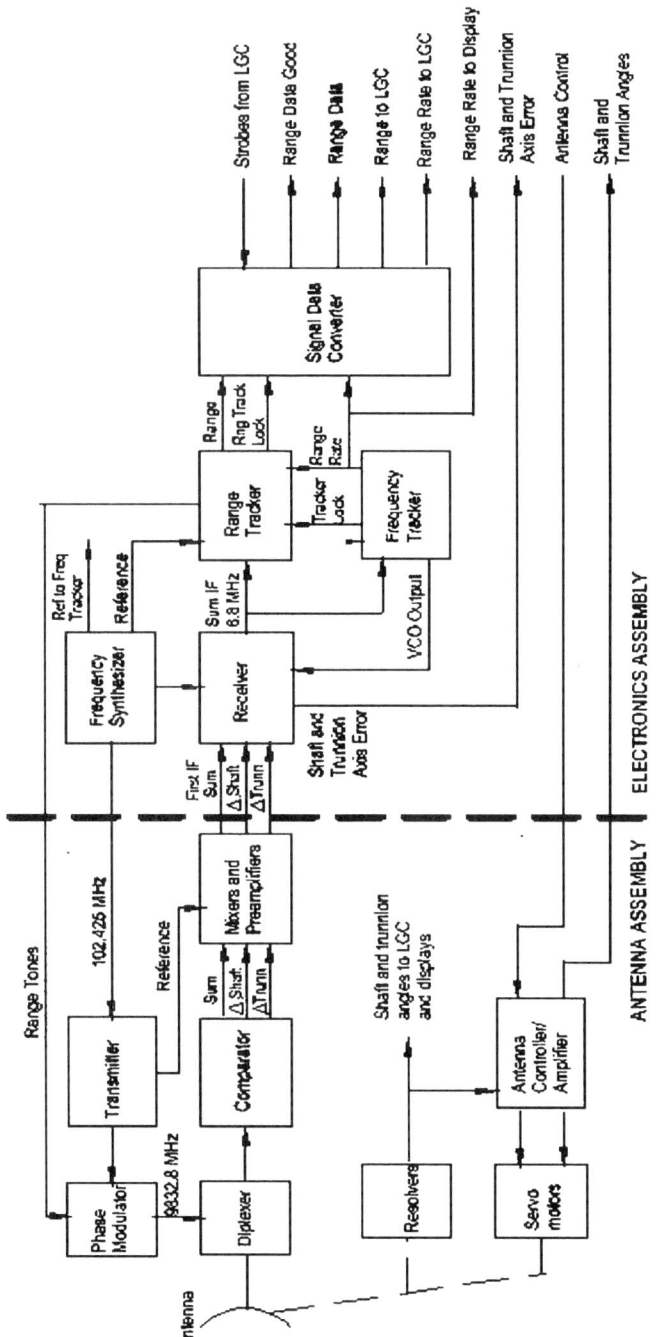

Fig. 8.24 Simplified block diagram of rendezvous radar (adapted from NASA graphics)

The shaft for the pitch axis of the antenna mount was parallel to the spacecraft Y-axis. The trunnion axis was perpendicular to the shaft axis. Drive motors for positioning were attached to the pitch and trunnion axes along with resolvers to provide angular position.

Antenna

The antenna was a Cassegrain type with a 24-inch parabolic reflector and a 4.7-inch hyperbolic subreflector. A four-port feed horn was located at the focus of the subreflector. The four ports of the feed horn allowed two-axis monopulse tracking of a target. The transmitter and monopulse receivers were mounted on the back of the parabolic antenna. The gain of the sum pattern was 32 dB and the beamwidth of the sum pattern was four degrees.

The antenna beam could be directed by one of three modes: automatic tracking, manual slew, or LM guidance computer control. The mode desired could be set by the astronauts by the rendezvous radar selector switch on control panel 3. There were two antenna orientations. Antenna orientation no. 1 was used for tracking during rendezvous. Antenna orientation no. 2 was used while on the lunar surface for overhead operation. The astronauts selected an orientation by inputs from the DSKY.

The angular coverage of the line-of-sight relative to the +Z-axis when in antenna orientation no. 1 encompassed shaft coverage of −70 degrees to +60 degrees and trunnion coverage from −55 degrees to +55 degrees. When in antenna orientation no. 2, the line-of-sight shaft coverage with respect to the +Z-axis was +45 degrees to +155 degrees, and the trunnion coverage was −55 degrees to +55 degrees with respect to the –Z axis.

Transmitter

The transmitter consisted of a varactor frequency multiplier that accepted a 102.425 MHz continuous wave signal from the electronics assembly and multiplied that input signal by a factor of 96 to obtain the transmitted frequency of 9,832.8 MHz. The transmitter output was phase modulated by three range tones of 200 Hz, 6.4 kHz, and 204.8 kHz. The lowest frequency range tone resulted in an unambiguous range in excess of 400 nautical miles, and the highest frequency range tone allowed good range measurement accuracy. The nominal radiated power was 300 milliwatts.

The output of the transmitter was fed through a diplexer to the sum port of the antenna. The diplexer allowed simultaneous transmission and reception by the antenna.

Transponder

The transponder in the Command and Service Module received the signal from the rendezvous radar by a horn-type antenna with wide beamwidth to accommodate expected look angles to the Lunar Module. The transponder was configured to search over a ± 104 kHz band for the signal from the rendezvous radar. The frequency search range accommodated the expected Doppler shift of the signal. Once acquired in frequency, the transponder activated a phase-locked loop to phase lock with the rendezvous radar carrier term. The phase-locked carrier was upshifted by 40.8 MHz and retransmitted with the range tones phase coherent with the tones received. The nominal transmitted power from the transponder was 300 milliwatts. Transmission was through the same wide-beamwidth antenna that had received the signal from the rendezvous radar antenna.

Received signal processing

The signal received from the transponder by the rendezvous radar antenna was directed to the four feed ports of the antenna. The signal from each of the feed ports was applied to a comparator. The comparator developed the sum pattern by adding the signal from all four feed ports (A + B + C + D). The difference patterns for shaft axis were obtained by the combination (A + D) – (B + C), and the difference pattern for the trunnion axis was obtained by the combination (A + B) – (C + D). The shaft and trunnion axes would be known as the elevation and azimuth axes in usual radar terms.

The sum and the two difference signals were applied by mixers along with a reference signal from the transmitter. The difference frequency (first intermediate frequency) was 40.8 MHz plus the Doppler shift due to range rate. The three intermediate frequency signals were conditioned by preamplifiers and applied separate channels in the receiver that was located in the electronics assembly.

Electronic Assembly

Referring to the block diagram of the rendezvous radar shown previously, the sum and difference signals at an intermediate frequency (IF) of 40.8 MHz from the antenna assembly were down-converted to a second IF frequency of 6.8 MHz and further amplified. The 6.8 MHz sum signal was applied to the frequency tracker and to the range tracker. The frequency tracker compensated for the Doppler shift on the signal by comparing the sum signal with a reference frequency from the frequency synthesizer. A phase-locked loop in the frequency tracker adjusted the frequency of a voltage-controlled oscillator (VCO) so that when mixed with the 40.8 MHz sum signal, the difference was exactly 6.8 MHz. The difference between the VCO frequency and a 40.8 MHz reference frequency was the Doppler shift.

The Doppler shift was converted to a signal representing range rate that was fed to displays on the control panel and to the signal data converter.

The Doppler corrected 6.8 MHz sum signal was fed to the range tracker where the phase-modulated range tones of 200 Hz, 6.4 kHz, and 204.8 kHz were extracted. The phase of the range tones were compared with the phase of reference tones to determine time delay and hence range to the transponder in the CSM. The range output was applied to the signal data converter and to displays.

The sum signal and the difference signals for the shaft and trunnion axes were down converted from the second intermediate frequency of 6.8 MHz to a third intermediate frequency of 1.7 MHz in the receiver. The difference frequencies were demodulated by mixing with the sum signal. The results were a signal for each axis with amplitude representing angular error of the target from boresight of the antenna and sign representing the direction of the error. The error signals were used by the antenna positioners to null the angular errors when in the automatic angle tracking mode.

The shaft and trunnion axis angles of the antenna were determined by resolvers. The analog data from the resolvers were applied to a coupling data unit (CDU) that converted the information to digital form for input to the LM guidance computer. The CDU also formatted the angle data for display on the instrument panel in the cockpit.

Signals from the range tracker and the frequency tracker were sent to the signal data converter to be put in a form that could be accessed by the LM guidance computer. These signals included range, range tracker lock, range rate, and Doppler sense (range rate sense). The signal data converter also provided an interface between commanded functions by the astronauts via the control panels to the elements in the rendezvous radar.

The LGC requested range data by sending a series of strobe pulses on the range select strobe line and then sending a series of pulses to read out a 15-bit shift register with range data. Likewise, to read out range rate data a series of pulses were sent to the range rate select strobe line followed by pulses to read out the 15-bit shift register now loaded with range rate data.

A range scale factor shift occurred when the range decreased to 50.6 miles. For ranges greater than 50.58 nautical miles, the least significant bit of the 15-bit range word was 75 feet. For ranges of 50.6 miles or less, the least significant bit of the 15-bit range word was 9.4 feet. A signal designating the range scale factor was sent from the signal data converter to the LGC.

REACTION CONTROL SYSTEM

The reaction control system (RCS) consisted of four groups of thrusters mounted to the ascent stage that were used for attitude control and translation about the three principal axes of the Lunar Module. The thrusters were normally controlled by the guidance, navigation, and control subsystem (GN&CS), but they could also be controlled manually by the astronauts.

A photograph of one set of four RCS thrusters mounted on the ascent stage of the Apollo 14 Lunar Module is shown below (Fig. 8.25). The detail shown is an expanded view of a NASA photograph of the LM. There were four sets of quad thrusters, one set on each corner of the ascent stage for a total of 16 thrusters.

Fig. 8.25 Reaction control system thrusters on Lunar Module of Apollo 14 (NASA photograph cropped by author)

Each thruster was 13.4 inches long, and the nozzle exit diameter was 5.6 inches. Each generated 100 pounds of thrust when activated. The pair of thrusters that appear near vertical in the photograph were aligned parallel to the X-axis of the LM. One was pointed in the +X direction and the other in the –X direction. The other two thrusters were aligned parallel to the Y-axis and Z-axis of the LM.

The thrusters used a mixture of equal parts of hydrazine (N_2H_4) and unsymmetrical dimethylhydrazine as fuel. Nitrogen tetroxide (N_2O_4) was used as oxidizer. The fuel and oxidizer were hypergolic, igniting upon contact with one another. Solenoid valves in the fuel and oxidizer lines were opened to fire the thrusters. The valves could be opened for as short as 14 milliseconds, but normally the thrusters were fired for pulses lasting a few seconds. Each thruster contained solenoid valves for fuel and oxidizer, a combustion chamber, and an expansion nozzle.

There were two reaction control systems for redundancy. Each reaction control system had its own fuel and oxidizer supply and control arrangement. Each independent system fed eight thrusters. Complete control in all axes could be achieved in case of failure of the other system. Normally, both RCS system were used together. The fuel tank for each redundant RCS system was 12.5 inches in diameter and 32 inches long. It held 99.3 pounds of available fuel. The oxidizer tank for each RCS system was 12.5 inches in diameter and 38 inches long. It held 194.1 pounds of usable oxidizer.

Fuel and oxidizer were contained in Teflon bladders within the tanks. Helium gas was ported to the tanks to squeeze the bladders and force the fuel and oxidizer into feed lines to the thrusters.

Normally, the RCS operated in the automatic mode where the LM guidance computer within the GN&CS controlled the firing of the thrusters to achieve necessary attitude or translation operations. A semiautomatic mode referred to as the "attitude hold mode" allowed the astronauts to change attitude by means of the attitude controller assembly. The duration of firing was proportional to the displacement of the controller. The computer maintained the new attitude when the controller assembly was returned to the detent or zero position. When in the attitude hold mode, the astronauts could manually translate the vehicle by displacing the translation control assembly. When the translation control assembly was returned to the detent or zero position, the thrusters stopped firing.

ASCENT PROPULSION SYSTEM

The ascent rocket engine was mounted in the center of the bottom of the ascent stage. Unlike the descent engine, the ascent engine was not throttleable, and it was not mounted on gimbals for steering. The engine could be fired, shut down, and restarted up to 35 times. It was mounted with its thrust axis oriented 1.5 degrees from the LM X-axis in the +Z direction to place the line of thrust near the center of gravity of the ascent stage.

The nominal thrust of the engine was 3,500 pounds. It operated with a chamber pressure of 120 pounds per square inch. The ascent engine was about half the size of the descent engine at 47 inches long and a nozzle exit diameter of 34 inches. The engine weighed 180 pounds.

The ascent engine used a mixture of equal parts of hydrazine (N_2H_4) and unsymmetrical dimethylhydrazine as fuel. Nitrogen tetroxide (N_2O_4) was used as oxidizer. The fuel and oxidizer ignited upon contact with one another. Fuel was contained in a spherical tank with volume of 36 cubic feet. The inside diameter computed from the volume was about 49 inches. It held 2,011 pounds of fuel. The oxidizer tank was also spherical with the same diameter. It held 3,211 pounds of oxidizer.

Fuel and oxidizer were forced out of the tanks into lines leading to the ascent engine by porting helium gas into each tank. Helium gas was contained in a spherical tank with a volume of 3.35 cubic feet. The computed internal diameter is about 22 inches. The gas was stored in the tank at a pressure of 3,050 pounds per square inch at room temperature. The weight of useable helium in the tank was 5.1 pounds. While firing, the engine used 4.3 pounds per second of fuel and 7.0 pounds per second of oxidizer. The ratio by weight of oxidizer to fuel was 1.6:1.

Temperature of the fuel and oxidizer tanks were displayed on the FUEL and OXID TEMP indicators on the control panel. Pressures were displayed on the FUEL and OXID PRESS indicators. The gross weight of the ascent stage with propellants was about 10,500 pounds. However, the gravity of the moon is about 16.5% of the gravity on earth so the effective weight of the ascent stage on the moon was about 1,733 pounds. The 3,500 pound thrust of the ascent engine easily lifted the ascent stage from the lunar surface and propelled it into lunar orbit.

Normally, firing of the ascent engine was controlled by the LM guidance computer using the computer program P12. In case of failure of the computer, engine firing was controlled by the abort guidance system. The engine could also be fired manually by setting the ENG ARM switch on the control panel to ASC and pressing the START push button.

If all went well during the mission and after exploration of the lunar surface by the astronauts, the ascent engine was used to launch the ascent stage from the moon and rendezvous with the Command Module. Should a problem occur that required an immediate abort of the mission, a series of events would be undertaken to put the ascent stage into a safe orbit about the moon. If abort was called during the powered descent, for example, LM guidance computer program P71 would be used to turn off the descent engine, separate the ascent stage from the descent stage, orient the ascent stage so that thrust from ascent engine would be in the proper direction, and then fire the ascent engine to put the ascent stage in orbit around the moon.

Ascent from the moon

The Apollo 11 astronauts made ready to leave the moon by first aligning the inertial platform in the PGNCS. This was accomplished by using one star sight and the gravity vector of the moon as reference. The position of the LM on the surface of the moon and other information needed by the computer to manage the ascent was received from MSFN and downloaded into the PGNCS. The time to launch the ascent stage was included in this information. The optimum time of launch was calculated to be several seconds after the Command and Service Module had passed overhead of the LM. The state vector in the PGNCS was transferred to the abort guidance system that provided backup to the PGNCS.

When the astronauts were ready to leave the moon, program P12 (Powered Ascent) was selected on the DSKY, and a countdown to the time of launch was displayed. In the case of Apollo 11, the time of launch was to be at 124:22:00 mission time. At 5 seconds before time of ignition, the DSKY began flashing Verb 99 to allow the astronauts to decide whether to proceed. Upon seeing the flashing 99 and by prior agreement, Mission Commander Armstrong pressed the ENGINE ARM and the ABORT STAGE buttons and Aldrin pressed the PRO (proceed) button on the DSKY keyboard. The ABORT STAGE button fired explosive bolts that freed the ascent stage from the descent stage and also fired guillotines that cut electrical cables and plumbing between the stages. The ascent engine was fired by the computer when the countdown reached zero.

The ascent stage rose vertically for about 10 seconds, and when the vertical velocity reached 40 feet per second, a pitchover to a flight path of 50 degrees from the vertical was initiated. The pitchover allowed the spacecraft to pick up horizontal velocity needed for lunar orbit. Once orbital velocity was reached, the computer turned off the ascent engine. The crew of Apollo 11 announced cutoff of the engine at 124:29:17 after a burn of about 7 minutes. The altitude at insertion into lunar orbit was 60,300 feet, vertical velocity was 32 feet per second, and horizontal velocity was 5,537 feet per second. The orbit had a apolune of 47.3 nautical miles and a perilune of 9.5 nautical miles.

The ascent engine had performed its intended function of placing the ascent stage into lunar orbit, and it was not used again. Subsequent thrusting to set up a rendezvous with the Command and Service Module was performed using the reaction control system (RCS) thrusters. Fine-tuning the orbit for rendezvous was best accomplished by the RCS with its much lower thrust and more easily controlled duration of thrusting. The PGNCS used measurements by the rendezvous radar of range, range rate, angle, and angle rate to the CSM to compute burn times for various maneuvers leading to rendezvous with the CSM.

After rendezvous and docking of the LM to the CSM, the LM crew transferred into the CSM with their collection of lunar soil and rocks. The ascent stage of Apollo 11 was undocked from the CSM at 130:09 mission time. The ascent stage remained in lunar orbit, but it eventually crashed into the moon. The descent stage for Apollo 11 still stands on the moon at the Sea of Tranquility site.

LM COMMUNICATIONS SUBSYSTEM

An effective telecommunications system was essential to the success of the Apollo mission. A highlight of that system was transmission of Neil Armstrong's famous words: "Houston, Tranquility Base here. The Eagle has landed." Those words came from Lunar Module Eagle that had just made a soft landing on the moon some 208,560 nautical miles from earth.

The communications subsystem in the Lunar Module was similar to that described previously for the Command Module. Major elements of the communications subsystem were the S-band communications equipment and the VHF communications equipment.

The S-band link was used to communicate between the Lunar Module and the Manned Spaceflight Network (MSFN) Center on earth. Communications through the MSFN was conducted by the Manned Spaceflight Center (MFC) in Houston. The VHF link was used to communicate between the LM and the CSM and between the LM and astronauts exploring the surface of the moon. VHF was also used for communications between the astronauts when they were outside of the LM. The LM served as a relay for communications between the astronauts exploring the surface of the moon and MFC in Houston. The astronauts communicated with equipment in the LM via the VHF link, and this communication was relayed to MSFN via the S-band link and then to MFC.

A simplified block diagram of the communications subsystem in the Lunar Module is shown on the next page (Fig. 8.26). The diagram is a composite of various NASA diagrams assembled by the author to show the various functions in simplified form.

VHF Communications

VHF communications in the Lunar Module included two separate receivers and two separate transmitters. One receiver-transmitter pair, channel A, operated at 296.8 MHz and the other pair, channel B, operated at 259.7 MHz. These operating frequencies corresponded to those used in the VHF receiver-transmitters in the CSM. Each VHF transmitter in the LM had an output power of 5 watts.

A particular receiver and a particular transmitter could be powered on by switches on control panel 12 that was located to the right of the Lunar Module pilot (LMP) station. The switches allowed turning either receiver on or off. Each transmitter was controlled by a three position switch with the center position of each switch being off. Transmitter B could be switched between VOICE, DATA, or OFF. Data consisted of low bit rate information such as status of the LM and biomedical data in pulse code modulation (PCM) format. Transmitter B could be switched between VOICE, VOICE/RNG, or OFF. The VOICE/RNG position allowed the received range tone to be retransmitted and also allowed voice transmission.

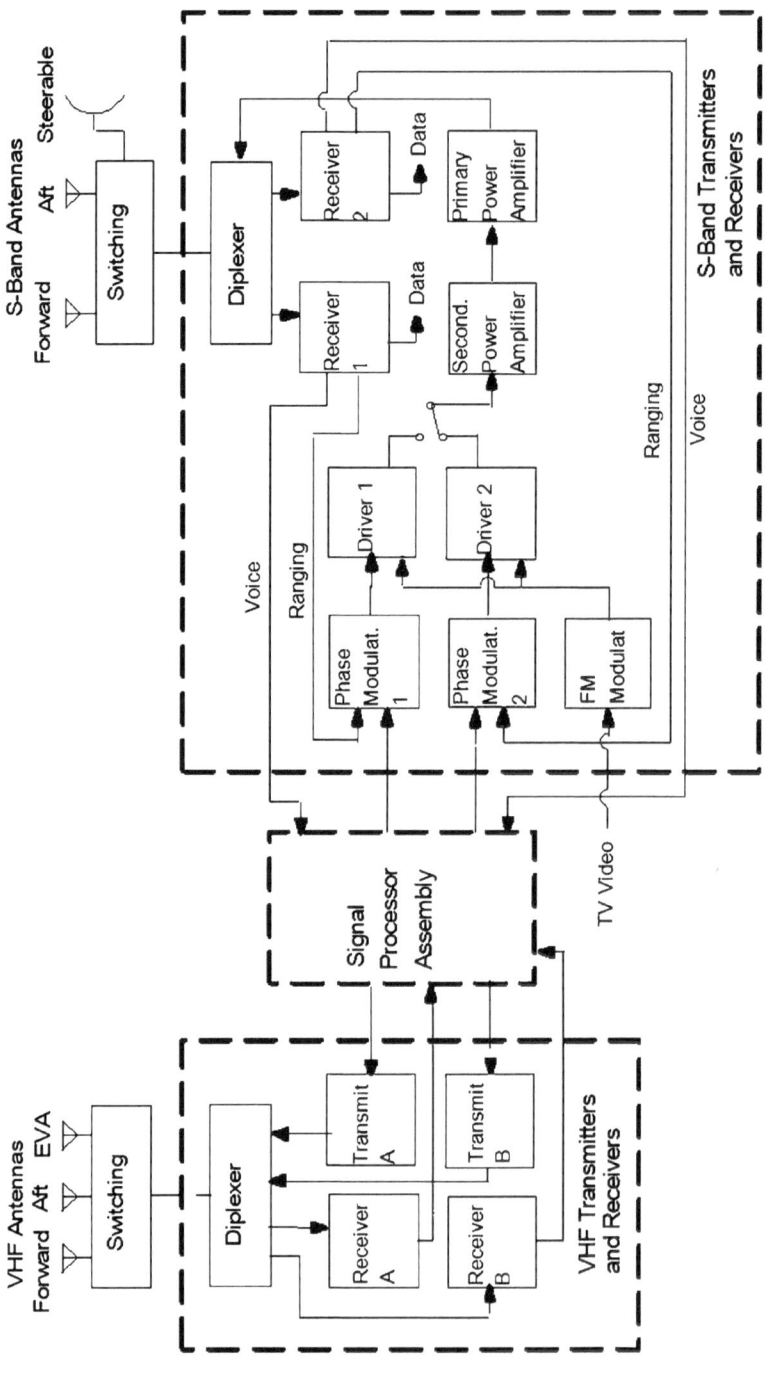

Fig. 8.26 Simplified block diagram of communications subsystem in Lunar Module

The signal processor assembly shown on the block diagram contained separate audio sections for the commander and Lunar Module pilot. Control panel 8 included switches to allow the commander to select which receiver-transmitter to be used and to set the volume for that receiver-transmitter. Likewise, control panel 12 included switches to allow the LMP to select which receiver-transmitter to be used and to set the volume for that receiver-transmitter.

Similar to the VHF communications channel for the CSM, the VHF communications equipment in the LM used on-off amplitude modulation to convey information. This type of modulation, although subject to some distortion, had a benefit of increasing the root-mean-square (rms) modulation level. This increased intelligibility for weaker signals.

The VHF communications subsystem in the LM could be switched to one of three antennas. Two of the antennas, referred to as in-flight antennas, were mounted on the side of the ascent stage near the top. One, labeled the forward antenna, was mounted on the forward side of the vehicle and the other, labeled the aft antenna, was mounted on the aft side. These antennas had near omnidirectional coverage, and either could be selected by a switch on control panel 12. One or the other of these antennas was used to communicate with the Command Service Module, depending on which had the best view of the CSM.

The third VHF antenna was referred to as the EVA antenna. The term EVA stands for extra vehicular activity. It allowed the astronauts exploring the surface of the moon to communicate with equipment in the LM which then relayed the communication to earth via the S-band communications equipment. The EVA antenna was a conical type with omnidirectional coverage in azimuth and located on top of the ascent stage. The VHF communications receiver-transmitter could be switched to this antenna by the same switch on control panel 12 used to select the in-flight antennas.

The portable life-support system (PLSS) worn by the astronauts while exploring the lunar surface contained a small VHF receiver-transmitter and a whip antenna to allow communication with each other and with VHF equipment in the LM via the EVA antenna. The equipment in the LM relayed these communications to the MSFN on earth via the S-band link.

The VHF communications subsystem in the Command Service Module and the Lunar Module contained a ranging function to measure range between the LM and the CSM. The range information was used during rendezvous of the ascent stage of the LM with the CSM as the ascent stage returned from the moon. During ranging, the 259.7 MHz transmitter in the Command Module was amplitude modulated with three range tones: 247 Hz, 3.95 kHz, and 31.6 kHz. These range tones were received by VHF receiver B in the LM, demodulated, and retransmitted back to the CSM by VHF transmitter A on 296.8 MHz. The range tones were demodulated by the receiver in the CSM and processed to determine time delay and hence range. The three tones gave an unambiguous range of 327 nautical miles and an accuracy of about 100 feet.

S-Band Communications

The S-band communications equipment was similar to that described previously for the Command Module. It provided uplink and downlink of data and communications between the Lunar Module and the Manned Spaceflight Center in Houston via the Manned Spaceflight Network (MSFN). The process of uplink of data and communications from earth to the LM and downlink of communications and data from the LM to earth was similar to that described for the Command Module.

The S-Band equipment included a primary and a secondary receiver and a primary and secondary transmitter. The transmitter configuration included a phase modulator, driver, and power amplifier. Either the primary or secondary receiver and either the primary or secondary transmitter could be selected. The dual configuration provided backup in case of failure of an item.

Voice communication between the astronauts in the LM and MSFN on earth was conducted through their headphones and microphones via the audio sections of the signal processor assembly and the S-band receivers and transmitters. When the astronauts were out of the LM exploring the lunar surface, they communicated to the LM via a VHF link and equipment in the LM relayed the communication to the MSFN over the S-band link.

The uplink signal from the MSFN to the LM had a carrier frequency of 2,101.8 MHz. Data uplinked from the MSFN, including the state vector of the LM, was frequency modulated on a 70 kHz subcarrier that in turn was phase modulated on the uplink carrier. The data was extracted by the receiver, decoded, and fed to the LM guidance computer. Uplink voice communication was frequency modulated on a 30 kHz subcarrier that was phase modulated on the uplink carrier. The voice signal was extracted by the receiver and fed to audio sections of the signal processor assembly for the commander and Lunar Module pilot.

The uplink carrier was also phase modulated by a long pseudorandom noise (PRN) code that was used to determine range from the MSFN to the LM. The PRN code was demodulated by an S-band receiver in the LM and fed to a phase modulators for a transmitter channel. This signal was transmitted down to the MSFN, and the round trip delay of the received code at the MSFM was measured and used to determine range to the LM.

Range rate from the MSFN to the LM was very accurately determined by measuring the two-way Doppler shift of the S-band carrier of the uplink to the LM which was shifted in frequency and retransmitted back to the MSFN.

Video from the TV camera was applied to a frequency modulator at the input to the transmitter chain. The frequency modulator also served as backup for telemetry and voice communication.

The primary power amplifier of the transmitter had a nominal power output of 18.6 watts at 2,282.5 MHz. The secondary power amplifier output, which passed through a path in the primary power amplifier, was 14.8 watts at 2,282.5 MHz.

Those power amplifiers were a microwave device called an amplitron, which was a crossed field microwave amplifier with the unique property that with power off there was very little signal loss in passing through the device. The pair of series amplifiers worked very well for redundancy since if the primary amplifier failed, its power could be turned off and the power of the secondary amplifier turned on without having to switch inputs and outputs.

S-band antennas

The S-band communications subsystem in the LM could be switched to one of three antennas. Two of the antennas, referred to as in-flight antennas, were mounted on the side of the ascent stage. Antenna 1 was mounted on the forward side of the vehicle, and antenna 2 was mounted on the aft side. These were conical spiral antennas with omnidirectional antenna patterns. The in-flight antennas served as backup in case of failure of the high-gain steerable parabolic antenna.

The third S-band antenna was a steerable parabolic reflector type. The parabola was 26 inches in diameter, and the antenna gain was 20 dB. The high gain of that antenna allowed good-quality voice and data communication with earth and facilitated tracking and ranging to the LM from the MSFN on earth. The antenna was steerable over a range of 174 degrees in azimuth (yaw) and 330 degrees in elevation (pitch). The antenna could be positioned manually using PITCH and YAW controls on panel 12. A TRACK MODE switch on panel 12 was set to SLEW to allow positioning the antenna. When the antenna was pointed within about 12.5 degrees to the line of sight to the MSFN, the switch could be set to AUTO whereupon the antenna would automatically track the line of sight to the MSFN.

A fourth S-band antenna used on Apollo 12 and Apollo 14 was a large, lightweight, high-gain, erectable parabolic antenna. That antenna was folded and stored in a compartment in the descent stage. It could be removed by the astronauts and erected on the lunar surface. The antenna was about 9.8 feet in diameter when erected and had a gain of 34 db. This high gain allowed transmission of high-quality color television data to earth. A cable connected the antenna to equipment in the LM.

APOLLO LUNAR SURFACE EXPERIMENTS PACKAGE

The Apollo Lunar Surface Experiments Package (ASLEP) was designed to gather information about the moon from instruments placed on the lunar surface by the astronauts. The package consisted of a central station that communicated with earth and a collection of experiments deployed around the central station and connected to it. The instruments were designed to operate for a year, but some operated until the monitoring program was shut down in 1977.

Electrical power for the ASLEP was provided by a SNAP-27 radioisotope thermoelectric generator (RTG). SNAP-27 was required to provide at least 63 watts of power at 16 volts for 1 year for early Apollo missions and 69 watts for 2 years for Apollo 17.

The collection of instruments carried on individual Apollo missions varied. Instruments were selected from the following list.

- Lunar Passive Seismic Experiment

  Measured natural seismic activity and response to objects hitting the moon

- Lunar Active Seismic Experiment

  Measured seismic response to small explosions set off a distance away from the instrument

- Heat Flow Experiment

  Measured temperatures at different depths in bore holes up to 2.3 meters deep

- Lunar Surface Magnetometer

  A three-axis fluxgate magnetometer measured magnetic field components in three axes over a long period of time.

- Lunar Dust Detector

  A series of photocells mounted on the central station determined lunar dust collection by measuring response to illumination over time.

- Solar Wind Spectrometer

  Seven sensors arranged around the instrument measured the energy and direction of travel of the solar wind.

- Superthermal Ion Detector and Cold Cathode Ion Gauge

  These instruments measured hydrogen and helium ions on the lunar surface derived from the solar wind.

- Charged Particle Lunar Environment Experiment

  Measured energies of protons and electrons in the solar wind at the lunar surface. It was only flown on Apollo 14.

- Lunar Surface Gravimeter

  Determined variations in the gravitational field of the moon by measuring gravitational field at several locations. It was only flown on Apollo 17.

- Lunar Ejecta and Meteorites

    Designed to investigate particles striking the moon and ejecta from their impact on the surface. It was only flown on Apollo 17.

- Lunar Atmospheric Composition Experiment

    Sensitive mass spectrometer that measured the composition of the lunar atmosphere. It was only flown on Apollo 17.

A photograph showing deployment of some of the ALSEP items during the Apollo 16 mission is shown below. The passive seismic experiment is in the foreground, and the central station is behind it. SNAP-27 is to the left and behind the central station (Fig. 8.27).

Fig. 8.27 Deployment of ALSEP during the Apollo 16 mission (NASA photograph)

Bibliography

Apollo 11 Mission Report, NASA document NASA SP-238, 1971

Apollo 17 Mission Report, NASA document JSC-07904, March 1973

Apollo Operations Handbook Lunar Module, NASA document LMA790-3-LM, April 1971

Bennett, Floyd V., *Apollo Experience Report – Mission Planning for Lunar Module Descent and Ascent*, NASA Technical Note NASA TN D6846, June 1972

Brooks, Courtney, G., Grimwood, James, M., Swenson, Loyd, S., *Chariots for Apollo,* Dover Publications, Mineola, New York, 2009

Dietz, Reinhold H., Rhoades, Donald E, and Davidson, Louis J., *Apollo Experience Report – Lunar Module Communications System*, NASA Technical Note NASA D-6974

Eyles, Don, *Tales from the Lunar Module Guidance Computer*, Presented at the 27th annual Guidance and cntrol Conference of the american Astronautical Society, February 2004

Farkas, Andrew J., *Apollo Experience Report –Lunar Module Display and Control subsystem*, NASA Technical Note NASA TN D-6722

Hooper, John C., *Performance Analysis of the Ascent Propulsion system of th Apollo Spacecraft*, NASA Technical Note NASA TN D-7400, December 1973

Kurten, Pat M., *Apollo Experience Report – Guidance and Control Systems: Lunar Module Aboart Guidance System*, NASA Technical Note NASA TN D-7990

LEM Guidance, Navigation, and Control Subsystem Course No. 30315, Grumman Aircraft document, March 1966

LM/AGS Operating Manual Flight Program 6, TRW document 11176-6033-T000, July 1969

Lunar Excursion Module Familiarization Manual, Grumman Document LMA 790-1, July 1964

Lund, Thomas J., *Radar Velocity Sensors and Altimeters for Lunar and Planetary Landing Vehicles*, Presented at Firt Western Space Congress, October 1970

NASA Apollo Lunar Module News Reference, Prepared by Grumman Aircraft Corp., undated

Reel, L. S., and Poehls, V. J., *The Apollo Lunar Landing Radar*, Presented at Institute of Navigation National Space Meeting, April 1969

Rozas, Patrick and Cunningham, Allen R., *Apollo Experience Report - Lunar Module Landing Radar and Rendezvous Radar*, NASA Technical Note NASA TN D-6849

Shelton, D. Harold, *Apollo Experience Report – Guidance and Control Systems: Lunar Module Stabilization and Control System*, NASA Technical Note NASA TN D-8086

9

Apollo Crew Personal Equipment

The Apollo program elevated safety and well-being of the crew to the highest priority. All of the equipment was designed with unprecedented levels of reliability. This attention was carried through to the garments worn by the astronauts who undertook the complex and potentially dangerous mission to explore the moon. Exporing the surface of the moon was especially hazardous because temperatures of objects in direct sunlight could reach 240°F, while the temperature of objects in the shade could drop to −140°F. Radiation and micrometeroids were also of concern since the moon does not have the protective atmosphere that shields us on earth. The suits were required to perform this shielding function.

The crew personal equipment included equipment required for life support and safety in the LM cabin as well as during extravehicular activity on the surface of the moon. The set of closed circuit pressure vessel garments that enveloped the astronauts while providing life support was called the extravehicular mobility unit (EMU).

The EMU interfaced with the environmental control subsystem (ECS) of the LM for oxygen and water hookups when the astronauts were in the cabin, and it interfaced with the Portable Life Support System (PLSS) when outside of the cabin.

A photograph of Buzz Aldrin standing on the surface of the moon during the Apollo 11 mission and wearing the full extravehicular mobility unit is shown on the next page (Fig. 9.1).

© Springer Nature Switzerland AG 2018   291
T. Lund, *Early Exploration of the Moon*, Springer Praxis Books,
https://doi.org/10.1007/978-3-030-02071-2_9

BACKGROUND OF EMU AND PLSS

NASA issued a request for proposals to develop and build PLSS and EMU equipment in April 1962. The Hamilton Standard Division of United Aircraft Corporation was announced the winner of the competion to develop the equipment in July 1962.

Fig. 9.1 Buzz Aldrin standing on the moon and wearing the extravehicular mobility unit (NASA photograph)

The contract, amounting to $1.55 million was signed on 5 October 1962. The contract required that the Inernational Latex Corporation be subcontracted to provide the pressure garment. The International Latex Corporation (ILC) was best known at the time for making form-fitting, flexible rubber undergarments with the trade name Playtex.

The first prototype EMU suits produced by ILC were incompatible with the cabin of the Command Module. Hamilton Standard canceled the contract with ILC citing poor performance. NASA then conducted competitive bidding for the suits. ILC was one of the bidders. When designs for the suits from competitors were evaluated, it was determined that suits from ILC were better than the competitors. In the end, NASA contracted directly with ILC for the EMU suits. Hamilton Standard built the PLSS.

The lead engineer at NASA for the PLSS was Maurice Carson. Hamiliton Standard's manager for the PLSS program was Cal Beggs and Earl Bahl was the lead engineer. Leonard Shepard was program manager at International Latex for the Apollo space suits and George Durney was senior development engineer. Another key member of the International Latex team was Eleanor Foraker, seamstress and group leader.

EXTRAVEHICULAR MOBILITY UNIT (EMU)

The extravehicular mobility unit consisted of three garmets worn together. They were:

- Liquid cooling garment
- Pressure garment assembly
- Integrated thermal micrometeroid garment

Apollo EMU garments were personalized for each astronaut. Great care and attention to detail were required to assemble the garments, which in the case of the thermal micrometeroid garment required stitching 14 layers together. Most of the stitching was done by hand to exacting standards for stitch length rather than using a sewing machine.

Liquid Cooling Garment

The liquid cooling garment contained a netwook of fine water tubes that allowed circulating water through the garment. The flow of water through the suit absorbed body heat, and a liquid transport loop transferred the warmed water to an external sublimator where it was cooled before returning to the suit. The sublimator in the environmental control subsystem of the LM was used when the astronauts were in the cabin, and the sublimator in the PLSS was used when outside of the cabin. A sketch of the liquid cooling garment is shown on the next page (Fig. 9.2).

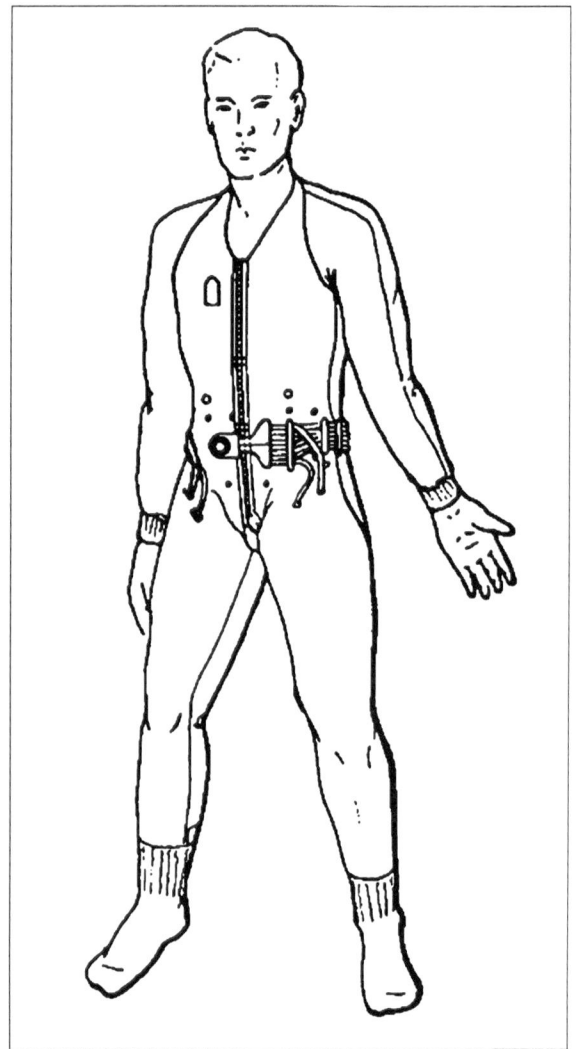

Fig. 9.2 Liquid cooling garment (NASA graphic)

Pressure Garment Assembly

The pressure garment assembly was a mobile life-support garment that could be worn in the cabin in case of loss of cabin pressure. It included a flexible torso and limb suit that covered the entire body except for the head and the hands. A separate helmet and intravehicle gloves completed the garment.

The torso and limb suit had four oxygen gas connectors, two for oxygen inlet and two for oxygen outlet. It also had a multiple water connector for inlet and outlet and an electrical connector. These connectors interfaced with tubes and cables that connected to the the environmental control subsystem when the astronauts were in the LM cabin and to the PLSS when they were outside of the cabin. A sketch of the pressure garment is shown below (Fig. 9.3).

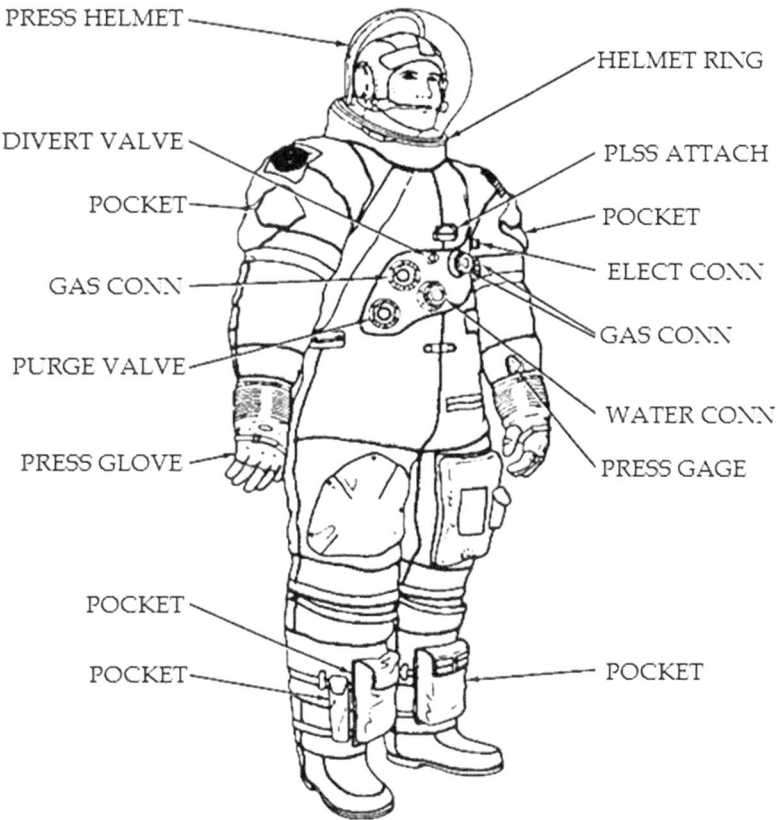

PRESS HELMET

HELMET RING

DIVERT VALVE

PLSS ATTACH

POCKET

POCKET

GAS CONN

ELECT CONN

GAS CONN

PURGE VALVE

WATER CONN

PRESS GLOVE

PRESS GAGE

POCKET

POCKET

POCKET

POCKET

Fig. 9.3 Pressure garment assembly (NASA graphic with larger captions)

Integrated Micrometeroid Garment

The integrated micrometeroid garment was worn over the torso and limb suit to protect the astronaut from radiation, heat transfer, and micrometeroid events when outside of the LM cabin. It was laced over the torso and limb suit. It was a multi-layer garment, with a protective outer layer followed by a micrometeroid shield-ing layer, followed by several layers of aluminized mylar that formed a thermal blanket, followed by a protective inner liner.

A sketch of an astronaut suited up for an extravehicular excursion with the inte-grated micrometeroid garment is shown below. The sketch shows him wearing the portable life support system and its control unit (Fig. 9.4).

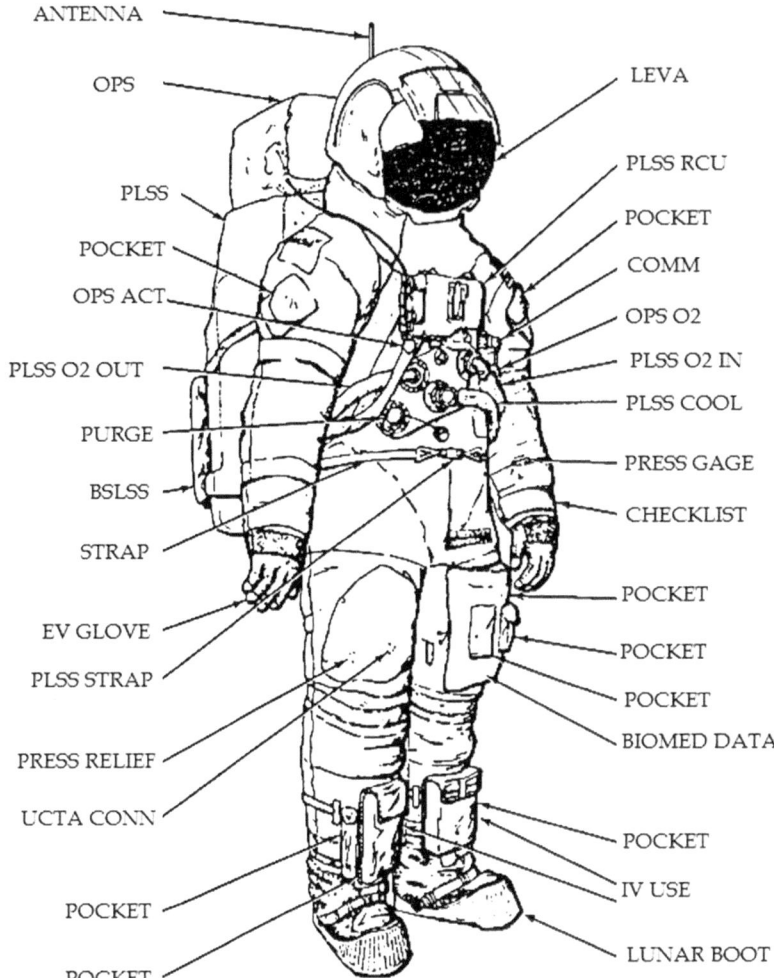

ANTENNA

OPS

PLSS

POCKET

OPS ACT

PLSS O2 OUT

PURGE

BSLSS

STRAP

EV GLOVE

PLSS STRAP

PRESS RELIEF

UCTA CONN

POCKET

POCKET

LEVA

PLSS RCU

POCKET

COMM

OPS O2

PLSS O2 IN

PLSS COOL

PRESS GAGE

CHECKLIST

POCKET

POCKET

POCKET

BIOMED DATA

POCKET

IV USE

LUNAR BOOT

Fig. 9.4 Astronaut suited up for extravehicular excursion (NASA graphic)

The gloves of the pressure garment assembly were replaced by extravehicular gloves for protection against heat and cold. The helmet of the pressure garment assembly was augmented by attaching an extravehicular visor assembly to it. The extravehicular visor assembly provided added protection against solar glare, ultraviolet radiation, space particles, and solar heat. Extravehicular boots were worn over the pressure garment boots. They provided protection against the extreme temperatures of the lunar surface.

PORTABLE LIFE SUPPORT SYSTEM

The portable life support system (PLSS) was a self-contained, rechargeable environmental control system that allowed the astronauts to spend several hours outside of the LM during their exploration of the moon. The PLSS was developed and built by the Hamilton Standard Division of United Aircraft Corporation.

The PLSS attached to the astronaut's back over the integrated micrometeroid garment by means of four straps. A photograph of Buzz Aldrin wearing the PLSS while descending from the Lunar Module Eagle during the Apollo 11 mission is shown on the next page (Fig. 9.5). A backup oxygen source called the oxygen purge system was contained in the package with a flag on it at the top of the PLSS in the photograph.

The PLSS supplied regulated oxygen at 3.9 pounds per square inch for breathing and suit ventilation, and it removed CO_2 and moisture from the oxygen circulating through the pressure suit. The ventilation flow rate, driven by fan, was about 5.5 cubic feet per minute. CO_2 was removed by passing the flow through a lithium hydroxide canister. Moisture was removed from the flow by passing it through passages in a cold sublimator and collecting the water that condensed. Fresh oxygen from the regulator made up for leakage from the suit and from conversion of oxygen to carbon dioxide by breathing of the astronaut.

The PLSS circulated cooled water through the liquid cooling garment by means of a water pump at a rate of about 240 pounds per hour (28.8 gallons per hour). The amount of water that passed through passages in a cold sublimator was controllable by the astronaut and that allowed him to control temperature of the water flowing through the liquid cooling garment.

An impotant element of the PLSS was the sublimator that cooled both the oxygen ventilation loop and the water that circulated through the liquid cooling garmet. The sublimator contained a porous plate of sintered nickel that was exposed to the vacuum of space. Water was applied to the plate and the fine pores in the plate allowed a small amount of water to seep through. The water turned to ice on the outer surface whereupon it sublimated to vapor. Sublimtion is a powerful cooling process that cooled the entire structure of the sublimator. A separate expendable feedwater source provided a continuous source of water to the pourous plate.

Water in the liquid transport loop that had passed through the liquid cooling garment was cooled by flowing through passages in the sublimator. The flow of oxygen passing through the pressure suit was cooled by flowing through other passages in the sublimator. Moisture picked up by the oxygen flow from the astronauts breath and skin condensed out and was gathered by a water separator associated with the sublimator.

Fig. 9.5 Buzz Aldrin wearing the PLSS while descending from Lunar Module Eagle (NASA photograph cropped by author)

The PLSS could operate in a space environment for 4 hours during early missions (Apollo 9 through 14) before replenshing the feedwater and oxygen sources and replacing the battery and lithium hydroxide canister. The operating time was extended to 8 hours for Apollo 15 through 17. Those later missions included the Lunar Roving Vehicle to traverse the lunar surface, and longer extravehicular activity (EVA) times were planned. The feedwater and oxygen were replenshed from supplies of water and oxygen in the Lunar Module. A new battery and a new lithium hydroxide canister were installed before each EVA.

The PLSS stored oxygen in a tank about 6 inches in diameter and 17 inches long. The pressure at time of filling was 1,020 psi for Apollo 9 through 14 and 1,410 psi for Apollo 15 through 17. Expendable feedwater for the sublimator was stored in bladders within a primary reservoir tank that held 8.5 pounds of water (about 1 gallon) for early missions. An auxiliary tank, which held 3.3 pounds of water, was added to accommodate the longer missions of Apollo 15 through 17. Condensate from a water separator in the oxygen flow path was forced into the space between the tanks and the feedwater bladders.

The PLSS was required to accommodate metabolic loads from the astronaut of 1,600 BTU/hour average and 2,000 BTU/hour peak. Heat was removed by a liquid transport loop that circulated water through the liquid cooling garment and through the sublimator. A three-position diverter valve located at the bottom right corner of the PLSS and accessible by an astronaut selected the amount of water passed through the sublimator. The three positions were marked MAX, INT (intermediate), and MIN. The associated temperature ranges of the water in the liquid transport loop were 45–50 degrees Fahrenheit, 60–65 degrees, and 75–80 degrees, respectively. The minimum position of the diverter valve was usually used by the astronauts, and they changed to the intermediate position when they were expending more energy.

The PLSS contained a VHF communications system and whip antenna for voice communications and telemetry. The system provided voice communications between the two astronauts, communications between the astronaut and Manned Spaceflight Center in Houston via the S-band link in the LM, and continuous telemetry from both astronauts.

The weight of the PLSS together with its remote control unit was 104 pounds. The reduced gravity of the moon resulted in an effective load on the astronauts of 17.2 pounds. The PLSS was 26 inches high, 20.5 inches wide, and 10.5 inches deep. The power for the PLSS was obtained from a 16.4-volt silver-zinc battery. The capacity of the battery was about 279 watt-hours for early missions and about 360 watt-hours for later mssions.

REMOTE CONTROL UNIT

The remote control unit (RCU) for the PLSS allowed an astronaut to activate and monitor critical functions of the PLSS. The RCU was mounted on the chest area of the outer suit.

A sketch of the remote control unit is shown on the next page (Fig. 9.6). Most of the controls are located on the upper side. The front side of the RCU contained a mount for a camera. The right side contained a mechanism to activate the oxygen purge system.

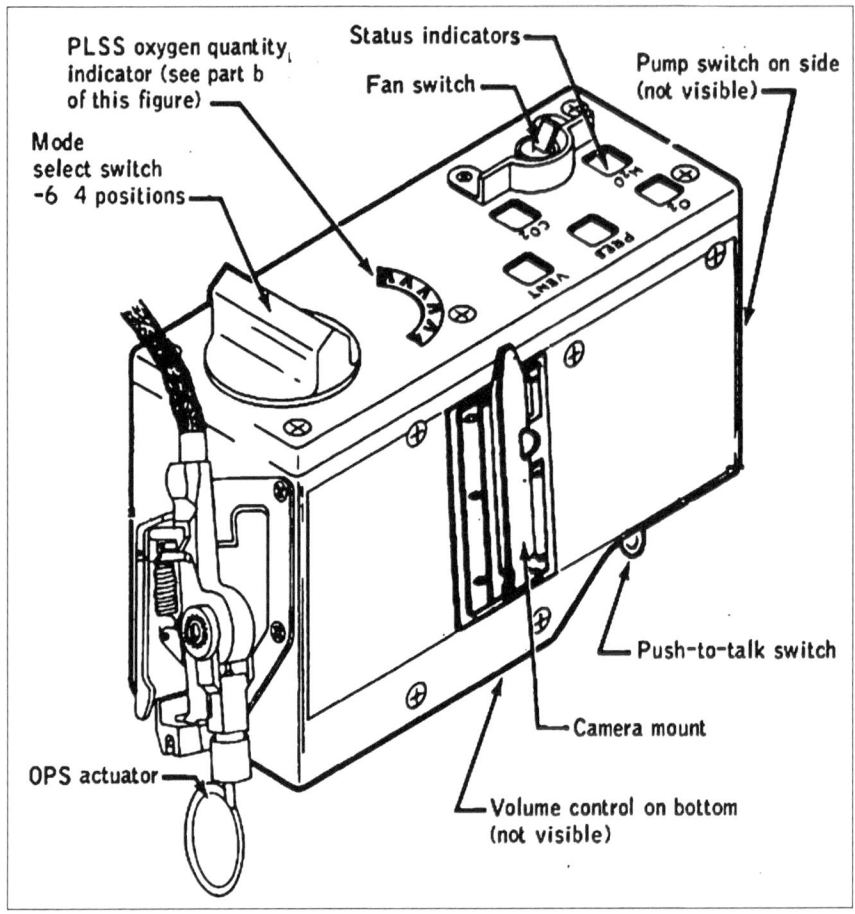

Fig. 9.6 Remote control unit for PLSS (NASA graphic)

The controls consisted of a fan switch, a pump switch, a communications mode selector switch, a push-to-talk switch, and a communications volume control. Displays included a PLSS oxygen quantity gauge and five status indicators. The status indicators and their functions are shown in the Table 9.1 below.

Table 9.1 Status indicators on RCU

Function	Label
High oxygen flow	O_2
Low PGA pressure	PRES
Low vent flow	VENT
Low feedwater pressure	H_2O
High CO_2	CO_2

A problem in any of these functions would illuminate a warning symbol in the particular warning indicator.

OXYGEN PURGE SYSTEM

An emergency oxygen supply, referred to as the oxygen purge system (OPS), was mounted on top of the PLSS during EVA. The OPS was contained in the package with a flag on it in the picture of Buzz Aldrin descending from the LM shown earlier. The OPS could supply breathing oxygen for about 30 minutes in case of failure of the PLSS.

The OPS contained two spherical tanks that each held about 2.1 pounds of useable oxygen. The outlet of the two tanks were connected in parallel and led to a shutoff valve that was remotely controlled by an actuator on the remote control unit. The output was regulated to 3.7 pounds per square inch at the outlet hose. The outlet hose connected to an oxygen inlet port on the pressure garment assembly. The OPS weighed 35.1 pounds. It was 18.4 inches long, 10 inches high, and 8.0 inches deep.

BUDDY LIFE SUPPORT SYSTEM

Special provisions called a Buddy Life Support System enabled an astronaut with a working PLSS to provide cooling water from his PLSS to the cooling circuit of the nonoperating PLSS. This was devised to allow an astronaut with a failed PLSS to return to the safety of the LM should his PLSS fail while out on the surface of the moon. His oxygen purge system would supply breathable oxygen during the emergency. A hose to connect the water circuits of the two PLSS units was carried in a separate small bag.

An illustration from the Apollo 14 press kit of the hookup between the two PLSS units is shown on the next page (Fig. 9.7).

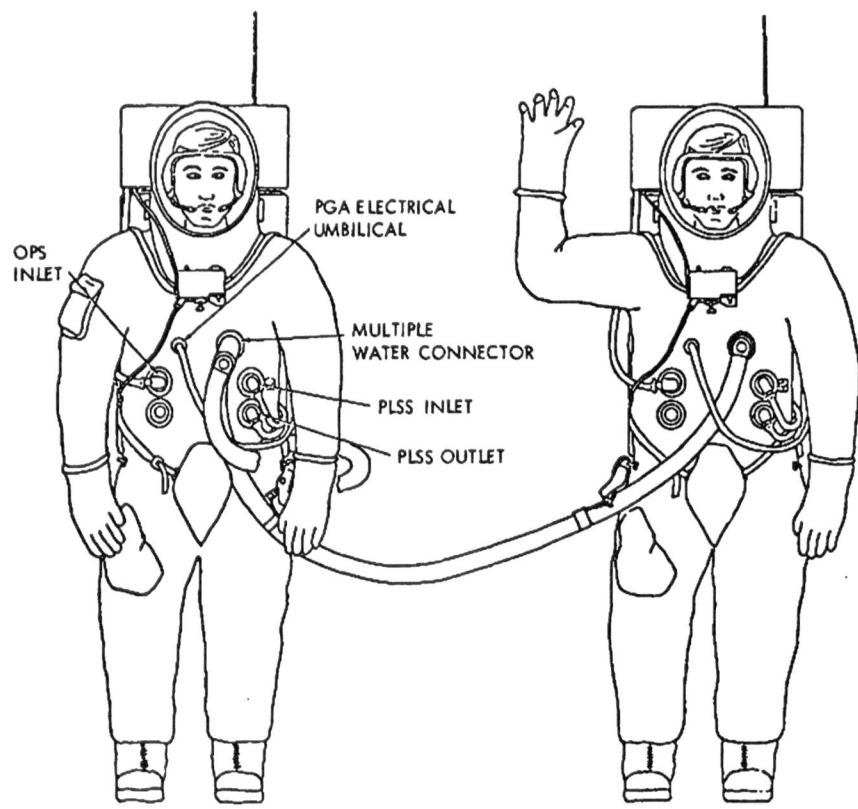

OPS INLET

PGA ELECTRICAL UMBILICAL

MULTIPLE WATER CONNECTOR

PLSS INLET

PLSS OUTLET

Fig. 9.7　Apollo Buddy Life Support System (Apollo 14 press kit)

Bibliography

Apollo Operations Handbook Lunar Module, NASA document LMA790-3-LM, April 1971
-- 2.11.2 Extravehicular Mobility Unit

Apollo Space Suit, A Historic Mechanical engineering Landmark, American Society of Mechanical Engineers, ILC Dover, Inc., 2013

Space Suit Evolution, NASA document, ILC Dover, Inc., 1994

10

Lunar Roving Vehicle and Exploration of the Moon

The Lunar Roving Vehicle (LRV) was an electric-powered, dune buggy-looking vehicle that greatly expanded the ability of the astronauts to explore the surface of the moon. A photograph of the LRV used on Apollo 15 on the next page (Fig. 10.1). The photograph was taken after its third and final sortie carrying David Scott and Jim Irwin on their exploration of the Hadley Rille site at the foot of the Apennines Mountains on the moon.

Three Apollo missions took LRVs along to the moon. The three missions and the LM crew for each were Apollo 15 (David Scott and Jim Irwin), Apollo 16 (John Young and Charles Duke), and Apollo 17 (Gene Cernan and Harrison Schmitt).

BACKGROUND OF LUNAR ROVING VEHICLE

The Marshall Space Flight Center (MSFC), led by Wernher von Braun, began studying concepts for a large mobile laboratory on the moon called MOLAB in 1964. That project proved to be too ambitious for the time, and studies of a vehicle that could transport astronauts on the moon were begun at MSFC in May 1969. The vehicle would be known as the Lunar Roving Vehicle (LRV). The MSFC program manager for the LRV was Saverio Morea.

MSFC issued a request for proposals for the Lunar Roving Vehicle in July 1969. The Boeing Company was announced to be the winner of the competition in October 1969. The contract for $19 million included development and production of four vehicles to be used on the moon and several test vehicles. Delivery of the first moon-capable vehicle was required by 1 April 1971.

© Springer Nature Switzerland AG 2018
T. Lund, *Early Exploration of the Moon*, Springer Praxis Books,
https://doi.org/10.1007/978-3-030-02071-2_10

Fig. 10.1 Lunar Roving Vehicle for Apollo 15 on the moon (NASA Photograph)

The Boeing effort was managed by Henry Kudish at Boeing's Huntsville, Alabama, manufacturing facility. The chassis was built at that facility and the vehicle was assembled there. The LRV engineering manager was Albert Fied. Boeing subcontracted to the Delco Division of General Motors for the running gear, and that included wheels, motors, and suspension. Ferenc Pavlics led that work at Delco.

Boeing delivered the first moon-capable model to MSFC in March 1971, 2 weeks ahead of the required date. The cost of the Boeing LRV program at the end was about $38 million.

PERFORMANCE OF LUNAR ROVING VEHICLES ON THE MOON

The LRV was a very capable vehicle, lauded by Apollo crews for its performance and reliability. Missions of Apollo 15, 15, and 17 were each planned to operate the LRV on three sorties, one per day. The LRV Operations Handbook, prepared by Boeing, states that the LRV was designed for a normal profile consisting of a maximum of 3 hours driving and 3 hours of station time for a total of 6 hours per sortie. Driving speed from 0 to 8.7 miles per hour was at crew's discretion. The LRV navigation system and console displays were to remain energized throughout the 6-hour sortie. Operation of the high-gain antenna and television camera was to be conducted only when the LRV was stopped.

The actual driving time during the missions was somewhat shorter than the specified maximum. A summary of performance for the three Apollo missions is tabulated below. Most of the values given were converted from NASA Technical Note TN D-7469, *Lunar Roving Vehicle Navigation System Performance Review*. Slightly different values are found in other documents (Table 10.1).

Table 10.1 LRV performance on the moon

Statistic	Apollo 15	Apollo 16	Apollo 17
Total driving time (hr:min)	3:00	3:19	4:29
Total distance (miles)	17.3	16.5	21.6
Average speed (mph)	5.8	5.0	4.8
Max range from LM (miles)	3.1	2.8	4.7
Longest traverse (miles)	7.8	7.0	12.6
Rock samples (pounds)	170	213	249

The maximum allowable range from the Lunar Module was limited to a "walk-back limit" which was a distance that the astronauts could walk back in case of failure of the LRV. This was a function of the time available from their portable life-support equipment. The limit was about 6 miles. The upper limit of speed of the LRU as given by the LRV Operations Handbook was 8.7 miles per hour. Gene Cernan set a lunar land speed record of 11.2 miles per hour during the Apollo 17 mission.

MECHANICAL CONFIGURATION OF THE LRV

The Lunar Roving Vehicle was foldable to allow it to fit in a compartment of the descent stage of the LM. The aluminum chassis was formed in three hinged sections. When the folded LRV was let down from the stowage compartment of the LM by lines and pulleys, the spring-loaded sections snapped into place and locked. The wheels were folded, and they also snapped into place and locked as the chassis was lowered from the descent stage.

When assembled, the LRV was 122 inches long, 77 inches wide, and 45 inches high. The wheelbase was 90 inches, and spacing between wheel centers was 72 inches. The LRV weighed 463 pounds, and it could carry a payload of 1,080 pounds on the moon. The vehicle remained stable on slopes up to 45 degrees. A sketch of the LRV with captions is shown on the next page (Fig. 10.2).

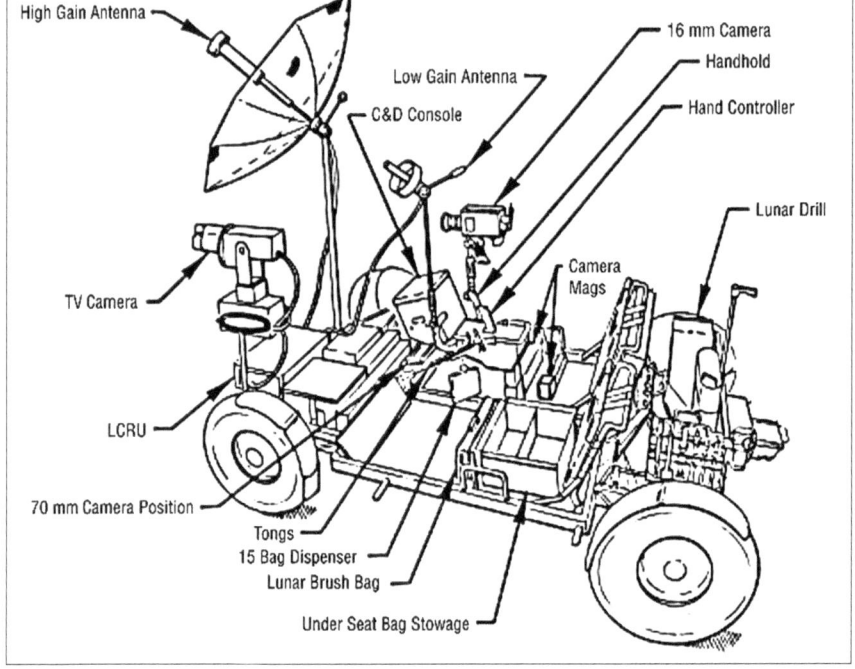

Fig. 10.2 Sketch of Lunar Roving Vehicle (NASA graphic cropped by author)

The wheels were 32 inches in diameter. What would be a tire on a vehicle on earth was formed from a woven mesh of steel wire.

The tread was formed from a chevron pattern of titanium strips riveted to the wire mesh. The wire mesh and the chevron strips can be seen in the photograph of a LRV wheel given on the next page (Fig. 10.3). A solid inner wheel 25 inches in diameter provided a bump stop should the wheel run over a large rock.

Each wheel was fitted with a traction drive assembly consisting of a 0.25 horse-power DC electric motor and a reduction drive with an effective gear ratio of 80:1. A mechanically actuated drum-type brake was provided for each wheel. Each wheel could be manually disconnected from the traction drive so that the wheel would rotate freely in case of failure of the drive. The suspension system included an upper and a lower triangular support, each connected to the chassis through a torsion bar. A shock absorber on each wheel provided damping of the suspension.

The electric drive motors were controlled from a Drive Control Electronics package. That package converted speed commands from the operator to a pulse width modulated, 36-volt rectangular waveform to the motors. The pulse repetition rate was 1,500 Hz.

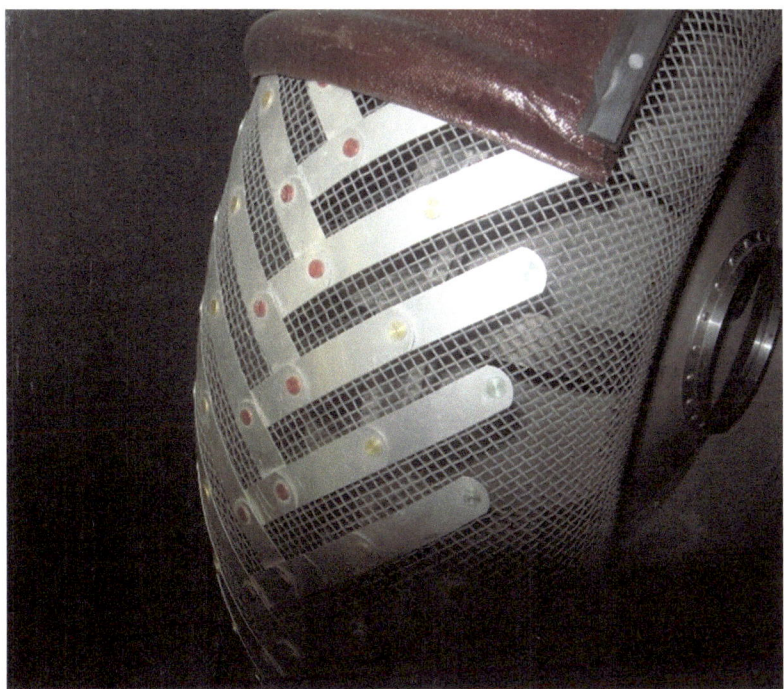

Fig. 10.3 Wheel of LRV showing wire mesh construction and titanium chevron tread (NASA Photograph)

The pulse width increased as higher speeds were commanded by the operator, and that resulted in higher average power provided to the motors. Slower speed commands resulted in shorter pulse widths and less average power.

The front and rear wheels could be independently steered by small electric motors. The turn radius of the LRV was 122 inches when both front and rear steering were used. Front wheel steering only or rear wheel steering only could also be selected.

A photograph of Apollo 17 Commander, Gene Cernan, driving the LRV at the Taurus-Littrow highlands landing site is shown on the next page (Fig. 10.4). The picture shows the wheels riding nicely on the lunar surface.

ELECTRICAL POWER

Electrical power for the electric motors that propelled and steered the vehicle and that operated the electronics and displays was obtained from two silver-zinc batteries. The batteries each had a nominal voltage of 36 volts and a capacity of 115 ampere-hours. The batteries were not rechargeable. Each battery was about 9.12 x

Fig. 10.4 Gene Cernan driving the LRV at the Taurus-Littrow highlands landing site (NASA photograph)

10.0 x 7.9 inches in size. They were mounted on a platform between the front wheels of the LRV and enclosed by a thermal blanket.

Normally, the electrical load was split approximately equal between the two batteries. Battery 1 powered Bus A and Bus B and Battery 2 powered Bus C and Bus D. The entire load could be switched to one battery in case of failure of the other. Meters on the display and control console allowed monitoring battery voltage, current being drawn, remaining ampere-hours, and temperature of each battery.

CONTROL OF THE LRV

A photograph showing the control and display console of the LRV is given on the next page (Fig. 10.5). The upper portion of the panel displayed navigation information. The lower section contained switching and metering for the motors and batteries. The size of the control and display panel was 10 by 12 inches.

A T-shaped hand controller located on a pedestal between the two astronauts allowed either of them to drive the LRV. The hand controller is the white T-shaped

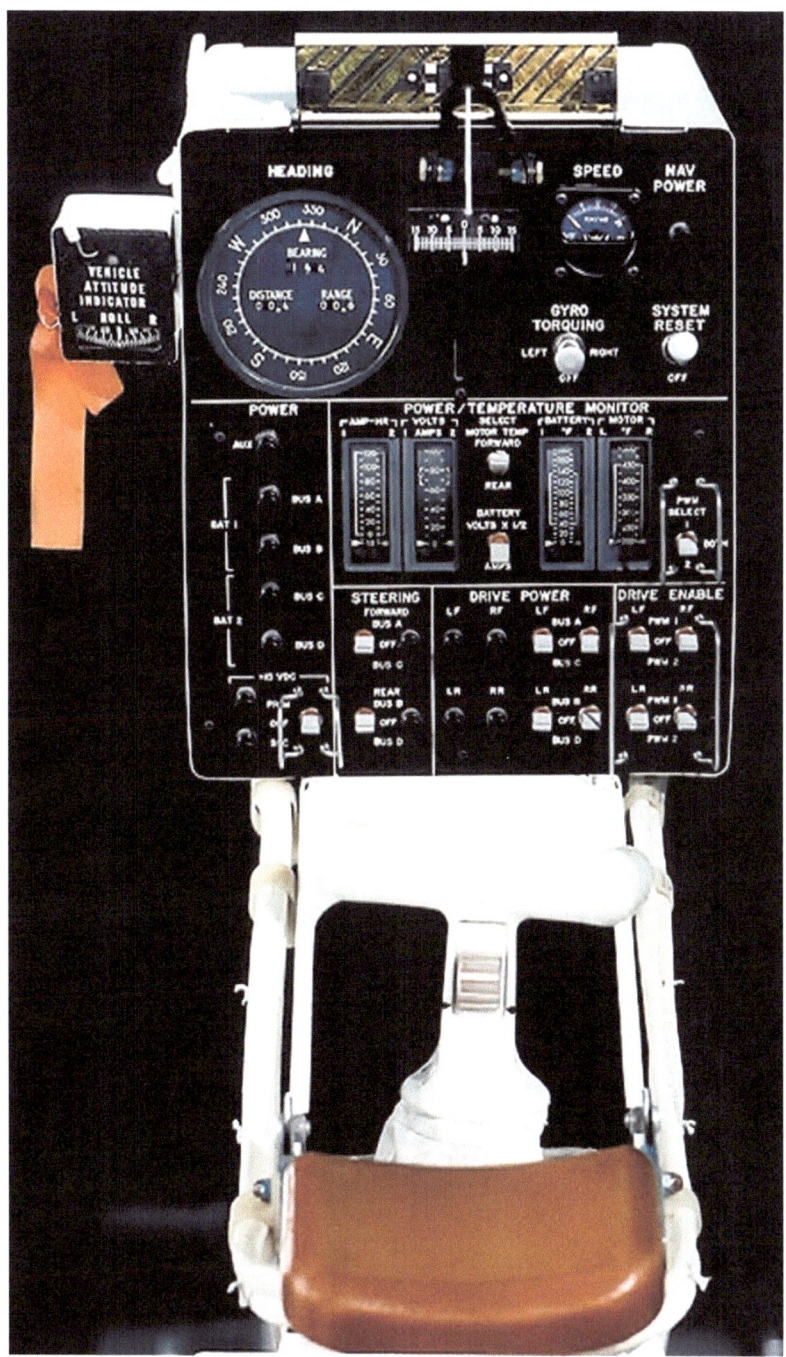

Fig. 10.5 Control and display module and hand controller for LRV (NASA photograph)

control in the lower foreground of the photograph. An armrest was provided below the hand controller.

The hand controller interfaced with the Drive Control Electronics package. The Drive Control Electronics in turn drove the various functions of the vehicle.

Pushing the hand controller forward caused the LRV to move forward at a speed proportional to the forward position of the controller. Moving the controller to the left caused the vehicle to steer to the left and moving the controller to the right caused the vehicle to steer to the right. Pulling the hand controller back either caused the LRV to move in reverse or it applied the brakes, depending on the position of a switch on the stem of the controller.　Pushing the switch up permitted travel in reverse. Pushing the switch down allowed the brakes to be applied when the hand controller was pulled backward. A spring-loaded catch was engaged when the controller was pulled nearly all the way back, and that locked the brakes in the parking position.

There were two switches under the label STEERING on the lower section of the control panel. The upper of the two switches switched power for front wheel steering from OFF to Bus A or to Bus C. The lower switch switched power for rear wheel steering from OFF to Bus B or to Bus D.

There were four switches under the label DRIVE POWER that controlled power to the four wheels. The switches were labeled LF, RF, LR, and RR. The LF and RF (left front and right front) switches could be switched from OFF to Bus A or to Bus B. The LR and RR switches could be switched from OFF to Bus B or Bus D. Lastly, there were four switches under the label DRIVE ENABLE that allowed switching each wheel drive circuit from OFF to PMW 1 or to PMW 2 where PMW was the acronym for pulse width modulation.

There were four vertical meters on the panel. The meter on the left displayed ampere-hours remaining for each battery. The display range was −15 to 120 Ah. The next meter displayed either volts or amperes being drawn from each battery. A switch next to the meter selected either volts or amperes. The display range was 0 to 100. The meter to the right of the switches displayed temperature of each battery. The display range was 0 to 180 °F. The last meter displayed temperature of the four drive motors. Temperature of either the front left and right motors or temperatures of the rear left and right motors were selected by a MOTOR TEMP switch. The display range was 200°F to 500°F.

LRV NAVIGATION SYSTEM

The navigation system provided important situational data including distance and bearing to return to the LM, distance traveled, and vehicle speed. The distance and bearing readouts to the LM could also be used to navigate to a predetermined location.

The navigation system included a directional gyro that was initially aligned to display vehicle heading with respect to lunar true north. It subsequently provided vehicle heading relative to true north essential to navigation as the vehicle maneuvered. The navigation system monitored odometer pulses from all four wheels and used information from the wheel with the third highest pulse rate (speed) to compute distance traveled. A small analog/digital computer gathered heading and odometer information and computed distance traveled, vehicle speed, range to the LM, and bearing to the LM.

Navigation during travels of the LRV required proper initial setting of the heading indicator relative to true lunar north. This was accomplished by first parking the LRV approximately on the down sunline and reading the shadow of a needle on the scale on the Sun Shadow Device (SSD). The SSD was a simple yet effective tool to help establish the heading of the vehicle relative to true north. A sketch of the control panel showing the SSD deployed is given below (Fig. 10.6).

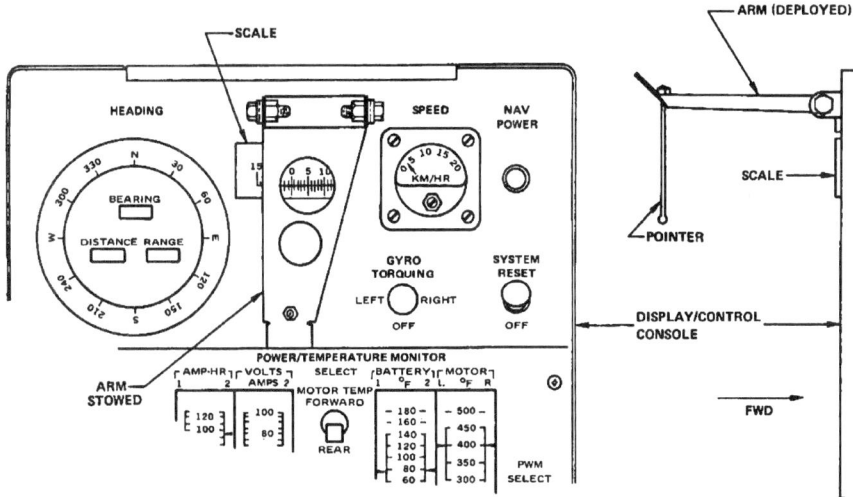

Fig. 10.6 Sketch of control panel showing SSD arm deployed (Bellcomm Graphic for NASA)

The reading on the SSD scale was communicated to the Manned Spaceflight Center in Houston along with pitch and roll of the LRV. Pitch and roll angles were read off of small pendulum-driven indicators in a box on the upper left side of the control and display panel.

The initial heading of the LRV was computed from the equation given below. The equation was solved by a computer at the Manned Spaceflight Center from measurements read down by the astronauts and from the azimuth and elevation angle of the sun obtained from ephemeris. The setting for the heading indicator (α) was then transmitted back up to the crew.

The setting for the heading indicator (α) was computed from the following equation.

$$\alpha = (\text{Sun Azimuth} - 180\,\text{deg}) - (\pm\text{SSD}) + \text{Roll Correction} + \text{Pitch Correction}$$

where:

SSD is the angle read on the SSD scale
Roll correction is equal to $\gamma(\sin\beta)/0.065$ where
$\gamma = 3.36\sin(\eta)$, $\beta = $ roll angle, and η is the elevation angle of the sun
Pitch correction is equal to $\text{SSD}[1 - 0.88(\eta - 26) + 0.46\sin(P)]$ where P is the pitch angle

The astronauts set the computed initial value for heading into the heading indicator by torqueing the gyro using the right-left gyro torqueing control on the control and display panel.

The LRV computer navigated by dead reckoning as the LRV traveled over the lunar surface. It used distance pulses from the odometer and heading from the heading indicator. The results were distance north-south and east-west from the starting point near the LM. The computer also continuously determined the range and bearing back to the LM. A reset button on the control panel allowed resetting all displays to zero at the beginning of each sortie.

Navigation information was displayed on the heading indicator on the control and display panel. That information included vehicle heading, distance traveled, range back to the LM, and bearing to the LM. A separate meter displayed speed of the vehicle from 0 to 20 km/hr.

The specified 3σ accuracy of the navigation data was as follows:

Navigation parameter	Accuracy
Heading relative to north	±6 degrees
Bearing to the LM	±6 degrees
Range to the LM	600 meters at 5 km
Distance traveled	2%
Velocity	±1.5 km/hr

COMMUNICATION WITH HOUSTON DURING SORTIES OF THE LRV

Communications between the astronauts on or near the LRV and the Manned Spaceflight Center (MSC) in Houston were carried out using VHF and S-band links via the Lunar Communications Relay Unit (LCRU) on the LRV. The LCRU subsystem consisted of a LCRU, a low-gain antenna, and a high-gain antenna. Communication functions with Manned Spaceflight Center via the Manned Spaceflight Network (MSFN) included:

- Duplex voice from astronauts via VHF AM link to the LCRU and LCRU to MSFN via S-band. Flight controller's voice from MSFN to LCRU via S-band and LCRU to astronauts via VHF AM
- Telemetry data to MSFN via S-band
- Television via the LCRU to MSFN via S-band
- Command data from MSFN to LCRU via S-band and hardline from LCRU to the Ground-Commanded Television Assembly

The low-gain antenna, high-gain antenna, and the LCRU Unit were installed on the LRV by the astronauts once the basic vehicle had been checked out on the moon.

Lunar Communications Relay Unit (LCRU)

The LCRU, built by RCA, was a rectangular assembly $22 \times 16 \times 6$ inches in size that weighed 55 pounds. A drawing of the control panel of the LCRU is shown on the below (Fig. 10.7).

The LCRU was mounted on a platform at the front of the vehicle such that the control panel of the unit was visible to an astronaut while positioning the high-gain antenna for best signal strength. The LRV would be parked, and the astronaut would be standing on the lunar surface to position the antenna.

The meter on the left side of the control panel gave an indication of signal strength when the switch just above and to the right of it was placed in the S-BD AGC (S-band automatic gain control) position. In the center position of the switch, the meter reads temperature of the heat radiator, and in the lower position, it reads battery voltage.

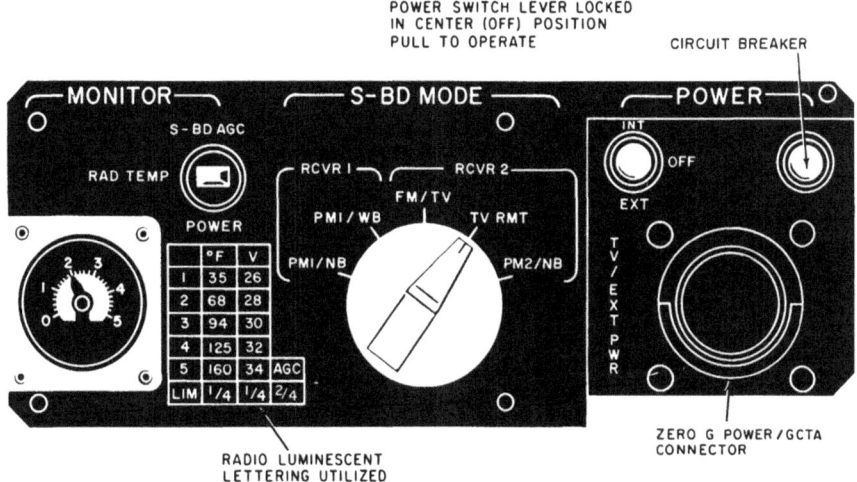

Fig. 10.7 Drawing of LCRU control panel (RCA/NASA graphic)

The temperature and voltage associated with meter readings were given on the table to the right of the meter. The S-band mode switch is discussed later.

The LCRU contained both VHF and S-band communications equipment. VHF equipment included a receiver transmitter and a whip antenna. The receiver operated at a frequency of 259.7 MHz. It received amplitude modulated voice and four subcarriers from the astronaut wearing the PLSS containing VHF communications system EVA-1. Voice and data from the PLSS containing EVA-2 was transmitted by FM modulation of its VHF carrier to EVA-1 where it was combined with voice and data from EVA-1. The combined voice and data from the two astronauts was picked up by the VHF receiver in the LCRU and transmitted to Houston over the S-band link.

The flight controller's voice from Houston was received over the S-band link and retransmitted by amplitude modulation of the VHF transmitter of the LCRU. The transmitter operated at a frequency of 296.8 MHz and the output power was 320 milliwatts. The signal transmitted by the LCRU was received by both astronauts.

The LCRU contained two S-band transceivers. Transceiver #1 consisted of a phase-modulated transmitter operating at 2,265.5 MHz and a phase-modulated receiver. The nominal output power of the transmitter was 6 watts.

The S-band receiver in Transceiver #1 received the 2,101.8 MHz uplink carrier from MSFN and extracted the flight controller's voice signal. The voice signal was used to amplitude modulate the VHF transmitter, and that transmission was received by both astronauts. Transceiver #1 was connected to the low-gain antenna.

Transceiver #2 consisted of a transmitter capable of either phase or frequency modulation and a phase modulation receiver. Frequency modulation was used to transmit video from the television camera. The frequency of the transmitter was 2,265.5 MHz, and the nominal output power was 8 watts. Transponder #2 was connected to the high-gain antenna on the LRV.

The receiver in Transceiver #2 was the same as in Transceiver #1 except that it had provisions to decode a 70 kHz subcarrier that contained control signals from MSC for the Ground-Commanded Television Assembly.

The LCRU was powered by a replaceable silver-zinc-type battery. The battery delivered 29 volts DC at a nominal current of 3.1 amperes.

Antennas Associated with the LCRU

There were two S-band antennas associated with the LCRU referred to as the low-gain antenna and the high-gain antenna. The low-gain antenna was used for voice and data communication with Houston when the LRV was in motion as well as when it was parked. The antenna had a helical feed and a cup-like reflector. The antenna gain was 10 dB on boresight and at least 6 dB over a 60 degree cone. The antenna was mounted by inserting its shaft into the left handhold on the LRV.

Elevation and azimuth pointing angles could be changed by the astronauts while the LRV was on a sortie to keep the earth in the broad beamwidth of the antenna.

The high-gain antenna was furled for transport to the moon. When unfurled it was a mesh parabolic dish with supporting ribs. The deployed diameter of the reflector was 36 inches. The antenna gain on boresight was 24 dB, and the gain was 20.5 dB over a 10 degree cone. The antenna was mounted by fitting the lower end of the mast into a receptacle on the left front structure of the LRV. The high-gain antenna was needed to transmit television video to the earth.

The high-gain antenna was positioned in azimuth and elevation about a ball joint by a positioning handle. An optical sight was provided to help point the antenna beam toward earth. The earth subtends an angle of about 1.8 degrees when viewed from the moon. Fine positioning to obtain best signal strength was done by watching the signal strength meter on the control panel of the LCRU Unit. Once positioned, the antenna orientation was friction locked in place. The high-gain antenna was only positioned and used when the LRV was at a stop.

Ground-Commanded Television Assembly

The Ground-Commanded Television Assembly (GCTA) included a color television camera and a television control unit. The "ground-commanded" portion of the name reflects the fact that the camera could be remotely controlled and pointed by signals from the ground (earth). Thus, operators in Houston could pan, tilt, and zoom the camera to observe various regions while the astronauts were exploring other areas. It was also used to observe ignition and liftoff of the ascent stage as the astronauts left the moon.

The GCTA was installed on the LRV by the astronauts once the basic vehicle had been checked out on the moon. A photograph of the camera attached to the television control unit is shown on the next page (Fig. 10.8). The GCTA was developed and produced by the RCA Corporation.

The camera used a silicon intensifier target (SIT) monochrome imaging tube with a rotating color wheel synchronized with the frame rate to generate color video. Standard NTSC line and frame rates were used. There were 262.5 scan lines per field and two interlaced fields in a frame giving a total of 525 lines per frame. The field refresh frequency was 60 Hz.

The camera optics had a zoom ratio of 6:1 and iris control from f/2.2 to f/22. It also had automatic level control (ALC) to adapt to a range of scene brightness. It had sensitivity and dynamic range of luminance of 1 to 10,000 foot-lamberts, and it could accommodate a scene dynamic range of 32:1.

The television control unit accepted commands sent up from MSC, and it could pan the camera up to 214 degrees to the right and 134 degrees to the left. It could

Fig. 10.8   Ground-Commanded Television Assembly (RCA photograph)

tilt the camera up to 85 degrees up and 45 degrees down. The camera could be removed from the television control unit and pointed by hand by the astronauts.

EXPLORATION OF APOLLO 17 USING THE LRV

Apollo 17 was the last Apollo mission to the moon and it was the most productive. The astronauts recorded the longest EVA time of all the missions at 22 hours and 5 minutes, and they traveled the longest total distance on the moon using the Lunar Roving Vehicle at 21.6 miles. A trained geologist, Harrison Schmitt, was one of the crew, and he carefully selected many of the rocks and regolith to be taken back to earth.

The landing site was in the Taurus-Littrow Valley between the South Massif and the North Massif. A picture of the landing site taken from the Lunar Module during the orbit preceding the landing is shown below (Fig. 10.9).

Fig. 10.9 Landing region for Apollo 17 (NASA photograph)

The South Massif is the mountain in the center of the picture, and the North Massif is across the valley from it to the right. The landing site was about 1 km to the right of the crater Camelot, which is the single crater above the cluster of small craters near the center of the picture.

The crew for Apollo 17 were Gene Cernan, Commander, Ron Evans, Command Module Pilot, and Harrison (Jack) Schmitt, Lunar Module Pilot. Informal photographs taken in the Command Module of the last Apollo crew to visit the moon are shown on the next page. The photographs were taken during the mission. Gene Cernan and Harrison Schmitt descended to the surface in the Lunar Module, while Ron Evans stayed behind to pilot the orbiting Command Service Module (Figs. 10.10, 10.11, and 10.12).

Fig. 10.10 Gene Cernan, Commander

Fig. 10.11 Ron Evans, Command Module Pilot

Fig. 10.12 Harrison Schmitt, Lunar Module Pilot

Cernan and Schmitt spent a little over 3 days (75 hours) on the moon. They performed three EVAs using the LRV, one each day. During the EVAs they deployed 11 experiments on the lunar surface. Eight of these experiments were carried for the first time on Apollo 17. Experiments carried for the first time were lunar seismic profiling, surface gravimeter, lunar atmospheric composition, lunar ejecta and meteorites, surface electrical properties, neutron probe, and the traverse gravimeter.

The lunar seismic profiling experiment was rather robust. It involved deploying four geophones in a pattern on the surface and then setting eight explosive packages at designated sites reached by traverses of the Lunar Roving Vehicle. Some of the explosive packages were located several kilometers from the geophone array. The explosive packages included an antenna and receiver, and the package responded to a detonate command sent from earth. The geophones sent the signals they received to a central processor which telemetered the signals back to earth.

The explosives were individually detonated by commands from earth after the astronauts had left the moon. The seismic signals produced from the detonations gave scientists insight into the lunar structure in the Taurus-Littrow region. Near the surface the seismic velocity was found to be about 250 meters per second. At depths below about 248 meters, the velocity was 1,200 meters per second, indicative of lava flow.

The traverse gravimeter was a portable, battery-operated gravimeter that allowed the astronauts to make measurements of local gravity at several locations. The gravimeter, which was about $6 \times 6 \times 12$ inches in size, was carried on the lunar rover and set on the surface at select locations. Gene Cernan usually operated the gravimeter, and he transmitted the reading displayed by the device at each location to Houston. The readings were an indication of the density of the underlying rock at each site.

The highlight of the Apollo 17 exploration was excursions by the astronauts in the Lunar Roving Vehicle (LRV). A photograph of Gene Cernan driving the LRV near the LM is shown below (Fig. 10.13).

Fig. 10.13   Gene Cernan driving the LRV near the Apollo 17 Lunar Module

The Traverses of the Apollo 17 LRV

The Taurus-Littrow highlands region of the moon explored by Apollo 17 was no "walk in the park." The rugged terrain is apparent in the photograph of Harrison Schmitt examining a large boulder on the surface shown on the next page (Fig. 10.14). The landing site was in smoother terrain. A total of 249 pounds of lunar material was gathered during sorties in the LRV.

One of the rock samples collected, assigned number 74255, is shown at its resting place on the moon in the photograph on the following page (Fig 10.15). The sample number was assigned by the Lunar Receiving Laboratory at the Manned Spaceflight Center when the rock had been returned to earth. The first digit, 7,

Fig. 10.14 Harrison Schmitt examining a boulder on the moon with the LRV in the foreground (NASA Photograph)

indicates that it was gathered by Apollo 17. The following digits represent location on the moon and size of the object.

A photograph taken by the Lunar Receiving Laboratory of a portion of the rock after it had been returned to earth is shown on the next page (Fig. 10.16). That portion of the rock was assigned number 74255,14.

The Apollo 17 astronauts made three traverses of the moon using the Lunar Roving Vehicle. The three traverses resulted in a total distance traveled of 21.6 miles. A plot of the paths taken on each traverse is shown on the map on the following page. **X** marks the landing location of the LM. The traverses are labeled EVA 1, EVA 2, and EVA 3 on the map where EVA refers to extravehicular activity. Traverse 1 traveled 1.5 miles, traverse 2 traveled 12.6 miles, and traverse 3 traveled 7.5 miles. Much of the time during EVA 1 was taken up by deploying and checking out the LRV and deploying and activating the experiments (Fig. 10.17).

Fig. 10.15 Location of rock 74255 on the moon (NASA photograph)

Fig. 10.16 Rock 74255,14 at the Lunar Receiving Laboratory

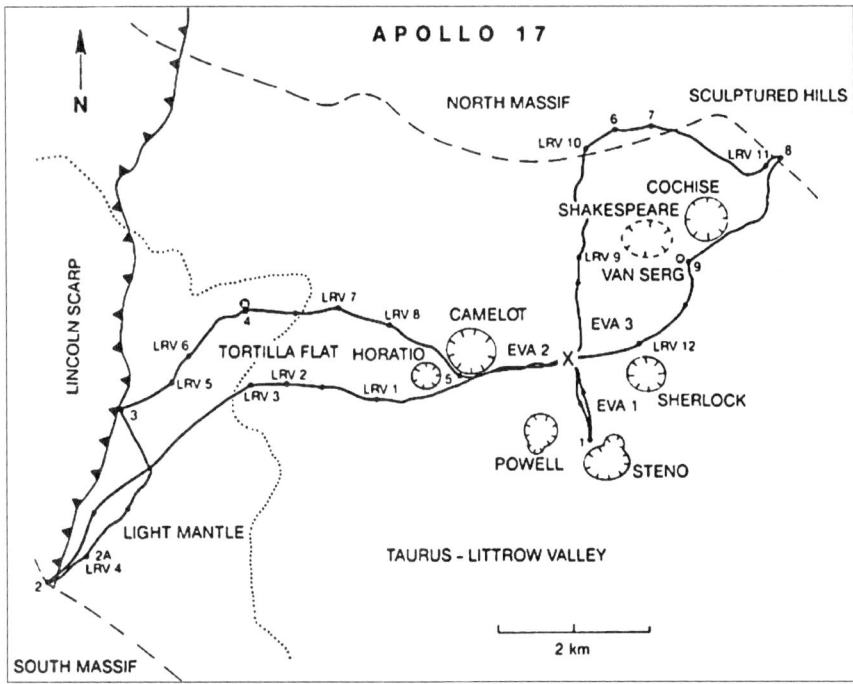

Fig. 10.17 Map showing the traverses of Apollo 17 (NASA graphic)

The numbers near the dots on the paths refer to station numbers where the astronauts stopped to explore. The photograph of Harrison Schmitt examining a boulder shown previously was taken at Station 6 of EVA 2. The line with the multiple arrowheads on the left side of the map denotes a steep escarpment. Major craters such as "Camelot" and "Shakespeare" are identified on the map.

The Ground-Commanded Television Assembly on the LRV was commanded to record the liftoff of the ascent stage of Apollo 17 as it departed the moon. A frame of the television image showing the descent stage after departure of the astronauts is given on the next page. That artifact of an amazing human adventure is a fitting end to this chapter and to the end of the remarkable saga of Apollo (Fig. 10.18).

Fig. 10.18   Apollo 17 Lunar Module descent stage rests on the moon for eternity (NASA image)

Bibliography

Apollo 17 Mission Report, NASA document JSC-07904, March 1973

Lunar Roving Vehicle Operations Handbook, Boeing document LS006-002-2M, April 1971

Lunar Roving Vehicle Systems Handbook, NASA document FC044, Prepared by Flight Control Division, June 1971

Lunar Roving Vehicle, Boeing Press document, 1972

Morea, Savero E., *The Lunar Roving Vehicle – Historical Perspective*, resented at 2nd Conference on Lunar Bases and Space Activities, April 1988

The Navigation System of the Lunar Roving Vehicle, Bellcomm Technical Memorandum, TM-70-2014-8, Decembert 1970

11

Russian Launch Vehicles, Lunar Impactors, and Flybys

BACKGROUND OF RUSSIAN EXPLORATION OF SPACE

The 1920s and early 1930s were a period of great theoretical and practical thinking about space travel in Russia. Rocket societies flourished, and experimental small rockets were launched by people who would one day conduct the Russian space program. Important papers on space flight were published, and flights to the moon and Mars were explored in detail. Leading scientists interested in space travel at the time included Mikhail Tikhonravov, Konstantin Tsiolkovsky, Alexander Shargei, Sergei Korolev, and Valentin Glushko.

This bright era came to an end in 1936 by cruel purges instituted by Premier Joseph Stalin. Several technical leaders were arrested and shot. Korolev and others were thrown into the gulag prison camp, and Glushko was placed under house arrest. World War II intervened, and the scientists in the gulag were put to work designing airplanes and military rockets.

After the war, Stalin relented and directed the scientists to work on military rockets and ballistic missiles. This work led to Soviet intercontinental ballistic missiles (ICBMs) in 1953. After Stalin's death in 1953, it again became thinkable to consider space flight. It was not much of a stretch to use the basic launch vehicles of ICBMs to launch earth satellites. Additional stages could be added to make flights to the moon.

The Soviet Union (Russia) conducted a robust space program starting in the mid-1950s. They were the first country to orbit a satellite of the earth (Sputnik 1 in October 1957), and they were the first to put a man in earth orbit (Yuri Gagarin in April 1961). They also had a very ambitious lunar exploration program during

© Springer Nature Switzerland AG 2018
T. Lund, *Early Exploration of the Moon*, Springer Praxis Books,
https://doi.org/10.1007/978-3-030-02071-2_11

the time period corresponding to the Ranger through Apollo lunar exploration programs in the United States. Russia was the major entity of the Soviet Union and spacecraft development, and space exploration was largely a Russian endeavor.

Russian unmanned lunar exploration programs scored several impressive successes along with many failures. Unfortunately, hardware for the more demanding manned lunar exploration did not rise to the occasion, and their manned lunar landing program floundered. The unmanned spacecraft programs included lunar impactors, lunar flyby, lunar landing, lunar orbiting, lunar sample return, and lunar rovers.

The most successful lunar exploration program involved the Lunokhod lunar rovers. The operation of the rovers demonstrated that Russian space engineering was very good. Had there been better program management and timely and clear Soviet Union government direction, the manned programs could have turned out more favorably.

Soviet practice was to only give names to spacecraft that made it out of earth orbit. Most unsuccessful launches were kept secret from the outside world and from their citizens as well. As a result, the outside world formed an early impression of considerable prowess of Soviet space program with select, achieved objectives. It was not until 20 years later that details of the troubled Soviet space programs emerged.

Management and development of space programs in the Soviet Union were organized in a series of experimental design bureaus known as OKB (Opytno Konstruktorskoye Buro). The lead design bureau was OKB-1 headed by Sergei Korolev. Korolev was both the technical and political leader for space programs. Several other design bureaus with expertise in particular areas supported OKB-1. Rocket engine design was assigned to OKB-456 led by Valentin Glushko. There was rivalry and considerable animosity between Korolev and Glushko and that impeded spacecraft development.

Other design bureaus involved with early space systems included OKB-52 headed by Vladimir Chelomei who designed ballistic missiles for the military; NII AP, headed by Nikolai Pilyugin, specializing in guidance systems; and NII-885, headed by Mikhail Ryazansky, specializing in radio communications systems. NII was an acronym for scientific test institute (Nauchno-Issledovatel'skly Institut).

The Soviet military wielded great influence on programs to be funded, and they sought intercontinental ballistic missiles (ICBMs) to assert Soviet military power. Sergei Korolev at OKB-1 was tasked with developing the first Soviet Union ICBM, to be known as the R-7, in 1953. The R-7 ICBMs were very successful.

The first artificial satellite of earth, Sputnik 1, was launched in October 1957 by an R-7 missile arranged for earth orbit.

RUSSIAN LAUNCH VEHICLES FOR LUNAR SPACECRAFT

Russian lunar exploration spacecraft were launched from the Baikonur Cosmodrome in the Republic of Kazakhstan. That classic launch site, first used for Soviet ICBMs, has been in constant use for space exploration since the launch of Sputnik 1 in 1957. In modern times, launches to the International Space Station are made from the extensive facilities at the cosmodrome.

Early Russian launch vehicles for lunar exploration were based on military launchers for ICBMs. The launch vehicles used for lunar exploration are described below. As with US early launch vehicles, there were many failures as well as successes.

Luna 8K72 Launch Vehicle

The Luna 8K72 launch vehicle lofted the first nine Russian lunar exploration spacecraft. Three of those missions were successful. Development of Luna 8K72 was begun in 1958 by OKB-1, headed by Sergei Korolev.

Luna 8K72 was an adaptation of the R-7 intercontinental ballistic missile with an upper stage added. Another launch vehicle based on the R-7 was the Vostok that launched Russian cosmonauts into space. A photograph of a Vostok 8K72K on display at the All-Russia Exhibition Center in Moscow is shown on the next page (Fig. 11.1). The Luna 8K72 was similar in appearance.

The R-7 had a central core first stage and four strap-on booster rockets. The Luna 8K72 launch vehicle used essentially the same core stage and strap-on booster rockets as the R-7. The third stage of Luna 8K72 was mounted above latticework near the top of the rocket. Propellants used for all three stages of Luna 8K72 launch vehicle were liquid oxygen and kerosene.

The four boosters that made up stage 1 consisted of cylindrical enclosures with propellant tanks and an RD-107 engine with thrust of 815 kilonewtons (183,200 pounds) at the end. The RD-107 engine had four combustion chambers feeding four nozzles and two smaller vernier engines. The vernier engines were steerable.

The booster enclosures were 2.68 meters (8.8 feet) in diameter and 19 meters (62 feet) long. The burn time was about 120 seconds. The boosters were jettisoned after their fuel was exhausted.

The core first stage of Luna 8K72 had a diameter of 2.99 meters (9.8 feet) and length of 28 meters (92 feet). It used a RD-108-8D55 engine with a thrust of 740.7 kilonewtons (166,500 pounds) at sea level. The engine had four combustion chambers and four nozzles. There were also four steerable vernier engines on the stage. The burn time was about 310 seconds. The core stage and boosters fired together at liftoff. Total thrust at liftoff was about 899,300 pounds.

Fig. 11.1 Soviet Union Vostok rocket (Wikimedia photograph posted by Sergei Arssenev)

The second stage of Luna 8K72 had a diameter of 2.56 meters (8.4 feet) and length of 9.31 meters (30.6 feet). It used a RD-105 engine with a thrust of 628 kilonewtons (141,000 pounds) in a vacuum. The burn time was 440 seconds.

Luna 8K72 carried a third stage mounted above the latticework at the upper end of the second stage. A Block E stage served as the third stage for lunar impact missions. The Block E stage, 2.4 meters in diameter and 5.2 meters long, was powered by a RD-105 rocket engine with a thrust of 49.4 kN (11,100 pounds).

Molniya 8K78 Lunch Vehicle

The Molniya 8K78 launch vehicle expanded the capability of the Luna 8K72 by replacing the third stage with two new upper stages. The first new upper stage produced 30,000 kg (66,150 pounds) of thrust in a vacuum and had a burn time of 200 seconds. It had a diameter of 2.6 meters (8.5 feet) and length of 2.8 meters (9.2 feet). The second upper stage had a thrust of 6,670 kg (14,700 pounds) and a burn time of 192 seconds. Its diameter was 2.6 meters (8.5 feet), and the length was 2.8 meters (9.2 feet). The two stages used liquid oxygen and kerosene for propellants.

The Molniya 8K78 launch vehicles were used to launch a total of 21 lunar exploration vehicles from January 1963 through February 1969. Of those, two made successful soft landings on the moon, and five orbited the moon. The other attempts were unsuccessful for various reasons.

Proton-K Launch Vehicle

What became known as the Proton launch vehicle was also known as the UR-500. It was developed by design bureau OKB-52 under Vladimir Chelomey to loft heavy military payloads into earth orbit. With the UR-500, OKB-52 became a rival of OKB-1 for launch vehicles for lunar probes.

The first assignment of the UR-500 was to launch a scientific satellite called Proton into orbit. Three out of four launches were successful. The launch vehicle was referred to as Proton after the satellite and that name stuck.

The Proton-K vehicle could be configured as either a three-stage or a four-stage launch vehicle with the fourth stage consisting of the capable Block D stage. The Block D stage served as a space tug for many different lunar missions. The first launch of a Proton-K with a Block D fourth stage was in March 1967.

The Proton-K was used to launch several lunar probes as well as probes to Venus and Mars. It was also used to loft modules to the Russian Mir space station and modules to the International Space Station. Proton-K had made 311 launches before it was supplanted by the Proton-M launch vehicle. A photograph of a Proton-K launch vehicle carrying the Zvezda Service Module to the International Space Station is shown on the next page (Fig. 11.2).

The first stage of Proton-K contained six booster rockets attached to the end of individual fuel tanks. The fuel tanks with rocket engines attached were clustered about a center oxidizer tank. The diameter of the oxidizer tank was about 4.1 meters (13.5 feet), and the diameter of each of the fuel tanks was about 1.6 meters (5.3 feet). The total length of the first stage was 21.2 meters (69.6 feet). The rocket engines were each gimballed to allow steering the first stage.

Fig. 11.2 Russian Proton-K launch vehicle (photograph from NASA files)

The engine for the boosters was the legendary RD-253 designed by Valentin Glushko, head of Design Bureau OKB-456. It was an advanced engine for the time developing 1,470 kilonewtons (330,500 pounds) of thrust at sea level. The total thrust of the six booster engines was about 8,820 kilonewtons (1.98 million pounds). The burn time was about 120 seconds. The first stage was jettisoned after completion of the burn.

Fuel for the engines was unsymmetrical dimethylhydrazine (UDMH), and the oxidizer was nitrogen tetroxide. This combination of fuel and oxidizer ignited upon contact (hypergolic). The second and third stages of the Proton-K were designed by Glushko, and they also used this combination of fuel and oxidizer.

The second stage contained four rocket engines mounted to a gimbal structure that allowed steering in flight. The total thrust of the four engines in a vacuum was about 2,400 kilonewtons (539,000 pounds). The burn time was 209 seconds. The diameter of the second stage was 4.14 meters (13.6 feet), and the length was 17.07 meters (55.9 feet). The second stage was separated from the third stage after its burn was completed.

The third stage was 4.15 meters (13.6 feet) in diameter and 4.11 meters (13.5 feet) long. It contained a single main engine and four gimballed vernier engines. The vernier engines provided steering and fine thrust control. The thrust of the main engine was 574 kilonewtons (129,000 pounds) in vacuum. The burn time was about 236 seconds. The third stage could inject a spacecraft into an earth parking orbit about 200 km (124 miles) high. The third stage also contained a flight control system that controlled the Proton-K vehicle during engine burn of the first three stages.

The fourth stage, referred to as Block D, was four meters (13.1 feet) in diameter and 5.5 meters (18 feet) long. Its RD-58M engine used kerosene fuel and liquid oxygen as oxidizer. The thrust was 85 kilonewtons (19,100 pounds), and the burn time was about 600 seconds. The engine could be started and stopped several times. The Block D served as a space tug for several lunar missions.

N1 Launch Vehicle

The N1 was a heavy-lift launch vehicle with a liftoff trust of 4,620 metric tons (10.19 million pounds). The N1 was planned to be the launch vehicle for a manned lunar landing program and other space explorations including missions to mars. Development of the N1 began in 1956 at a low level due to funding constraints. It was not until August 1964 when Soviet leaders approved a manned landing mission that priority was given to development of the N1. A great deal of time and effort was devoted to developing the N1, as recounted in Chapter 14 of this book, but it never achieved a successful launch.

LUNAR IMPACT MISSIONS

The lunar impact missions were managed, and hardware for the flights was developed by the Experimental Design Bureau, OKB-1, headed by Sergei Korolev. Korolev was the premier space systems designer in the Soviet Union. He was a respected leader of Soviet Union (Russian) space programs.

The impactor spacecraft were launched toward the moon by Luna 8K72 launch vehicles. They were put into a translunar trajectory by a Block E third stage. A picture of a mockup of a successful impactor, Luna 2, is shown on the next page. The impactor spacecraft were spherical in shape with a diameter of 0.9 meters (35.5 inches). They weighed about 390 kg (860 pounds) (Fig. 11.3).

The mission of the impactor spacecraft was to gather scientific information on the interlaying space between the earth and moon and scientific information on the moon before impacting the moon. The scientific measurements included magnetic field, radiation, cosmic particles, and micrometeoric impacts. An underlying purpose was to show the world that the Soviet Union had the technical expertise and wherewithal to send the first-ever spacecraft to the moon and guide it to impact at a particular location.

Six launch attempts were made to send lunar impact spacecraft to the moon between September 1958 and September 1959. Only one, Luna 2 launched in September 1959, impacted the moon. Luna 1, launched in January 1959, missed the moon but gathered data during the flyby.

The first three attempts to send a spacecraft to impact the moon failed at launch or shortly thereafter. The first attempt, launched in September 1948, failed due to vibration leading to structural failure of the strap-on booster rockets. The second attempt, launched in October 1958, again failed due to vibration, and the launch vehicle exploded shortly after launch. The third attempt, launched in December 1958, failed due to a failure of a hydrogen peroxide pump gearbox.

The fourth attempt launched in January 1959 was partially successful. The spacecraft missed the moon, but it returned a large amount of scientific data. The spacecraft velocity ended up too high because the burn of the Block E stage was not cut off in time due to a problem on the ground. As a result, the spacecraft crossed the path of the moon too soon and missed the moon by about 3,726 miles. This flight was given the name Luna 1.

The fifth impact flight attempt in June 1969 was unsuccessful due to a guidance system failure. The sixth attempt to impact the moon launched in September 1959 was completely successful. After a successful trajectory and operation of spacecraft systems, it impacted the moon in the Palus Putredinus region. Its sensors and telemetry systems operated until impact, and good scientific data was received back on earth. The flight was given the name Luna 2.

Fig. 11.3 Mockup of Luna 2 (NASA graphic)

The Soviet Union's custom was to only assign names to lunar spacecraft that made it out of earth orbit. Thus, only the flights of Luna 1 and Luna 2 were publicized.

Instruments aboard the Luna 2 spacecraft included:

- Fluxgate magnetometer to measure magnetic fields
- Cherenkov detectors to measure cosmic particles

- Micrometeorite impact counters
- Ion traps to measure plasma in interplanetary space
- Scintillation counters to measure ionizing radiation
- Geiger counters to measure ionizing radiation

Referring to the photograph of Luna 2, the magnetometer was mounted on top of the slender probe on the top of the spacecraft to distance the magnetometer from influence of the spacecraft. One of the two micrometeorite impact counters can be seen as the small rectangular device divided into four spaces on the left side of the picture. There were four ion traps mounted on the surface of the spacecraft. One can be seen directly above the micrometeorite counter in the picture. Three Geiger counters with different shielding material were mounted along the lower end of the magnetometer support probe.

Scientific readings were encoded onto telemetry signals and sent back to earth. The transmitter operated at a frequency of 183.6 MHz, and the data was encoded as pulse time modulation. The transmitter fed the four whip-like antennas on the top side of the spacecraft in the photograph. The 183.6 MHz radio system was also used in conjunction with ground equipment to measure velocity and range of the spacecraft.

The information telemetered to earth by the 183.6 MHz radio system was also transmitted at 19.993 MHz. That transmission used a "V" antenna formed by two steel ribbons that unfurled when the Luna spacecraft separated from the Block E stage. The wound-up ribbons can be seen at the bottom of the photograph of Luna 2.

Electrical power for the Luna 1 and Luna 2 spacecraft was obtained from silver-zinc and mercury oxide batteries. The batteries were sized to operate the spacecraft for 40 hours. The flight time of Luna 2 until impact was 33.5 hours.

The instruments flown in Luna 1 and Luna 2 added important knowledge about the moon and interplanetary space. The sensitive magnetometer determined that there was no measureable magnetic field in the vicinity of the moon. As a corollary, the radiation detectors confirmed that there were no radiation belts around the moon. Micrometeorite impacts detected were the equivalent of 0.002 impacts per square meter per second. The ion traps detected a flow of plasma in space. This flow is now known as the "solar wind."

In total, six impactor spacecraft were launched toward the moon. One successfully impacted the moon.

LUNAR FLYBY MISSIONS

The lunar flyby missions were managed by Soviet Union (Russian) Experimental Design Bureau, OKB-1, headed by Sergei Korolev.

Three spacecraft were launched with a mission of looping around the moon and photograph the back side. The three launches were made between October 1959 and April 1960 from the Baikonur Cosmodrome using Luna 8K72 launch vehicles.

The first of these spacecraft, named Luna 3, was successful. It was launched in October 1959. A second lunar flyby spacecraft with improved cameras was launched on 15 April 1960. That mission was unsuccessful when an upper-stage engine cut off too soon. A third lunar flyby spacecraft was launched on 19 April 1960. The spacecraft was destroyed by a launch explosion.

The successful spacecraft, Luna 3, was launched on a trajectory that took it around the moon and looped back to earth. The spacecraft took a series of pictures of the far side of the moon on its loop. The photographs were taken on film that was developed in the spacecraft, scanned, and transmitted to earth in electronic form. The resulting pictures gave the world the first views of the far side of the moon. The spacecraft continued to orbit the earth-moon system and eventually burned up in the earth's atmosphere.

A photograph of the Luna 3 spacecraft resting in a stand is shown on the next page. Luna 3 was about 1.2 meters (47 inches) high and 0.95 meters (37.4 inches) in diameter. The spacecraft weighed 278 kg (614 pounds). The flange near the top of the spacecraft held solar cells on both the top and bottom surface. Solar cells were also arranged around the cylindrical middle section of the spacecraft and in a band around the lower portion. The camera was located in the cylindrical opening at the top of the picture. It took pairs of images simultaneously through a 200 mm focal length lens and through a 500 mm focal length lens. The imaging medium was special radiation resistant 35 mm film. There was enough film for 40 frames of images (Fig. 11.4).

The camera lenses were fixed in the spacecraft, requiring the spacecraft to be maneuvered to point the camera at the moon. Maneuvering was accomplished by nitrogen gas jets mounted on the lower portion of the spacecraft as it is oriented in the photograph.

The far side of the moon was in sunlight at the time the trajectory of Luna 3 took it around the far side. The direction to the sun was sensed by photocells, and their signals were used to orient the spacecraft so that one end was pointed at the sun and the end holding the camera was pointed at the moon.

As the spacecraft passed around the far side of the moon, photocells on the camera end sensed the bright moon and triggered the first pair of pictures. At that time, the spacecraft was 63,500 km (39,460 miles) above the moon. At that distance the 200 mm lens imaged the full disk of the moon, and the 500 mm lens imaged a smaller area with higher resolution. A series of 29 frames were exposed over the next 40 minutes with the last image taken at a distance of 66,700 km (41,450 miles) above the moon.

Fig. 11.4 Luna 3 spacecraft (photograph from NASA collection)

The exposed film was moved to a film processor that developed, fixed, and dried the film. The developed film was moved to a scanner by a command from the ground. The scanner consisted of a cathode ray tube (CRT) that generated a fine bright spot, a film transport mechanism, and a photoelectric cell on the opposite side of the film. The film frame was positioned next to the face of the CRT, while the CRT performed a raster scan of the frame. A raster frame consisted of 1,000 lines. A photoelectric cell on the opposide side of the film picked up the varying light intensity as the film was scanned and created a video signal.

A slow or a fast scan rate could be selected. The time to scan one line was 1.25 seconds for the slow scan rate, and the transmission time for one frame was about 30 minutes. The fast scan rate was appreciably faster with a frame scan time of 10 seconds and transmission time for the frame of 15 seconds. The slow scan rate had a much narrower video bandwidth and hence had a higher signal-to-noise ratio when received on earth.

The video signal resulting from the scanning process was used to frequency modulate a transmitter operating at 183.6 MHz. Transmission of the image to the ground station was concident with scanning. The transmitter fed a qudrapole antenna consisting of the four rods that can be seen at the top of the photograph of Luna 3.

Williams and Friedlander at NASA Goddard Space Flight Center have cataloged the images from Luna 3. They identify the first image of the far side of the moon taken through the 200mm lens as that shown below (Fig. 11.5).

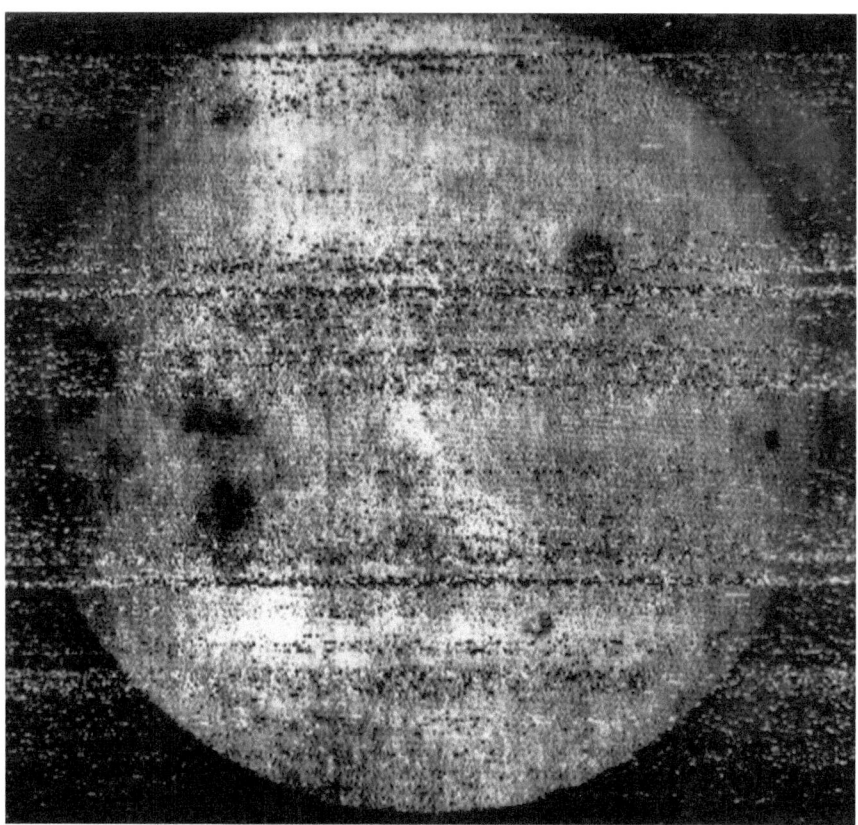

Fig. 11.5 First picture taken of the far side of the moon (NASA graphic)

The image is somewhat noisy, likely due to low signal-to-noise ratio on transmission to earth. The picture of the far side of the moon shows terrain more rugged than on the near side with only a few maria that appear as dark areas in the image. Close-up pictures were also taken with the 500mm lens. The pictures were quite noisy, and they will not be shown here.

The Russian lunar flyby spacecraft gave the world the first look at the far side of the moon, a view never seen from earth. The program pioneered circumlunar flight, although only one of the three spacecraft launched was successful. The pictures returned by Luna 3 were important in revealing previously unknown features of the moon.

Bibliography

Embry-Riddle Aeronautical University course "Russian Space Operations and Technology."

Hardesty, Von and Eisman, Gene, *Epic Rivalry – The Inside Story of the Soviet and American Space Race,* National Geographic, Washington, DC, 2007

Harvey, Brian, *Soviet and Russian Lunar Exploration*, Praxis Publishing, Chichester, UK, 2007

Kruse, Richard, *R-7 Family of Rockets / Proton (UR-500) Family of rockets,* http://historic-spacecraft.com/Rockets_Russian.html

Siddqi, Asif A., *Deep Space Chronicle*, NASA Report NASA SP 2002-4524

Zak, Anatoly, *Vostok Launch Vehicle,* russianspaceweb.com

12

Russian Soft Landers, Orbiters, and Rovers

Following exploration by impactor and flyby spacecraft, the Soviet Union (Russia) launched a series of soft landing spacecraft to the moon. That effort included spacecraft that landed on the moon and conducted photographic surveys; capable spacecraft that landed, gathered samples of soil, and returned the samples to earth; and spacecraft that orbited the moon. Lunar roving vehicles were also soft-landed on the moon. Driven from earth, these vehicles conducted sophisticated scientific experiments during lengthy sorties on the lunar surface.

LUNAR LANDING MISSIONS

A total of 13 spacecraft were launched toward the moon with the intent of making a soft landing and taking a series of photographs from the landing site. Two of the spacecraft made successful soft landings, two missed the moon, three crashed into the moon, and the other spacecraft did not make it out of earth's orbit. The spacecraft were launched by Molniya 8K78M launch vehicles from the Baikonur Cosmodrome. A summary of flights of the 13 lunar landing spacecraft is given in the table on the next page (Table 12.1).

Russian engineers at OKB-1 and their managers were certainly persistent. Eleven failures in a row could try men's souls.

The twelfth attempt, launched in January 1966, was a complete success! The spacecraft was given the name Luna 9. The thirteenth and final attempt, launched in December 1966, was also a complete success. It was given the name Luna 13.

A photograph of a replica of one of these successful spacecraft, Luna 9, is shown on the following page (Fig. 12.1). The photograph shows a replica of Luna 9 spacecraft attached to an E-6 bus. The spacecraft is on display in the Museum of Air and Space in Paris (Musée de l'air et de l'espace).

© Springer Nature Switzerland AG 2018   339
T. Lund, *Early Exploration of the Moon*, Springer Praxis Books,
https://doi.org/10.1007/978-3-030-02071-2_12

Table 12.1 Fights of Russian lunar landing spacecraft

Launch date	Name	Results
January 1963	–	Electrical power was lost and spacecraft remained in earth orbit
February 1963	–	Attitude control was lost and spacecraft did not achieve earth orbit
April 1963	Luna 4	Missed the moon when attitude control loss prevented midcourse correction
March 1964	–	Third stage cut off too soon and the spacecraft did not achieve earth orbit
April 1964	–	Third stage cut off too soon and the spacecraft did not achieve earth orbit
April 1965	–	Upper stage did not fire stranding the spacecraft in earth orbit.
April 1965	–	Third stage did not fire and spacecraft did not reach earth orbit
May 1965	Luna 5	Crashed on the moon due to failure of a gyro coupled with human error
June 1965	Luna 6	Missed the moon when retrorocket didn't shut down after midcourse correction
October 1965	Luna 7	Crashed on the moon after losing attitude control during landing
December 1965	Luna 8	Airbag punctured during landing and spacecraft crashed on the moon
January 1966	Luna 9	Successful landing and photo taking
December 1966	Luna 13	Successful landing and photo taking

The two successful spacecraft, Luna 9 and Luna 13, landed in the Oceanus Procellarum region of the moon. Luna 9 landed at 7.08° N and 64.37° W in the lunar coordinate system and Luna 13 landed at 18.87° N and 62.06° W. Both conducted television camera surveys around their landing area. Luna 9 returned the first pictures ever taken from the surface of the moon.

The Luna 9 and Luna 13 series of lunar landers were two-part spacecraft consisted of an E-6 bus and a landing capsule. The landing capsule was contained in the bulbous structure at the top of the spacecraft in the photograph of Luna 9. The bus carried the landing capsule down to just above the lunar surface and ejected the capsule when a probe under the landing vehicle contacted the surface. The spherical landing capsule alighted on the moon, bounced a few times, righted itself, and extended pedals to stabilize the capsule for television surveys.

An artist's drawing of the landing capsule is shown on the following page (Fig. 12.2). Pedals to stabilize the capsule are extended as they would be when the capsule was resting on the lunar surface.

Landing Sequence

As Luna 9 approached the moon, the spacecraft oriented itself with the retrorocket pointed toward the local vertical in preparation for the braking burn. A radar altimeter on the bus made a mark at an altitude of about 77 km (46.6 miles). The mark

Fig. 12.1 Replica of Luna 9 (Wikipedia posting by Pine)

was used to fire the retrorocket, inflate airbags that protected the landing capsule, and jettison equipment that would no longer be used, including the radar altimeter.

The velocity of the spacecraft was about 2.6 km/s (8,530 feet per second) at the time of firing the retrorocket. Spacecraft velocity during the descent was computed by integrating acceleration from the last known velocity. The retrorocket was shut off when the velocity had decreased to a preset value. This occurred at about 250 meters (820 feet) altitude. Four vernier engines continued to thrust as

Fig. 12.2 Drawing of Luna 9 landing capsule with pedals extended (NASA graphic)

the spacecraft descended. The vernier engines were shut off and the landing capsule was ejected when a contact probe extending 5 meters (16.4 feet) below the bus contacted the lunar surface.

The landing capsule, protected by airbags, hit the surface at about 22 km/h (13.7 mph), bounced a few times and came to rest. The airbags were deflated and the capsule came to rest with the heavier pedal side down. The pedals opened to stabilize the lander, the antennas deployed, and the lander was ready to take photographs.

Description of Luna 9 Spacecraft

The overall Luna 9 spacecraft, including the E-6 bus and the landing capsule, was 2.7 meters (9.5 feet) long and weighed 1538 kg (3391 pounds). The landing capsule was spherical in shape, 0.58 meters (22.8 inches) in diameter, and weighed

99 kg (218 pounds). The bus held a retrorocket, four vernier engines, fuel and oxidizer tanks, and a guidance and navigation system that included a radar altimeter.

The retrorocket used amine fuel and nitric acid oxidizer. The combination ignited upon contact. The retrorocket generated 45.5 kilonewtons (10,200 pounds) of thrust and it could be shut down and restarted. A pressurized cylindrical section of the bus just below the landing capsule contained communication equipment, control and navigation system components, and batteries.

The spherical landing capsule contained communication equipment, antennas, a camera system, a radiation sensor, and batteries. The antennas automatically deployed when the pedals of the capsule opened. There were strings with weights at the end attached to the end of the probe antennas. These were used to give the camera images a reference for the local vertical.

The camera used a scanning mirror to reflect the image through an objective lens covered by a diaphragm with a pinhole. The image from the pinhole, which represented a scanned spot on the lunar surface, was applied to a photomultiplier tube. The objective lens was focused at the hyperfocal distance, and images at distances from about five feet to infinity were acceptably sharp. As an aside, "point and shoot" cameras are focused at the hyperfocal distance.

The scanning mirror was contained in the cylindrical object at the top of the capsule. The mirror scanned rapidly in elevation over an included angle of 29 degrees. Don Mitchell in his treatment of *Soviet Space Cameras* indicates that there were 6,000 vertical lines scanned in a panoramic sweep of 360 degrees. The mirror was rotated in the horizontal plane by a pixel width for each vertical sweep.

A complete scan of 29 degrees by 360 degrees took about 100 minutes. This equates to one vertical scan per second. The horizontal resolution based on 6,000 lines in 360 degrees would be 0.06 degrees. Mitchell reports that the analog video signal was 250 Hz which would correspond to 500 pixels. The vertical resolution would then be 0.058 degrees.

The images were transmitted to earth using standard facsimile encoding used on earth to send pictures by wire or radio. The transmission from Luna 9 was picked up by the large antenna at the Jodrell Bank Observatory in England, and the Jodrell Bank engineers noticed that transmission of images was in standard facsimile format. They borrowed a facsimile receiver from the Daily Express newspaper and downloaded the images. The Daily Express published the pictures long before they were available in the Soviet Union to the dismay of Soviet leaders and scientists.

The first photograph ever taken of the lunar landscape with a camera on the moon is shown on the next page (Fig. 12.3). The picture was taken shortly after landing on 3 February 1966. The tilt of the horizon in the picture indicates that the capsule came to rest at an angle of about 15 degrees from the vertical. The sun was at a very low angle for this first picture as is apparent from the long shadow cast by the rock in the foreground.

A full 360 degree panoramic series of pictures was taken on 4 February. Another panorama was taken on 5 February and again on 6 February. Radiation measurements made by Luna 9 indicated a dose of about 30 millirads per day. The batteries were depleted by the end of the day 6 February and communications came to an end.

Fig. 12.3 First picture of the lunar surface taken by Luna 9 (NASA graphic)

Description of Luna 13 Spacecraft

Luna 13 was launched on 21 December 1966 and it soft-landed on the moon on 24 December.

The Luna 13 landing capsule was similar to the Luna 9 capsule except that a second camera was carried to obtain stereoscopic images, and it carried equipment for additional scientific experiments. The weight of the Luna 13 capsule was 150 kg (330 pounds), about 51 kg heavier than Luna 9.

The cameras were contained within two cylindrical objects on top of the spacecraft. The capsule had two deployable booms 1.5 meters (59 inches) long. A soil penetrometer was attached to the end of one boom and a backscatter radiation densitometer was attached to the other.

The penetrometer consisted of a cylindrical fixture that held a shaft about 3.5 cm (1.4 inches) in diameter with a conical tip along with a small explosive charge. The explosive charge when ignited gave the shaft a thrust of five to seven kg (11 to 15 pounds) for 0.6 to 1.0 seconds. The depth of penetration of the shaft into the lunar soil was measured and used to estimate mechanical properties of the soil.

The radiation densitometer assembly included a cesium-137 source of gamma rays and three radiation detectors. The radiation detectors measured the level of backscattered gamma rays from the surface. This backscattered data, transmitted to earth, was used to estimate the composition of the soil.

A three-axis accelerometer assembly located in the capsule measured deceleration upon landing of the capsule. These readings along with penetrometer data were used to determine mechanical properties of the surface.

The Luna 13 experiment collection also included an infrared radiometer to measure heat transfer from the lunar surface. The instrument had four radiometer sensors on the capsule for that purpose. The capsule also contained a radiation detector that was mounted near the top of the capsule.

One of the two cameras did not function on Luna 13 and stereoscopic images were not obtained. However, good images were obtained from the other camera. A total of five 360 degree panoramic images of the surface around the landing areas were made at different sun angles.

Soft landing a spacecraft on the moon was difficult in those pioneering years and only two of the thirteen spacecraft launched accomplished a soft landing. Yet, the project returned the first-ever pictures taken from the surface of the moon, and it sent back information on properties of the lunar soil.

LUNAR ORBITER MISSIONS

The lunar orbiter program originated in Russian Experimental Design Bureau OKB-1. The program was transferred to the Design Bureau, OKB Lavochkin in 1965.

Five spacecraft were launched with a mission of orbiting the moon between March 1966 and April 1968. Four of these were successful. The successful spacecraft were given names Luna 10, Luna 11, Luna 12, and Luna 14. All of the launches were made from the Baikonur Cosmodrome by Molniya 8K78M launch vehicles.

The spacecraft used for Luna 11, Luna 12, and Luna 14 were essentially the same and consisted of a single orbiting vehicle. Luna 10 was different. It used a bus similar to that used in Luna 9 and Luna 13 to enter lunar orbit and then ejected a sizeable instrument capsule into orbit.

Luna 10

The Luna 10 spacecraft, which included a bus and an instrument capsule, is pictured on the next page (Fig. 12.4).

The instrument capsule is positioned above the smooth cylindrical section of the bus. The capsule was roughly cylindrical with a diameter of 0.75 meters (29.5 inches) and length of 1.5 meters (59 inches). It weighed 245 kg (540 pounds). The capsule held seven instruments: a gamma ray spectrometer, a three-axis magnetometer, a micrometeoroid detector, gas discharge counter to detect ionizing radiation, X-ray detector, sensor to measure infrared radiation from the moon, and charged particle detectors. It also contained batteries to operate the experiments and the communication system. It did not contain a camera.

Luna 10 entered lunar orbit on 3 April 1966. The instrument capsule was separated from the bus shortly after entering lunar orbit. The orbit had a perilune of 349 km, an apolune of 1,015 km, and an inclination from the equator of 71.9 degrees. The Luna 10 instrument capsule operated for 56 days and completed 460 orbits of the moon before its batteries were depleted. A total of 219 transmissions were made from the spacecraft to relay scientific information to earth.

Data telemetered to earth showed no measureable magnetic field and no atmosphere. Measurement of cosmic radiation indicated five particles per square centimeter per second. Gamma ray spectrometer measurements of the lunar surface were representative of basalt. An important discovery from tracking of the spacecraft in orbit was that there were regions of mass concentrations (mascons) that resulted in a varying gravitational field.

Luna 11, Luna 12, and Luna 14 Spacecraft

The Luna 11, 12, and 14 spacecraft were all essentially the same. A picture of the spacecraft is shown on the following page (Fig. 12.5). The lower portion of the spacecraft consisted of the E-6 bus. Equipment for the science mission was contained in the conical portion of the spacecraft above the bus.

Luna 11 was launched on 24 August 1966 and it entered lunar orbit on 28 August 1966. The orbit was 160 km by 1,200 km oriented 27 degrees from the equator. It had made 277 orbits of the moon and made 137 transmissions of data to the ground by the time the batteries were depleted on 1 October 1966. The spacecraft carried a set of equipment similar to that of Luna 10 for measurements of gamma ray and X-ray emissions from the moon, micrometeorite flux, and solar charged particles. The spacecraft included two cameras but no useable pictures were returned because of problems with the spacecraft attitude control system.

Luna 12 was launched on 22 October 1966 and it entered lunar orbit on 25 October. The orbit was 100 km by 1,740 km at 10 degrees inclination to the equator. By the time that the batteries expired on 19 January 1967, it had made 602 orbits of the moon and had made 302 transmissions of data to earth. Scientific measurement equipment on board was similar to that in Luna 10.

Fig. 12.4 Luna 10 lunar orbiter resting on a stand (from NASA database)

Luna 12 carried two cameras, one with a 500 mm lens and the other likely with a 200 mm lens. It returned 40 images per camera. Scenes were photographed on film 25.4 mm wide. The film was developed and dried and then scanned to produce an analog video signal. Scanning could be done at either 1100 or 550 lines per image.

Fig. 12.5 Luna 11 lunar orbiter (Graphic from NASA collection). The captions on the drawing are as follows: 1 Gas supply for attitude control system. 2 Phototelevision system. 3 Thermal control system radiator. 4 IR radiometer. 5 Instrument compartment. 6 Battery. 7 Attitude control system sensors. 8 Antennas. 9 Attitude control system electronics. 10 Vernier engines. 11 Retrorocket housing

Luna 14 was launched on 7 April 1968 and it entered lunar orbit on 10 April. The orbit was 160 km by 270 km, inclined 42 degrees from the equator. Luna 14 carried instruments and cameras similar to those on Luna 12. Communication with the spacecraft continued until 4 July 1968.

Only a few of the photographs taken by the orbiters have been published. Some of the pictures show good detail of the surface. The pictures taken later by Zond 8 were better quality.

LUNAR SAMPLE RETURN MISSIONS

Russia developed a unique series of spacecraft that would land on the moon, gather a sample of the lunar soil, and transport the sample back to earth. The spacecraft was developed and the missions were managed by OKB Lavochkin.

A total of 10 sample return spacecraft were launched between June 1969 and August 1976. Three of the spacecraft were successful in returning samples to earth. The successful spacecraft were named Luna 16, Luna 20, and Luna 24. All of the launches were made by Proton 8K82K launch vehicles.

A summary of flights of the sample return spacecraft is given in the table below.

Table 12.2 Flights of sample return spacecraft

Launch date	Name	Results
June 1969	–	Electrical failure prevented Block D firing and did not reach earth orbit
July 1969	Luna 15	Crashed during landing on moon
September 1969	–	Block D stage failed and spacecraft was stranded in earth orbit
February 1970	–	Second stage of Proton K shut down before reaching earth orbit
September 1970	Luna 16	Successful mission. 106 grams of soil returned to earth
September 1971	Luna 18	Ran out of fuel and crashed during landing
February 1972	Luna 20	Successful mission. 50 grams of soil returned to earth
October 1975	–	Block D stage failed and spacecraft did not reach earth orbit
November 1975	Luna 23	Toppled over on landing
August 1976	Luna 24	Successful mission. 170 grams of soil returned to earth

Luna 15 was launched three days before the launch of Apollo 11 in July 1969. It was an attempt by the Soviet Union to return a sample of lunar soil to earth before samples were returned by Apollo 11. Signals from Luna 15 ceased four minutes after initiation of the lunar deorbit burn and the spacecraft was presumed to have crashed on the moon.

Luna 23 toppled over upon landing. High-resolution images from the US Lunar Reconnaissance Orbiter published in 2012 pictured Luna 23 lying on its side.

An illustration of the Luna 16 spacecraft is shown on the next page (Fig. 12.6). The spacecraft consisted of a descent stage and an ascent stage. The ascent stage launched from the descent stage and traveled to earth carrying the lunar soil sample. The overall spacecraft stood about four meters (13.1 feet) high and weighed 5,750 kg (12,679 pounds).

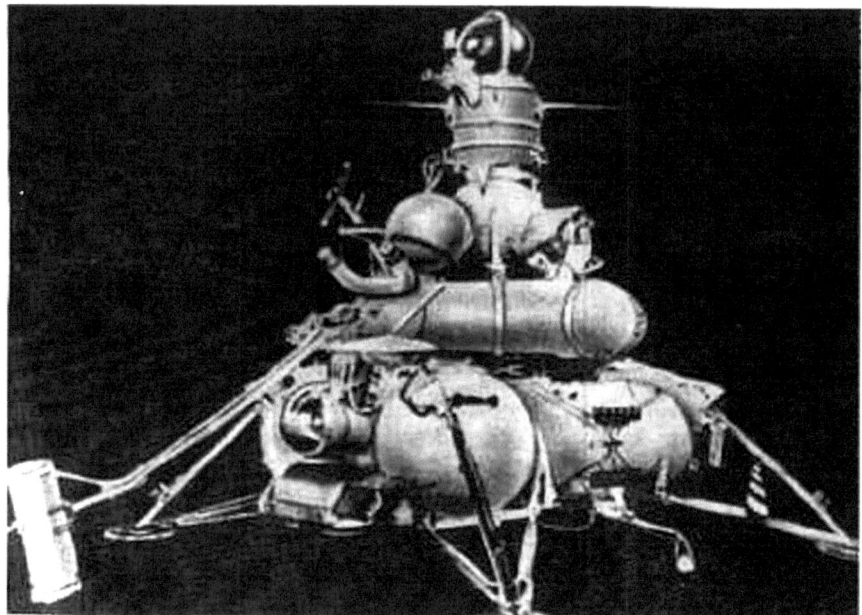

Fig. 12.6 Luna 16 spacecraft (graphic from NASA collection)

The ascent stage was made up of the structure located above the horizontal cylindrical object in the drawing. It included the spherical fuel tank shown in the drawing. The spherical structure at the top of the ascent stage was the earth entry capsule that held the lunar soil sample. The earth entry capsule was 50 cm (19.7 inches) in diameter and weighed about 39 kg (86 pounds). The total weight of the ascent stage was 520 kg (1,147 pounds). The ascent stage was propelled by a KRD-61 rocket engine that produced 1.9 tonnes (4,190 pounds) of thrust.

The descent stage held tanks for propellants and the landing legs were attached to the stage. The stage contained an 11D417 rocket engine that was used to decelerate the spacecraft out of lunar orbit and reduce its velocity during descent to the surface. The engine was throttleable over a range of 750 kg to 1929 kg (1,654 to 4,253 pounds) of thrust.

The landing stage also contained a radiation sensor, temperature sensor, landing radar, and communications equipment. The illustration of the spacecraft does not show four large cylindrical propellant tanks that were attached to the descent stage but jettisoned before landing. Propellants in those tanks had been used to establish orbit around the moon and modify the orbit.

After landing softly on the moon, a drill in the descent stage was used to penetrate the lunar surface to obtain a core sample. It is the appendage on the left side of the illustration. After gathering the core sample, the arm pivoted up and deposited the sample in the earth entry capsule. The capsule was then sealed in the vacuum of space to preserve the contents as it returned to earth.

A photograph of a model of Luna 16 displayed in the Moscow Museum of Cosmonautics is shown below.

Fig. 12.7 Model of Luna 16 spacecraft in Moscow Museum of Cosmonautics (Wikimedia posting by Bembmv)

Sample Return Operations on the Moon

Luna 16 was placed in a nearly circular orbit 111 km above the moon on 17 September 1970. Data on properties of the orbit was accumulated for several days to observe variations in lunar gravity. On 20 September the descent engine was fired to set up an elliptical orbit with perilune of about 15 km. The engine was fired again to begin a descent to the surface and continued to fire to slow the spacecraft for landing.

A landing radar provided altitude and altitude rate data to the control system during the descent. The descent engine was cut off at an altitude of 20 meters, and vernier engines controlled the spacecraft down to about 2 meters above the surface when they were cut off and the spacecraft free-fell to the surface. The velocity had been slowed to about 2.4 meters per second (7.9 feet per second) by the time the vernier engines were cut off.

Luna 16 landed in Mare Fecunditatis (the Sea of Fertility) on 20 September 1970. The landing was at 0.518° S, 56.364° E in the lunar coordinate system.

Shortly after landing, the spacecraft deployed the drill assembly onto the lunar surface. It had drilled down about 35 mm (1.4 inches) when the drill struck something hard and could drill no further. The drill assembly with the core sample was swung up to deposit the sample into the earth return capsule and the capsule was sealed.

The ascent stage launched from the descent stage toward earth on 21 September 1970. The entry capsule contained a heat shield for entry into the atmosphere. After the capsule was slowed by the heat shield, a parachute was deployed to further slow the capsule for a survivable landing.   The earth entry capsule returned to earth intact on 24 September. The core sample from the lunar surface weighed 105 grams. The major component of the core sample was found to be basalt.

Luna 20

The configuration of the Luna 20 spacecraft was the same as Luna 16. Luna 20 was launched from the Baikonur Cosmodrome on 14 February 1972 and made a soft landing in the Apollonius Highlands region of the moon on 21 February 1972. The landing was made on a plateau between two mountain peaks at 3.786° N, 56.624° E in the lunar coordinate system.

The drill was deployed soon after landing. A television camera monitored the drilling operation and transmitted pictures of the process to earth. The drill worked its way down to 25 cm (9.8 inches) below the surface. The core sample was deposited in the earth return capsule on 21 February.

The ascent stage launched from the moon on 22 February 1972, and the earth return capsule was returned to earth on 25 February. The core sample contained 30 grams of lunar soil. The major component of the soil was anorthosite. The sample contained the largest percentage of aluminum oxide and calcium oxide of all the lunar samples. It also contained a small amount of pure iron.

Luna 24

The Luna 24 spacecraft was similar to Luna 16, but it carried a new drill assembly with a new deployment arm. The new drill was capable of obtaining a core sample 2.5 meters deep into the lunar soil. The drill was capable of both rotary drilling operation and impact penetration. It preserved the layers as it drilled.

Luna 24 was launched from the Baikonur Cosmodrome on 9 August 1976 and made a soft landing on Mare Crisium on 18 August 1976. Coordinates of the landing were 12° N, 62.12° E in the lunar coordinate system. Mare Crisium is a nearly circular basin about 555 km in diameter filled with basalt. It is located northeast of Mare Tranquillitatis.

The drill was deployed on 18 August and it drilled 2.25 meters into the lunar surface. The core sample passed through an 8 mm diameter passage in the hollow bit and was transferred to a flexible tube 12 mm in diameter that preserved the layering. The tube was rolled up in a flat spiral fashion by a core holder. The core holder was transferred to the earth return capsule on the ascent stage after the drilling was complete.

The ascent stage was launched from the moon on 19 August. The earth return capsule returned to earth and was recovered on 22 August 1976. There was 170 grams of material in the core sample. The composition was largely different types of basalt.

The recovery of the capsule from Luna 24 in 1976 marked the end of the Russian early lunar exploration program. That exploration program scored many firsts and added to knowledge about the moon. Russia's efforts were somewhat overshadowed by the immensely successful U.S. Apollo program.

Chronologically, the Luna 24 mission took place four years after the previous sample return mission and after completion of the lunar rover program. The largely successful lunar rover program is covered next.

EXPLORING THE MOON WITH THE LUNOKHOD ROVERS

The Soviet Union (Russia) had ambitious plans for manned missions to the moon, and the Lunokhod rovers were planned and designed to examine selected sites in preparation for manned landings. The name lunokhod in Russian is translated as moonwalker.

The very successful Lunokhod rovers demonstrated that Russian space engineering was very good. The rovers were driven from earth and conducted sophisticated scientific experiments during their travel over the lunar surface.

A photograph of a model of a Lunokhod-1 rover on display at the Moscow Museum of Cosmonautics is shown on the next page (Fig. 12.8). The lid is open and the solar cells on the inside of the lid are visible. A view with the lid closed is shown on the following page (Fig. 12.9).

Development of the Lunokhod rover was begun in 1963 by the Experimental Design Bureau, OKB-1. The design of the rover itself was subcontracted to the VNII Transmash Company who had designed and built tanks for the Russian army in World War II. The project was transferred from OKB-1 to Design Bureau OKB Lavochkin in 1965 to allow OKB-1 to concentrate on the manned lunar landing program.

The Lunokhod rovers with their carrying bus were launched to the moon from the Baikonur Cosmodrome by Proton 8K82K/Block D launch vehicles. The first attempt at launch in February 1969 failed when Proton exploded. Luna 17, carrying Lunokhod-1, was launched successfully in November 1970, and Luna 21, carrying Lunokhod-2, was launched successfully in January 1973.

Fig. 12.8 Model of Lunokhod-1 rover in Moscow museum (Wikimedia posting by Petar Milošević)

The rovers were carried to the moon by a spacecraft referred to as the bus that made a soft landing on the moon. The bus was similar in construction and used components common to the bus that carried the sample return spacecraft to the moon. The bus carrying the rover was put in orbit around the moon and then decelerated out of orbit and controlled to a soft landing on the moon. A drawing of the spacecraft with Lunokhod-1 on top of the bus is shown on the following page (Fig. 12.10).

After landing, ramps were unfolded from the bus to allow the rover to be driven off onto the lunar surface. The ramps are in the folded up position in the drawing. Ramps were provided on both the front and back sides of the rover in case a rock or crater would impede exit on one side.

The Luna 17 spacecraft entered orbit around the moon on 15 November 1970. On 17 November the bus fired its rocket engine and descended to a soft landing in

Fig. 12.9 Lunokhod-1 rover with lid closed (NASA collection)

Mare Imbrium (Sea of Rains). The landing was at 38.17° N, 35.0° W in the lunar coordinate system.

The Luna 21 spacecraft entered orbit around the moon on 12 January 1973. On 15 January the spacecraft descended to a soft landing in the LeMonnier crater. The crater is ancient with a collapsed wall and flooded with lava extending in from Mare Serenitatis. The landing was at 25.999° N, 30.408° E in the lunar coordinate system.

DESCRIPTION OF LUNOKHOD-1

A major structure of Lunokhod-1 was the tub-shaped instrument compartment. The compartment had a round top that was 2.15 meters (7 feet) in diameter. The top of the compartment was sealed by an airtight cover that functioned as part of thermal control for the compartment. A moveable lid that could be opened and closed fit over the interior cover.

The structure of the instrument compartment was made from a strong, light-weight magnesium alloy. The height of Lunokhod-1 to the top of the instrument compartment was about 1.3 meters (51 inches). The weight of Lunokhod-1 was 756 kg (1667 pounds).

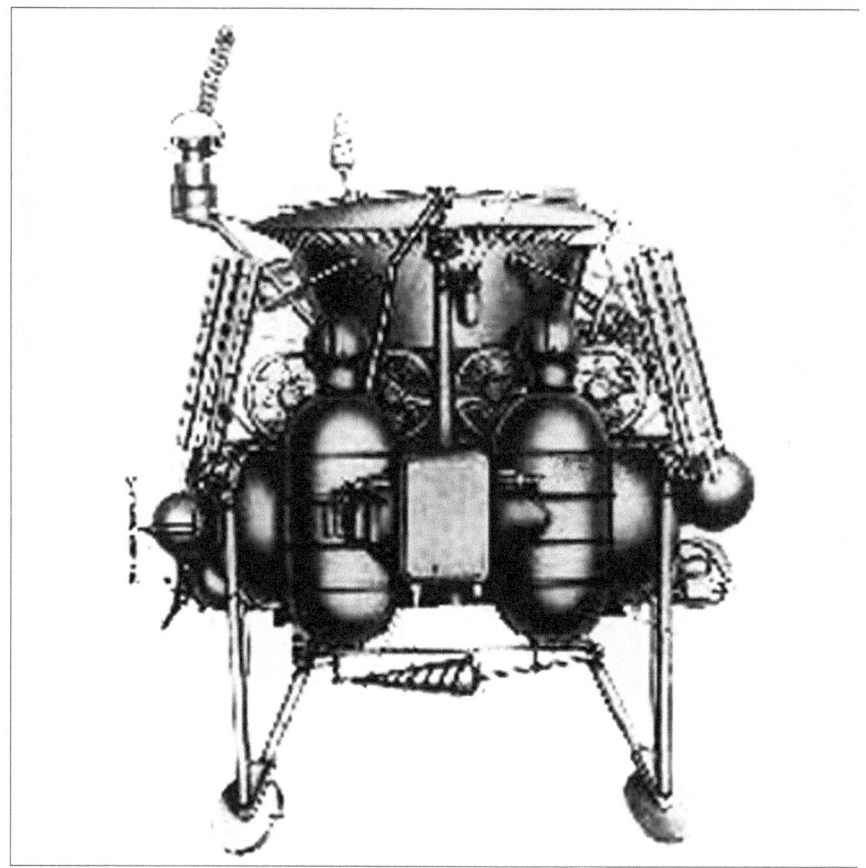

Fig. 12.10 Luna 17 spacecraft carrying Lunokhod-1 (NASA collection)

A moveable lid fit over the top of the cover of the compartment. The bottom side of the lid was filled with solar cells. When the lid was open and approximately directed toward the sun, ample electrical power was generated to keep the batteries charged. Battery power was used to operate and propel the rover. The lid could be positioned at selected angles up to 180 degrees from the closed position.

The lid was open during the lunar day (about 14 earth days) while the rovers moved about on the lunar surface and performed various measurements. At the end of the lunar day, the vehicle was parked so that the rising sun would illuminate the solar cells when the lid was opened. The lid was then closed for the very cold lunar night that lasted about 14 earth days. The equipment was kept warm at night by a polonium-210 radioisotope heat source in conjunction with a nitrogen gas circulation system.

The well-insulated instrument compartment contained a cooling system that operated during the lunar day and a warming system that operated during the lunar night. The cooling system made use of an efficient heat radiator that enclosed the top of the compartment. The special surface had a very low heat absorption coefficient from the sun's rays while serving as a good radiator.

The sealed instrument compartment was filled with nitrogen at earth's atmospheric pressure. A fan continuously circulated the nitrogen gas around the electronic components and across the radiator at the top of the compartment where the heat was radiated to space.

The lid was closed during the lunar night to minimize radiation. Ducts were opened to allow the circulating nitrogen gas to pass over a heater system deriving heat from a polonium-210 isotopic source.

The instrument compartment contained receivers and transmitters to communicate with earth and remote control equipment to allow driving the vehicle and managing of experiments from earth. It also contained two thermal control systems, electronic units for the experiments, batteries, and power converters.

The rover had eight wheels, four on each side. The wheels were 510 mm (20 inches) in diameter. They consisted of wire mesh formed around three rings supported from a hub by wire spokes. The wheels were 200 mm (7.9 inches) wide and they had cleats around the periphery to improve traction. The spacing between wheels (track) was 1.6 meters (63 inches). The distance from front of the front wheel to the back of the back wheel was 2.2 meters (86.6 inches). Torsion bars were used to provide independent suspension for the wheels.

Each wheel was independently powered and controlled. The hub for each wheel contained an electric motor, gear train, brake, temperature sensor, and a wheel revolutions counter. The vehicle was steered by varying the rotation speed of the wheels on the right and left side. The normal turn radius was about 3 meters (10 feet), but it could also turn in place by commanding the wheels on one side of the vehicle to move forward and the wheels on the other side to move backward. The vehicle had two speeds available for traveling: 0.8 km/h (0.5 mph) and 2.0 km/h (1.2 mph).

The Lunokhod rovers were driven and controlled remotely from the earth. The driver used inputs from television cameras and telemetry data to control the rovers. Two television cameras pointed forward were located in front of the instrument compartment. They can be identified in image of the rover with the cover closed by the lens caps that are hanging down. Two other television cameras were mounted on each side of the instrument compartment.

Other equipment on the Lunokhod rovers included a high-gain antenna with its positioning apparatus that was located forward and above the instrument compartment. It appears to be a helix-type antenna with narrow beamwidth and relatively high gain to give adequate signal-to-noise ratio on the ground for wide bandwidth television transmission. A conical antenna, mounted above and to the side of the instrument compartment, had a broad beamwidth with attendant low gain and handled lower bandwidth communication between Lunokhod and earth.

Experiments Carried by Lunokhod-1

The Lunokhod rovers served as pathfinders for planned manned lunar exploration. To that end, determining the characteristics of the lunar soil was very important. These characteristics included bearing strength and traction properties.

Soil penetration tests

Bearing strength was determined by means of a penetrometer that applied force to a conical tipped probe and measured penetration as a function of force applied. The probe was a shaft 5 cm (2 inches) in diameter with a conical tip with an apex angle of 60 degrees.

Some results of penetrometer testing by Lunokhd-1 for various terrain types have been published. The data were given as penetration in millimeter as a function of vertical load in kilogram. At a vertical load of 16 kg, for example, penetration varied from 30 mm to 80 mm at different sites. Lunokhod-1 made about 500 penetration tests.

A measure of slippage of the wheels of the rovers was determined by using a ninth wheel that was not powered and that freely turned on the lunar surface. The ratio of the number of turns of the ninth wheel to the driven wheels gave a measure of slippage of the powered wheels.

RIFMA spectrometer

The chemical composition of the soil was determined by means of a RIFMA spectrometer where RIFMA stands for Roentgen Isotropic Fluorescent Method of Analysis.

The spectrometer was contained in the box-like package that located between the front wheels in the photograph of the rover. It emitted high-energy X-rays toward the surface that ionized atoms in the regolith. Electrons in the outer shells of the atom then moved to the open spot in the next orbit down and so doing gave up energy that was accompanied by radiation. The discrete line spectrum of the secondary radiation identified a particular atom. Lunokhod-1 made chemical analysis in 25 different locations on the lunar surface.

X-ray telescope

A telescope sensitive to X-ray radiation was mounted such that it pointed to the local zenith when the rover was on level ground. The wavelengths of interests were about 1.0 to 6.0 angstroms (0.1 to 0.6 nanometers). The telescope had two channels, each with a field of view of 3.5 degrees. The telescope incorporated a simple X-ray photon counter for each channel. One channel had a filter that blocked radiation in the 1.0 to 6.0 angstrom band. The other channel was unfiltered. The channel with the filter yielded the background measurement. That count was subtracted from the count in the unfiltered channel to get the photon count in the band of interest.

Laser retroreflector

An array of laser retroreflectors was mounted on the front top of the vehicle. A property of those retroreflectors was to reflect laser energy back to the source from a wide range of angles. It was used by earth-based lasers to obtain precise measurements of the distance from the earth to a particular point on the moon.

DESCRIPTION OF LUNOKHOD-2

Lunokhod-2 was similar to Lunokhod-1 except for a few improvements and additions. The changes brought the weight of Lunokhod-2 up to 840 kg (1852 pounds). A photograph of a model of Lunokhod-2 on display in the Russia in Space exhibition at the airport in Frankfurt, Germany, is shown on the next page (Fig. 12.11).

A third television camera was added to Lunokhod-2 to aid navigation of the vehicle by operators on earth. The added camera was located in front of the vehicle at about eye level of a walking person. It is the device with the square aperture on the upper right side of the photograph.

Apparatus for a new experiment to measure sky brightness during the lunar day and night was added. This experiment used a two channel astrophotometer with one channel sensitive to visible light and the other sensitive to ultraviolet light. The visible light channel had a field of view of 12.5 degrees, and the ultraviolet channel had a field of view of 17.4 degrees. The instruments were pointed toward the local zenith when the vehicle was on level terrain. A major finding was that in visible light the sky was about a factor of 20 brighter than expected during the day. It was deduced from those findings that there is a swarm of dust particles around the moon.

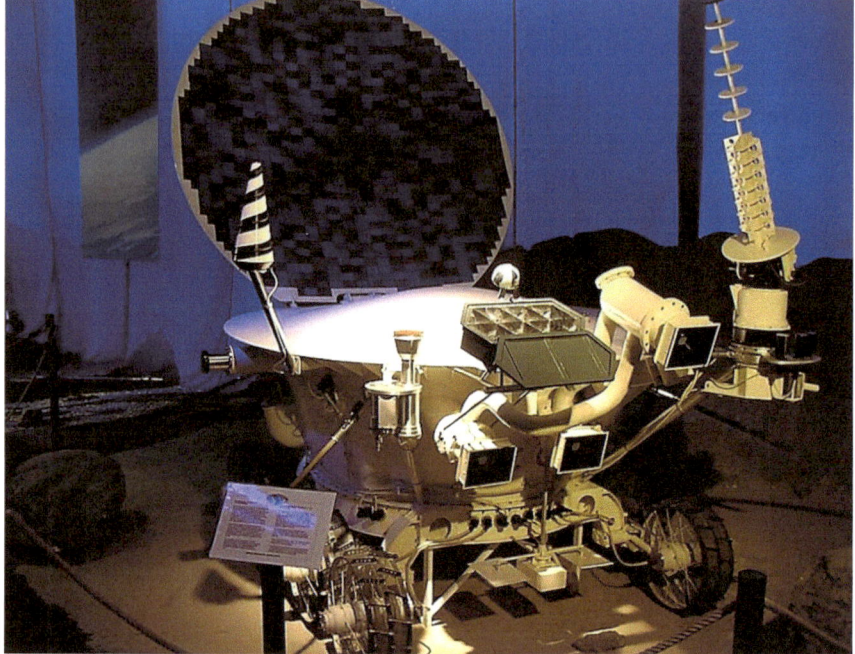

Fig. 12.11 Model of Lunokhod-2 at Frankfurt airport (Wikipedia posting by de:Benutzer:HPH)

A magnetometer was added to the experiments for Lunokhod-2. It was fastened to an extendable boom at the front of the instrument capsule. Moving it out on a long boom minimized magnetic interference from the rover. The boom is not attached to the model in the photograph.

Other experiments were essentially the same as carried by Lunokhod-1.

DRIVING THE LUNOKHOD ROVERS

The Lunokhod rovers were controlled remotely from the earth. The ground crew included five people: crew chief, driver, navigator, antenna operator, and engineer. The driver used inputs from television screens, telemetry data, and inputs from other crew members to control the rover. The crew chief directed the driver on which of the two forward speeds to use and when to stop and start. He also directed the course to be followed. The navigator kept navigation data on the vehicle up-to-date and recommended the course to be followed and course changes.

Television cameras on the vehicle provided situational awareness to the ground operators. Lunokhod-1 had two television cameras in front of the vehicle that provided stereoscopic images in the direction of travel. The frame rate of those cameras varied between 3 and 20 seconds per frame depending on vehicle speed and type of terrain being traversed. A third camera, raised to about eye level with a walking person, was added to the front of the vehicle for Lunokhod-2.

In addition, there were two television cameras on each side of the vehicle. One camera on each side was mechanically scanned to provide a panorama of 180 degrees in the horizontal plane and 30 degrees in the vertical plane. The other camera was mounted to allow it to scan in the vertical plane. One application of the cameras was to image the earth and sun simultaneously and that allowed operators on the ground to determine the heading of the vehicle. A directional gyro in the vehicle kept track of vehicle heading between updates.

In addition to visual cues transmitted to the earth, telemetered data included data from experiments, attitude angles and heading of the vehicle, RPM and temperature of each wheel, and electrical current to the drive motor for each wheel.

SUMMARY OF THE ROVER MISSIONS ON THE MOON

Lunkhod-1 operated through 11 lunar day/night cycles (322 earth days). It traveled a total of 10.5 km (6.5 miles) while driving on each of its 11 lunar days. It returned about 20,000 television pictures and 206 panoramic photographs. It conducted 500 soil penetration tests and 25 chemical analysis of lunar soil.

Lunkhod-2 operated through 4 lunar day/night cycles (about 113 earth days) and traveled 37 km (23 miles) while driving during its 5 lunar days. Its operation was cut short on the fifth lunar day when it brushed the side of a crater and the solar cells became covered with soil. The soil fell on the heat radiators when the cover was closed. That caused the machine to overheat and it quit working. Lunkhod-2 returned about 80,000 television pictures and made 86 panoramic photographs. It also made hundreds of soil penetration tests and chemical analysis of the lunar soil.

The Lunokhod rovers left distinct tracks in the lunar soil that were photographed by the U.S. Lunar Reconnaissance Orbiter spacecraft. Analysis of the photographs taken by the Lunar Reconnaissance Orbiter Camera (LROC) indicated that the actual length of travel of Lunokhod-2 was 39.16 km rather than 37 km as reported.

A Russian scientist, Rusian Kuzmin, who had participated in the Lunokhod rover program as a young planetologist, paid a visit to the LROC Science Operations Center in 2010. He had analyzed television photographs from the rovers at the time of their traverses of the moon. During his visit with U.S. scientists, he viewed pictures taken by the Lunar Reconnaissance Orbiter of the Lunokhod rovers and their distinct tracks on the moon. He wrote a nostalgic "thank you" note to the staff and enclosed a picture taken by Lunokhod-2 of the Fossa Recta area of the moon. That picture is reproduced below.

Fig. 12.12 Picture of Fossa Recta area of the moon taken by Lunokhod-2 (NASA/GSFC/ASU photograph)

Bibliography

Elshafie, Ahmed, *Subsurface Planetary Investigation Techniques and their role for Assessing Subsurface Planetary Composition*, University of Arkansas dissertation, 2012

Harvey, Brian, *Soviet and Russian Lunar Exploration*, Praxis Publishing, Chichester, UK, 2007

Kassel, Simon, *Lunokhod-1 Soviet Lunar Surface Vehicle*, ARPA report R-802, Rand Corporation, September 1971

NASA-NSSDCA-Spacecraft-Details, *Luna 17/Lunokhod 1*, NSSDCA/COSPAR ID: 1970-095A

NASA-NSSDCA-Spacecraft-Details, *Luna 21/Lunokhod 2*, NSSDCA/COSPAR ID: 1973-001A

Severny, A. B., Terez, E. I., and Zvereva, A. M. *The Measurements of Sky Brightness on Lunokhod-2*, Paper presented at Conference on Interactions of the Interplanetary Plasma with the Modern and Ancient Moon, September 1974.

13

Russian Manned Circumlunar Flight

The Soviet Union (Russia) conducted six different unmanned lunar exploration programs between 1958 and 1973. Each of those programs had successful missions along with several failures. Those unmanned explorations of the moon have been described in the previous two chapters of this book.

Soviet leadership under Nikita Khrushchev was reluctant to commit the considerable resources necessary for a manned landing on the moon. That goal had been stated as early as 1961 in the United States by President Kennedy. As late as the fall of 1963, the Soviet Union had only committed to a manned circumlunar flight without orbiting or landing.

Progress of the Apollo program in the United States was open knowledge, and prestige to be gained by placing the first man on the moon was seductive. Soviet leadership had a change in heart and decided to enter the race to land a man on the moon in August 1964. Pressure to enter the race had also come from Russian engineers and scientists.

There was considerable rivalry between the Soviet design bureaus, particularly OKB-1, headed by Sergei Korolev; OKB-52, headed by Vladimir Chelomey; OKB-586 headed by Mikhail Yangel; and OKB-456, headed by Valentin Glushko. All vied to be dominant in the manned space programs. Relations between Korolev and Glushko were particularly bitter.

The design bureaus OKB-1, OKB-52, and OKB-586 all individually submitted proposals for manned missions to the moon involving both a lunar flyby and a lunar landing. The Central Committee of the Communist Party and the Council of Ministers acted on the proposals and set forth a 5-year plan for space programs on 3 August 1964. Key points of the directive regarding manned lunar flights were reported by Asif Siddiqi in *Spaceflight*, May 2004, as summarized below.

© Springer Nature Switzerland AG 2018
T. Lund, *Early Exploration of the Moon*, Springer Praxis Books,
https://doi.org/10.1007/978-3-030-02071-2_13

Manned circumlunar flight – This would be managed and hardware developed by OKB-52, headed by Vladimir Chelomei. The UR-500 launcher would be used. The first flyby would take place in 1965.

Manned lunar landing – This would be managed and hardware developed by OKB-1, headed by Sergei Korolev. The N1 launcher would be used. The first manned landing would take place in 1967.

Awarding the lunar flyby mission to Chelomey involved politics at the highest level.

The N1 heavy-lift launch vehicle had been pushed by Korolev in OKB-1 for several years, but funding was not forthcoming and progress was slow. The decision by Soviet leadership to pursue a landing on the moon elevated development of the N1 to a high priority.

The Soviet decision to approach the moon in two different paths addressed the rivalry between leaders of the OKB design bureaus, but it diluted the effort. The United States concentrated on the manned landing and treated orbiting the moon as incidental.

In hindsight, the Soviet Union had little chance to beat the United States in the race to land a man on the moon. The commitment to land a man on the moon was not made by Soviet leadership until August 1964. In contrast, by the end of 1962, essential details of the Apollo mission had been worked out, and major contractors to develop Apollo spacecraft hardware had been selected. Further, the United States had a strong central agency for space activity, NASA, which was lacking in the Soviet Union.

The Soviet design bureaus were reorganized and renamed in 1966. OKB-1 became TsKBEM and OKB-52 became TsKMB. To avoid confusion, the original names will be used in this chapter.

FIRST SOVIET MANNED SPACEFLIGHT

We will digress a moment to give credit to early Soviet manned space flight. Twenty cosmonauts were selected in February 1959 to participate in manned Soviet space programs, and a cosmonaut training center was established in 1960.

To compete with the US space program, a major effort was made to place the first cosmonaut in orbit around the earth before the launch of the first US manned Mercury suborbital flight. The first Soviet manned flight was held in abeyance until two consecutive successful flights with dogs had been achieved. Finally, on 12 April 1961 a Luna 8K72 launch vehicle lofted a Vostok spacecraft with Lieutenant Yuri Gagarin aboard into earth orbit.

After Gagarin's planned one orbit of earth, the retrorocket engine was fired and the Vostok capsule reentered the earth's atmosphere. At an altitude of 7,000 meters

(22,970 feet), a hatch was jettisoned and an ejection seat with Gagarin wearing a space suit was ejected from the capsule. The ejection seat had descended on parachute to about 4,000 meters (13,130 feet) when Gagarin separated from the seat and descended to earth on his personal parachute. This landing approach seemed a little more sporting than the US approach of landing the capsule with the astronauts inside, but the US approach required a water landing.

Gagarin received worldwide acclaim for his pioneering flight. The first US Mercury suborbital flight with Alan Shepard aboard took place shortly afterward on 5 May 1961. John Glenn became the first US astronaut to orbit the earth in a Mercury capsule in February 1962.

CIRCUMLUNAR SPACECRAFT

The Soviet Union planned a series of manned circumlunar flights to the moon as a prelude to manned landings. The trajectory of these flights would launch from the earth, swing around the moon, and return to the vicinity of earth. The mission called only for a lunar flyby without going into orbit around the moon.

Soviet leadership assigned the circumlunar mission to Vladimir Chelomey, head of OKB-52, in August 1964. The UR-500, which OKB-52 had been developing to launch military payloads into orbit, would be the launch vehicle. The UR-500 with upper stages was later known as Proton-K. OKB-52 had begun design of a manned spacecraft called the LK-1 for circumlunar flight in 1963.

Khrushchev was deposed in 1964, and new Soviet leadership reevaluated the space programs. Sergei Korolev lobbied hard to use a modification of the 7K-OK spacecraft for the circumlunar flights instead of letting Chelomey develop his LK-1 spacecraft.

The 7K-OK spacecraft was an extension of the Soyuz spacecraft being developed for earth orbit testing of rendezvous operations and space walks. It was also to be the basis of the spacecraft for the manned landing program. The new Soviet leadership sided with Korolev, and development of the LK-1 spacecraft by OKB-52 was canceled. The circumlunar spacecraft derived from the 7K-OK by OKB-1 would be named the L1.

Korolev also pressed hard for control of the circumlunar program, citing technical and cost advantages in having one design bureau be responsible for both lunar programs. Once again, Korolev prevailed, and the circumlunar program was transferred from OKB-52 to OKB-1 in October 1965.

Korolev decided to use Chelomey's Proton launch vehicle for the circumlunar program with the addition of a Block D upper stage. The Block D was being developed by OKB-1 for the manned landing program.

Unfortunately, the new Proton-K launch vehicle had ongoing problems that resulted in several launch failures. Ten unmanned circumlunar flights were attempted. Two were fully successful and one was nearly successful. Of the seven unsuccessful flights, five involved problems with the Proton.

Leaders of the space program had decreed that four successful unmanned flights must be flown before circumlunar flights with cosmonauts could proceed. This bar was not reached, and placing cosmonauts in circumlunar flight was never attempted.

Description of L1 Spacecraft

The L1 spacecraft that would fly the circumlunar missions was based on the Soyuz spacecraft that first carried cosmonauts in earth orbit in November 1966. A sketch of the L1 spacecraft as given in NASA RP 1357 is shown below (Fig. 13.1).

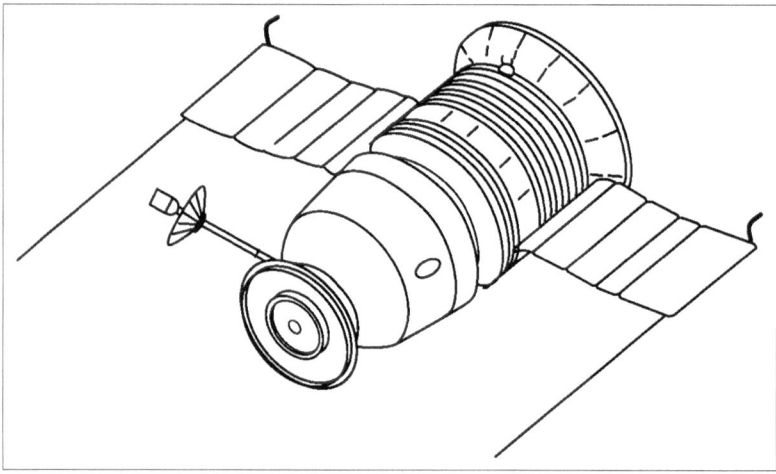

Fig. 13.1 Sketch of L1 spacecraft (NASA drawing)

The cabin in the forward portion of the L1 was designed for two cosmonauts. The appendages to the sides in the drawing are solar panels. The drawing shows them slightly bent when deployed. Sketches of later models of the L1 show straight solar panels. A high-gain communications antenna protruded from the side. The disk-shaped appendage at the front of the module was an adapter for attachment of the launch escape system. It was jettisoned before translunar injection.

The length of the spacecraft was about 4.9 meters (16 feet), and the diameter in the area of the solar panels was about 2.7 meters (8.9 feet). The weight of the spacecraft varied depending on configuration. The weight of the most successful L1 spacecraft, designated Zond 7, was 5,979 kg (13,184 pounds).

The forward portion of the spacecraft was referred to as the Descent Module. It contained the crew compartment and a heat shield for protection when reentering the earth's atmosphere. The habitable volume of the crew compartment was about 3.5 cubic meters (124 cubic feet). The Descent Module separated from the Instrumentation and Service Module before reentry into the earth's atmosphere at the end of the mission.

The aft portion of the spacecraft, including the solar panels, was referred to as the Instrumentation and Service Module. It contained the main rocket engine and propellants along with electrical, communications, and life support systems. It also had small thrusters for maneuvering the spacecraft.

The main rocket engine was a KTDU-53 type that developed 425 kg (937 pounds) of thrust. Fuel for the engine was UDMH and the oxidizer was nitric acid. The combination ignited upon contact. Sufficient propellants were provided for a burn time of 270 seconds.

A Soviet Union photograph of the L1 in the process of being assembled to the launch vehicle is shown below. The L1 spacecraft is the forward object to the right in the picture. The silver-looking object with the fairing is the Block D stage, and the white structure behind the Block D stage is part of the launch vehicle (Fig. 13.2).

Fig. 13.2 L1 spacecraft being assembled to launch vehicle (Wikipedia posting by RKA of Soviet Union photograph)

The guidance system of the L1 included a digital computer, a three-axis gyro platform, and a series of optical sensors. The optical sensors consisted of two star trackers, an earth sensor, and a sun sensor. The sun and earth trackers were sufficient for alignment for midcourse correction, but the star tracker was necessary to align the gyro platform to allow accurate alignment of the spacecraft for reentry.

The L1 was the first Soviet spacecraft to use a digital computer. The computer was referred to as Argon-11S. It was developed by the Scientific Research Institute of Electronic Machines (NIEM). It was a fixed point machine with word length of 14 bits. The command word length was 17 bits. The random-access memory (RAM) capacity was 128 14-bit words, and the read-only memory (ROM) capacity was 4,096 17-bit words. Addition time was 30 μsec and multiplication time was 160 μsec. The dimensions of the computer were $305 \times 305 \times 550$ mm ($12 \times 12 \times 21.7$ inches). It weighed 34 kg (75 pounds) and consumed 75 watts of power.

Photography of the lunar surface was an important task for the L1 spacecraft. The camera on Zond 7 used a 300 mm objective lens and photographs were made on 5.6×5.6 cm size film. Both panchromatic and color films were used. Panchromatic film is good-quality black-and-white film. Zond 8 used a camera with a 400 mm lens, and photographs were made on 13×18 cm panchromatic film. The Zond 8 photographs are among the best taken of the moon.

Other instrumentation on the L1 spacecraft included ion traps to measure solar wind and sensors to measure micrometeoroid flux, cosmic rays, and magnetic fields.

L1 Hardware Built and Flown

There were 3 prototype test models and 12 flight models of the L1 spacecraft built. The first prototype, 1P, was used for ground testing. The next two prototypes, 2P and 3P, were used for testing in earth orbit, primarily to verify engine operation of the Block D stage.

There were nine launches of flight models of the L1 spacecraft between March 1967 and October 1970 that were intended to circumnavigate the moon. The launch vehicle for these flights was the Proton-K with a Block D fourth stage. A summary of results of the nine flights is given in Table 13.1.

In 1969 Soviet leadership began questioning the purpose of continuing the manned circumlunar program. Apollo 8 had placed three astronauts in orbit around the moon in December 1968, and Apollo 11 landed two astronauts on the moon in July 1969. A manned circumlunar flight would pale in light of US accomplishments.

Following the successful unmanned circumlunar flight of Zond 7 in August 1969, Soviet leadership canceled further work on a manned circumlunar flight. They allowed one more unmanned flight. The successful flight of Zond 8 was the last circumlunar mission. Zond 8 returned some of the best pictures ever taken of the moon.

Table 13.1 Fights of L1 circumlunar spacecraft

Launch date	Name	Results
September 1967	–	One of six engines of first stage of Proton failed. Vehicle was destroyed
November 1967	–	Second stage failed, and the spacecraft did not achieve earth orbit
March 1968	Zond 4	Successful test flight in direction away from moon. Entered atmosphere on ballistic trajectory and was destroyed
April 1968	–	Second-stage engine shut down too soon, and spacecraft did not achieve earth orbit
September 1968	Zond 5	Successfully circumnavigated the moon and photographed back side. Capsule made a high-G ballistic entry before being recovered at sea
November 1968	Zond 6	Circumnavigated the moon, but parachute was jettisoned during recovery, and return capsule crashed
January 1969	–	Third stage of Proton-K stopped firing too soon and so did not attain earth orbit
August 1969	Zond 7	Successfully circumnavigated the moon, photographed back side, and was recovered on earth
October 1970	Zond 8	Successfully circumnavigated the moon, photographed back side, and was recovered on earth

Bibliography

Hardesty, Von and Eisman, Gene, *Epic Rivalry – The Inside Story of the Soviet and American Space Race,* National Geographic, Washington, DC, 2007

Harvey, Brian, *Soviet and Russian Lunar Exploration*, Praxis Publishing, Chichester, UK

Johnson, Nicholas, L., *The Soviet Reach for the Moon*, Cosmos Books, River Vale, NJ, 1995

Siddiqi, Asif, A., *A Secrete Uncovered*, Spaceflight, Vol 46 May 2004

Siddiqi, Asif, A., *Challenge to Apollo: The Soviet Union and the Space Race*, NASA History Series, NASA SP-2000-4408, published by aIc Books

Zak, Anatoly, *7K-L1: Soyuz for Circumlunar Mission*, russianspaceweb.com

14

Russian Manned Lunar Landing Endeavors

Spacecraft development within the Soviet Union was largely a Russian endeavor. Russia, the major entity within the Soviet Union, was the country around which the Soviet Union was formed in 1921, and it picked up the pieces after the Soviet Union dissolved in 1991.

A manned lunar landing program was pushed strongly by engineers and scientists. Finally, Soviet leaders realized the potential benefits of a manned lunar landing program and set priorities and awarded funding. A manned landing on the moon was a difficult endeavor, however, as demonstrated by the extraordinary expenditure of technical talent and treasure by the United States to make Apollo a success. Hardware for the Soviet manned program did not rise to the task.

There was considerable rivalry between the Soviet Union and the United States at the time and that rivalry extended to space accomplishments and exploration of the moon. Upper management of the Soviet Union space program was in some disarray during the "space race" with the United States. The political intrigue and troubled management of the Soviet space program would not come to light outside of the close-knit space program until years later.

As with most ambitious technical projects, there were failures as well as successes in the Soviet programs. In the end, with the failure of four launches in a row of their colossal N1 booster that they planned to use to place a cosmonaut on the moon, the Soviet Union withdrew from the manned lunar exploration space race. The overwhelming success of the U.S. Apollo program likely contributed to their decision to withdraw.

The Soviet Union was looked on as a formidable competitor in the race to place a man on the moon during the development of Apollo hardware. The secrecy of their program heightened the speculation. The author recalls philosophical discussions over dinner with my counterpart at Grumman after a day wrestling with

© Springer Nature Switzerland AG 2018

T. Lund, *Early Exploration of the Moon*, Springer Praxis Books,
https://doi.org/10.1007/978-3-030-02071-2_14

Apollo technical issues. I distinctly recall us coming to an agreement that the United States was going to win the race, not necessarily because of better hardware, but because ours was an open society with technical details, successes, and failures openly discussed throughout the country.

RUSSIAN MANNED LUNAR LANDING SPACECRAFT

Experimental Design Bureau OKB-1, headed by Sergei Korolev, received direction from the Central Committee of the Communist Party of the Soviet Union in August 1964 to develop spacecraft systems to land a man on the moon. The program would use the N1 heavy-lift launch vehicle then in development by OKB-1.

Like the United States, scientists in the Soviet Union considered several approaches for a manned lunar landing. The approach, favored up til 1964, would use two or more launches from earth to carry portions of the spacecraft to earth orbit. The final spacecraft would be assembled in earth orbit and dispatched to the moon. Initial design of the N1 had assumed this multiple launch approach.

The final configuration that evolved in 1965 would use a single heavy-lift launcher to send two joined spacecraft plus a Block D space tug to the moon. One of the spacecraft would be a small lunar lander crewed by one person. The other larger spacecraft would provide habitat for a two-man crew, and it would remain in lunar orbit. The lander would detach from the larger spacecraft and land on the moon. After exploration of the moon by the cosmonaut, the landing spacecraft would launch from the moon and rendezvous and dock with the orbiting spacecraft. The principles were similar to those chosen by the Apollo program three years earlier.

Development of the N1 launch vehicle was the most difficult task to be accomplished on the manned lunar landing program. The capability of the N1 would directly influence the design of the orbiter and lander. In the end, it was the unreliability of the very large and complex N1 launcher that doomed the Soviet manned moon landing program.

Sergei Korolev's N1 launch vehicle would have to be pushed to the extreme to loft the lunar orbiter/lander spacecraft into earth orbit. The final design of the first stage delivered about 4,620 metric tons (10.19 million pounds) of thrust. The N1 had the highest liftoff thrust of any booster at that time.

L3 LUNAR LANDING SPACECRAFT ASSEMBLY

The overall lunar landing spacecraft assembly was referred to as L3. It consisted of a Block G stage with 41 metric tons of thrust, a Block D stage with a restartable engine with 8.5 metric tons of thrust, the lunar orbiter spacecraft, and the lunar landing spacecraft.

The lunar orbiting spacecraft was called the Lunniy Orbitalny Korabl (LOK). The English translation is Lunar Orbital Ship. The LOK performed a function similar to that of the Command Service Module of Apollo. The LOK was a design adaption of a Soyuz 7 K spacecraft, and portions of it were similar to the L1 spacecraft used for circumlunar flight.

The lunar landing spacecraft was called the Lunniy Korabl (LK). The English translation is Lunar Ship. The LK performed a function similar to the Lunar Module of Apollo although it carried only one cosmonaut.

The Block G stage was essentially the fourth stage of the N1 launch vehicle. It was fired while in earth orbit to place the spacecraft assembly on a translunar trajectory. It was jettisoned after the translunar trajectory had been achieved.

The assembly that would be sent to the moon included the Block D stage, the Lunar Orbiter Ship (LOK), and the Lunar Ship (LK). The Block D stage performed course correction burns, translunar insertion burn, burns to correct the lunar orbit, and the initial portion of the burn for powered descent of the LK from lunar orbit.

LOK Lunar Orbiter Spacecraft

The LOK lunar orbiter spacecraft was an adaption of the successful Soviet Soyuz manned spacecraft. A drawing of the LOK spacecraft is shown below (Fig. 14.1).

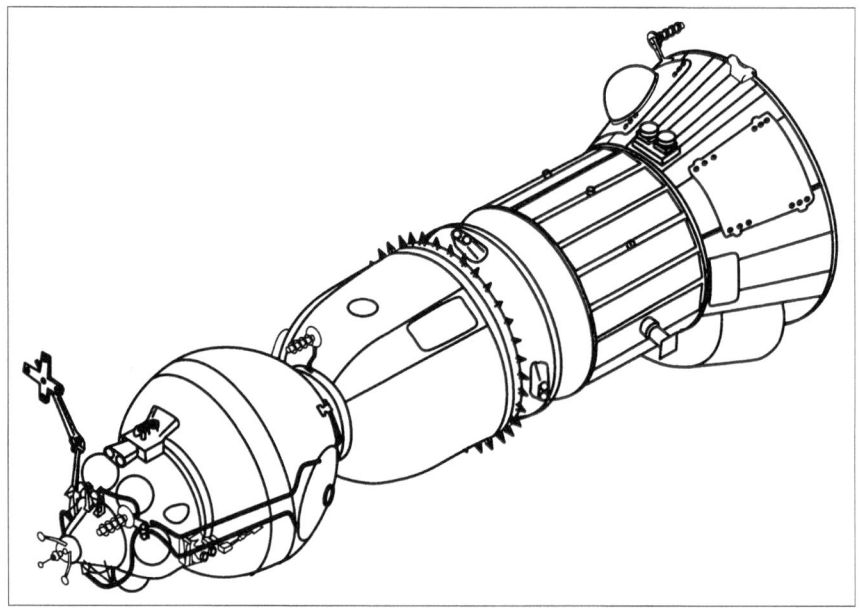

Fig. 14.1 LOK lunar orbiter spacecraft (NASA drawing)

The spacecraft was 10 meters (32.8 feet) long and the principal diameter was 2.9 meters (9.5 feet). The LOK weighed 9,850 kg (21,719 pounds) at launch.

The spherical-shaped portion at the front of the drawing was the Habitat Module for the two cosmonauts. The Habitat Module was composed of a front hemisphere with radius of 1.09 meters (3.6 feet), a transition 236 mm (0.77 feet), and a rear hemisphere with a radius of 1.14 meters (3.7 feet). The structure behind the Habitat Module contained the Descent Module for the crew that included a heat shield. The Descent Module would be detached from the spacecraft and used to reenter the atmosphere upon return to earth.

There was a hatch for passage of the cosmonauts from the Habitat Module to the Descent Module. The Habitat Module contained another hatch in the side of the module that opened to space to allow a cosmonaut to spacewalk to and from the LK lunar landing spacecraft. The LK lunar lander with the Block D stage attached was mounted to the flared portion of the LOK at the right side of the drawing above.

A control panel and manual controls in the forward portion of the Habitat Module allowed the cosmonaut to control the spacecraft as it rendezvoused and docked with the LK on the lander's return from the lunar surface.

A spacecraft orientation assembly was mounted at the front of the Habitat Module. It contained four sets of thrusters and six small spherical tanks holding N_2O_4 and UDMH propellants. These propellants ignited upon contact in the thrusters. The orientation assembly performed attitude control of the spacecraft and fine control for docking with the LK.

The rendezvous system used a radio frequency-based system to guide to LK to rendezvous with the LOK spacecraft. A laser-based optical system was used by the cosmonauts for close-in guidance to manually position the LOK in position to dock with the LK. There is very little information about the radio rendezvous system or close-in laser system in the literature. Swedish space researcher Sven Grahn has identified several antennas on the LOK and antennas on the LK that may be part of a radio frequency rendezvous system.

The docking system, referred to as Kontakt, was a relatively simple system that did not require precise alignment of the two spacecraft for docking as required by Apollo. The Kontakt system included a flat circular honeycomb structure about one meter in diameter on the top of the LK lander. The honeycomb structure consisted of 108 hexagon openings. Upon return of the LK lander to the LOK lunar orbiter, a set of three probes on the front of the LOK was inserted into any of the hexagon openings and then the three probes were displaced outward to lock the LK to the LOK. The cosmonaut then opened the hatch of the LK and spacewalked back to the Habitat Module of the LOK.

The guidance system for the LOK was developed by NII AP (Scientific research Institute of Automatics and Instrument Making) led by Nikolai Pilyugin. Because of development time constraints, first models of the LOK and the LK used an analog guidance system. Later versions used a digital system based on an S-530 digital computer developed by NII AP. The gyro platform was an existing design by NII-994.

The LOK had two sets of rocket engines. One engine, with two combustion chambers and two nozzles with total thrust of 417 kg (920 pounds), was used for lunar orbit maneuvers and for course correction during the return to earth. A larger engine, with two combustion chambers and two nozzles providing a total thrust of 3,388 kg (7,470 pounds), was used at the end of the lunar orbiting portion of the mission to initiate the trans-Earth trajectory.

Propellants for the engines were the hypergolic combination of unsymmetrical dimethyl hydrazine (UMDH) and nitrogen tetroxide (N_2O_4). The propellants were held in a spherical tank divided into two chambers. A total of 3,152 kg of propellants were carried.

Electrical power for the LOK spacecraft was provided by hydrogen-oxygen fuel cells supplemented by batteries. Water generated by the fuel cells was used by the crew. There were four fuel cells, and each could provide 1.5 kW of power at 27 volts.

A total of seven LOK orbiter spacecraft were built. One was launched as payload of the fourth N1 launch vehicle in November 1972. The payload included a mockup of the landing vehicle, LK. The N1 exploded shortly after launch. None of the other LOK spacecraft were flown.

LK Lunar Lander

A photograph of an engineering model of the LK lunar lander that is displayed in the Science Museum, London is shown on the next page (Fig. 14.2). The model shown may have been painted a gold color for display purposes. Most of the landers in other museums are natural aluminum color.

A drawing of the LK lander from NASA report RP 1357 is shown on the following page (Fig. 14.3). The drawing is of the right side of the lander as shown in the photograph. The sloped structure at the left side of the drawing is the recessed window area in the photograph.

The height of the lunar lander was 5.2 meters (17 feet), and the weight including rocket motor and fuel was 5,560 kg (12,269 pounds). For reference, the fully fueled Apollo Lunar Module weighed 33,200 pounds and stood 23-feet high.

The LK lunar lander was composed of the Lunar Landing Aggregate (Lunnyi Posadochnyi Agegat) (LPA) and the Lunar Ascent Vehicle (Lunnyi Vzletnyi Apparat) (LVA). To shorten the names, the composition of the lander will be referred to here in the same terms used to describe the Apollo Lunar Module. The Lunar Landing Aggregate will be referred to as the descent stage, and the Lunar Ascent Vehicle will be referred to as the ascent stage.

Fig. 14.2 Model of LK Lunar Lander in London Science Museum (Wikipedia posting by Andrew Gray)

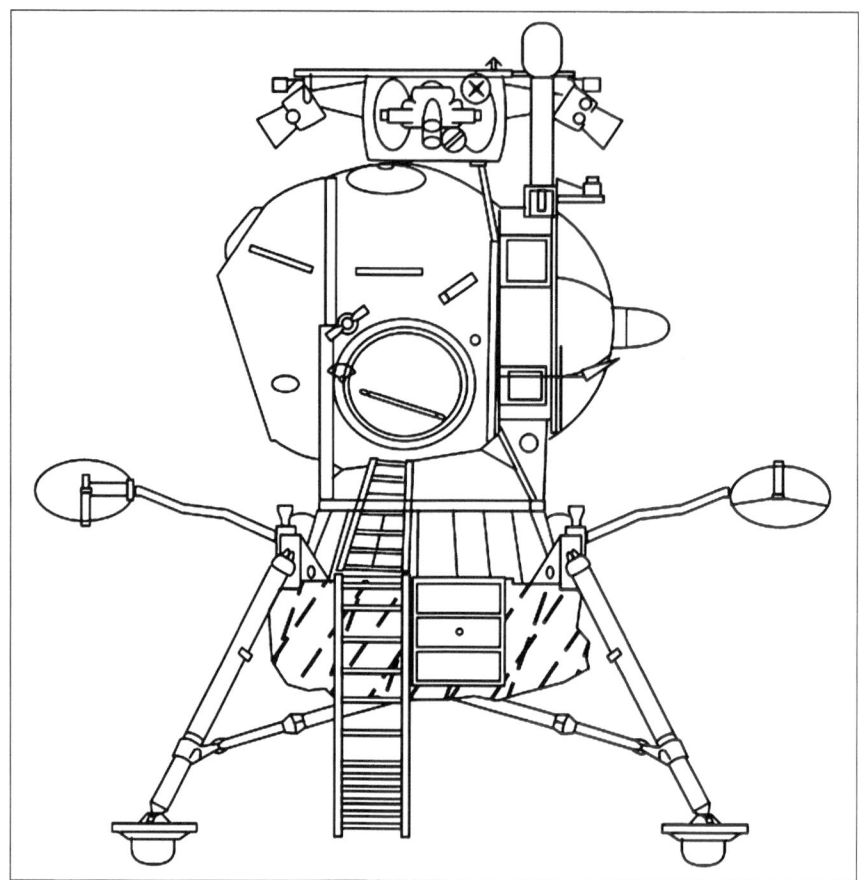

Fig. 14.3 Drawing of LK lunar lander (NASA drawing)

The ascent stage contained a pressurized crew capsule that was semispherical in shape with extent of about 2.3 meters (7.5 feet) by 3.0 meters (9.8 feet). The crew capsule is the dominant hemispherical looking structure in the photograph. The habitable volume of the curved crew capsule was about 4 cubic meters (140 cubic feet). The crew compartment was joined to a cylindrical instrument compartment.

The breathing atmosphere in the crew capsule was standard earth atmosphere of nitrogen and oxygen mix. The pressure was maintained at 0.74 atmospheres. The spacecraft designers were concerned with a fire hazard of using pure oxygen as used in the Apollo LM.

An attitude control assembly was mounted to the top of the crew compartment. That assembly had two 40 kg thrusters to control pitch, two 40 kg thrusters to control yaw, and four 10 kg thrusters to control roll. The thrusters were commanded by the guidance and control system. The circular metal plate that can be seen at the top of the attitude control assembly was part of the docking mechanism.

The landing stage contained four deployable legs and a crushable honeycomb structure that held the ascent stage. The landing legs had a maximum span of 5.4 meters (17.7 feet). Unlike the Apollo descent stage, the landing stage did not contain an engine. The rocket engine was mounted to the ascent stage, and it fired through the descent stage to slow the spacecraft for landing. It also provided thrust for ascent from the lunar surface.

The engine assembly, referred to as Block E, included a type 11D411 main engine with a single combustion chamber and a type 11D412 backup engine with two combustion chambers. The nozzle of the main engine was in the center of the spacecraft, and the two nozzles of the backup engine were located on each side of the main engine nozzle. Four vernier engines were located around the main and backup engine nozzles.

The main engine had a maximum thrust of 2,050 kg (4,520 pounds). The backup engine with two combustion chambers had a total thrust of 2,045 kg (4,510 pounds). The main engine was throttleable over a 525 to 1125 kg (1,819 to 2,480 pound) range. The engines burned the hypergolic combination of unsymmetrical dimethyl hydrazine (UMDH) and nitrogen tetroxide (N_2O_4).

Both engines were used during landing and during liftoff. If both engines were operating normally after liftoff, the backup engine was shut down. The throttleable engine was one of the more challenging items to develop. Flight testing of the three prototype models of the lander in earth orbit focused on engine performance. The Apollo program also found the throttleable descent engine in the Lunar Module to be very challenging.

The landing stage held two high-gain parabolic antennas for communications and transmission of television data. It also contained a set of batteries and four tanks for water used as part of a vaporization cooling unit for the spacecraft. An interesting feature was small rocket motors burning solid fuel that were pointed up and mounted on top of each landing leg. These rockets would be ignited at touchdown and served to hold the spacecraft fast to the lunar surface to minimize the chance of bouncing or falling over.

The LK was capable of either an automatic or a manual controlled landing using its guidance and control system. A set of instruments and a hand controller were provided to allow the cosmonaut to make a manual landing. He could also maneuver the spacecraft as it approached to dock with the lunar orbiter.

The guidance and control system was analog based for early models. A digital-based system with an S-530 digital computer was planned for later models when that digital system came available. A three-axis gyro platform provided attitude information. A landing radar provided velocity and range information to the lunar surface.

The landing radar, which was referred to as Planta, consisted of a three-beam Doppler velocity sensor with antenna beams splayed out like the corners of a pyramid and a radar altimeter with single antenna beam in the center of the velocity sensor beam arrangement. The radar altimeter was required to acquire the lunar surface at an altitude of 3,000 meters (9,845 feet). Presumably, the velocity sensor began tracking at about that same altitude. Shutdown and jettisoning of the Block D stage and igniting of the LK engine were cued from altitude data from the landing radar.

Landing Sequence

The landing procedure was similar to that of Apollo. One cosmonaut would descend to the lunar surface in the LK and explore while the other remained in lunar orbit in the LOK. The landing sequence was briefly as follows:

- The cosmonaut, wearing a space suit, opened the hatch of the Habitat Module of the LOK and spacewalked to the LK where he opened the hatch and entered the capsule.
- After checkout of the LK and Block D stage by the cosmonaut, the LK was separated from the LOK.
- The Block D stage engine was fired to cause the LK to leave lunar orbit and descent toward the lunar surface. It continued to fire to slow the spacecraft.
- At an altitude of about 2 km (6,560 feet) above the surface, as measured by the landing radar, the Block D stage was shut down and separated from the LK. The velocity of the spacecraft had been reduced to about 100 meters per second (328 feet per second) by this time.
- The rocket engine of the LK was ignited and the guidance system controlled the thrust of the engine and the attitude control thrusters to achieve a soft landing.
- At touchdown, upward-firing solid fuel rocket engines ignited to hold the spacecraft firmly to the surface and prevent bouncing or overturning.

The general plan was for the cosmonaut to spend about four hours on the moon on the first flight and up to 48 hours on the moon on later flights. During time on the surface on the first flight, the cosmonaut would exit the crew capsule and descend via the ladder to the surface. He would explore, gather samples, deploy scientific instruments, take pictures, and operate a television camera. At the end of the surface mission, he would return to the crew capsule.

At the proper time to effect rendezvous, the rocket engines would be ignited and the ascent stage of the LK would launch from the surface and rendezvous with the LOK. After docking, the cosmonaut would spacewalk from the LK back to the

Habitat Module of the LOK. The LK would be jettisoned and the rocket engine of the LOK would be fired to set the LOK spacecraft with the two cosmonauts on a trans-Earth trajectory.

Development and Test of the LK Lander

General management of the LK lander program was retained by Sergei Korolev's OKB-1 design bureau, but the detailed design and development of the lander was assigned to design bureau OKB-586 headed by Mikhail Yangel. Weight of the LK spacecraft was a crucial issue, reflecting the marginal weight-lifting ability of the N1 launch vehicle. Weight reduction was a constant driver during the development of the LK.

Three abbreviated versions of the LK, referred to as T2K, were built to evaluate the spacecraft in earth orbit. Particular emphasis was given to operation of the Block E propulsion system. Three T2K test models were launched into earth orbit, and all performed well. T2K #1 was launched in November 1970 as Cosmos 379. A series of burns of the throttleable engine were made simulating the descent and ascent portions of the mission. A series of fine burns were also made simulating rendezvous and docking.

T2K #2 was launched in February 1971 as Cosmos 398. The same series of engine burns as in Cosmos 379 were made and all systems performed well. T2 k #3 was launched in August 1971 as Cosmos 434. Again, all systems performed well. The backup engine was tested to simulate liftoff from the moon.

As a result of the T2K flight testing and ground testing, the LK lander was deemed ready for manned flight. Unfortunately, the LK never got a chance to carry a cosmonaut to the lunar surface because of development problems with the N1 launch vehicle.

At least five flight models of the LK have survived and are in museums and other institutions in Russia. In addition, there are a few engineering units and models on display.

DEVELOPMENT OF THE N1 LAUNCH VEHICLE

Just as the Apollo program relied on Saturn V to lift the very heavy Apollo space-craft off the earth and send it on a path to the moon, so did the Soviet manned lunar landing program rely on the N1 launch vehicle to do the same heavy lifting.

The books "Challenge to Apollo: The Soviet Union and the Space Race, 1945-1974" by Asif Siddiqi and "Soviet and Russian Lunar Exploration" by Brian Harvey give descriptions of the emerging technology, mismanagement, and political problems that plagued the Soviet manned lunar exploration programs. Some of the material given below on the development of the N1 launch vehicle was drawn from these books.

In February 1960, Sergei Korolev, head of OKB-1 Experimental Design Bureau, and his staff set forth goals for a new heavy-lift launch vehicle that would serve the following programs:

- Low earth orbit defense-related activity
- Global space-based communications and weather forecasting satellites
- Exploration of the moon and inner planets

OKB-1 defined parameters for the new heavy-lift launch vehicles as follows:

Booster	Time frame	Payload to low-earth orbit	Payload to the Moon
N1	1960–1962	40–50 tons	10–20 tons
N2	1961–1964	60–80 tons	20–40 tons

The Soviet space program was under control of the military. As a result, development of intercontinental ballistic missiles (ICBMs) and reconnaissance satellites proceeded while funding for such nondefense projects as manned exploration of the moon was limited. The ambitious schedule set out by the OKB-1 fell by the wayside.

Development of the N1 launch vehicle proceeded at a low level at OKB-1 due to limited funding. The N1 was intended to be a launch vehicle suitable for various missions, including a mission to Mars. It was not tailored for a moon landing mission as was the Saturn V.

Russian engineers were aware of the specific impulse advantage of using liquid hydrogen and liquid oxygen as propellants for the upper stages of the N1, but liquid hydrogen-liquid oxygen engines were still in development in the Soviet Union. The propellants selected for the first three stages of the N1 were liquid oxygen and kerosene. As a result, even though the first stage of the N1 had greater total thrust than the first stage of the Apollo Saturn V, the weight that could be lofted to the moon by the N1 was significantly less than by the Saturn V. Saturn V used liquid hydrogen and liquid oxygen as propellants for the second and third stages.

There was bitter animosity between Sergei Korolev, chief designer of the N1 and head of OKB-1, and Valentin Glushko, the foremost rocket engine designer in the Soviet Union. Glushko was head of design bureau OKB-456. Their unresolved differences sorely hindered successful development of the N1.

Glushko was adamant that the best propellant for the N1 was the hypergolic combination of nitrogen tetroxide (N_2O_4) and unsymmetrical dimethylhydrazine (UDMH), and he refused to design an engine for the N1 using liquid oxygen and kerosene. The engine design fell to a less-experienced design bureau, OKB-276, headed by Nikolay Kuznetsov. The result was a very fine engine, the NK-33, but it developed a modest 154 metric tons (340,000 pounds) of thrust.

The original design of the N1 contained 24 engines and was capable of lofting about 75 metric tons into earth orbit. At the time, the plan was to put two or more elements of the lunar spacecraft in earth orbit, assemble them, and then depart for the moon. After the decision in 1965 to put the spacecraft into earth orbit with a single launch, six engines were added to the N1 bringing the total to 30. Other changes increased the amount of propellant and chilling the propellants before launch. The trajectory would also be changed to change the launch azimuth to 51.6 degrees and lower the orbit from 300 km to 220 km.

The final configuration of the N1 contained 30 NK-33 engines giving a total liftoff thrust of 4,620 metric tons (10.19 million pounds). In comparison, the Apollo Saturn V launch vehicle contained five F-1 engines of 680 metric tons each for a total liftoff thrust of 3,400 metric tons (7.5 million pounds).

Korolev's optimistic analysis predicted that with the powerful booster, 95 metric tons could be placed in earth orbit. This put a severe weight constraint on weight budgets for modules of the L3 spacecraft assembly. Later, when liquid hydrogen-liquid oxygen engines were available for the upper stages, heavier payloads could be placed in orbit. The initial design to be fielded had no margin and could not accommodate the inevitable growth in weight as the detailed design of the L3 spacecraft assembly progressed. It appears to have been an untenable situation, but the N1 program went forward.

Development of the N1 and the entire Soviet space program suffered a setback in January 1966 when Sergei Korolev died. Asif Siddiqi writes in *Challenge to Apollo*: "Sergy Pavlovich Korolev's death ended one man's unprecedented twenty-year reign over soviet missile and space programs.... no (other) single person had expertise in managing the design bureau, dealing with Soviet politicians, brokering deals with other chief designers, and instilling a vision of space exploration among the thousands who worked at the firm." His influence reflected his larger-than-life personality.

Korolev was succeeded as the head of OKB-1 by his deputy, Vasili Mishin. From several accounts, Mishin was a capable engineer, but he did not have the drive, charisma, and high-level connections that Korolev possessed. To his credit, he inherited programs fraught with technical problems and unrealistic schedules, and he tried hard to make the programs successful.

An important accompaniment to development of the N1 launch vehicle was construction of a launch complex at the Baikonur Cosmodrome near the town of Tyuratam in what is now the Republic of Kazakhstan. The new launch complex included two launch pads, assembly building, and a transporter that rode on rails to carry the N1 from the assembly building to the launch pad. A partially operable mockup of the N1 was assembled, and several dry runs were made with the mockup to prove out the launch complex and procedures.

When finally assembled, the N1 launch vehicle was massive. Standing vertically with the payload of the L3 lunar orbiting and lunar landing spacecraft, the spacecraft was 105 meters (345 feet) high. It weighed 2,788 metric tons (9.15

million pounds) when fully fueled. The N1 launch vehicle without payload stood over 60 meters (195 feet) high.

A photograph of the N1 with payload on the launch pad at the Baikonur Cosmodrome is given below (Fig. 14.4). The photograph shows a mockup of the N1 on the second launch pad.

Fig. 14.4   N1 heavy-lift launch vehicle on launch pad (NASA photo collection)

The N1 launch vehicle consisted of the Block A first stage, Block B second stage, and Block V third stage. The three stages, separated by latticework, can readily be seen in the photograph. The L3 spacecraft group would be mounted at the top of the third stage. Likely, the payload in the photograph was a L1 spacecraft and mockup of the LK lunar lander.

The Block A first stage was 30 meters high and the diameter at the base was 16.8 meters (55 feet). It contained 30 NK-33 engines and two large spherical propellant tanks. One tank, 10.5 meters in diameter, held kerosene, and the other, 12.8 meters in diameter, held liquid oxygen. A photograph of the N1 in a horizontal position showing the placement of the 30 engines in the Block A stage is shown below (Fig. 14.5). Credit to Alex Panchenko at USSR-AIRSPACE for the photograph.

Fig. 14.5 Photograph of N1 launch vehicle showing placement of engines in Block A stage (USSR-AIRSPACE photograph)

An engine management system known as KORD (Kontrol Roboti Dvigvaeli) monitored operation of the 30 engines. In case of poor performance of an engine, it would shut it down and also shut down a good engine directly opposite to balance the thrust. It would shut down all engines upon command in case of an errant trajectory during early flight, for example.

The Block B second stage, mounted on latticework above the Block A stage, was 20.5 meters (67 feet) high, and it was 10.3 meters (34 feet) in diameter at its base. It contained eight NK-43 engines with 179 metric tons of thrust each, a spherical kerosene tank 7.0 meters in diameter, and a spherical liquid oxygen tank 8.5 meters in diameter.

The Block V third stage, mounted on latticework on top of the Block B stage, was 11.5 meters (38 feet) high, and the diameter at its base was 7.6 meters (25 feet). It contained four NK-39 engines with 41 metric tons of thrust each, a spherical kerosene tank 4.9 meters in diameter, and a spherical liquid oxygen tank 5.9 meters in diameter.

The L3 spacecraft complex was mounted on top of the third stage of the N1. The L3 complex consisted of a Block G stage with 41 metric tons of thrust, a Block D stage with a restartable engine and 8.5 metric tons of thrust, the LOK lunar orbiter spacecraft, and the LK lunar landing spacecraft.

The Block D stage, which was also used in conjunction with the L1 in the circumlunar program, was a workhorse space tug. It was 5.5 meters (18 feet) long and 4 meters (13 feet) in diameter. The restartable engine with 8.5 metric tons (18,740 pounds) of thrust burned kerosene and liquid oxygen. Propulsion also included a set of thrusters for fine control that used N_2O_4 and UDMH hypergolic propellants.

Flight Testing of N1 Launch Vehicles

First N1 flight

The first N1 launch vehicle was assembled and transported from the assembly building to the launch pad in May 1968. This first flight model of the N1 carried a modified L1 capsule that had been designed for circumlunar flight instead of the lunar landing L3 payload. The flight plan called for the Block D stage and the L1 capsule to orbit the moon for 2 or 3 days and then return the L1 capsule to earth.

Testing of the N1 systems on the launch pad revealed several problems and the vehicle was taken back to the assembly building for further testing. It was deemed ready for flight in February 1969 and moved back to the launch pad.

Launch occurred on 21 February 1969. Two of the 30 engines shut down just after liftoff. All the rest of the engines shut down 69 seconds after launch. The vehicle had reached an altitude of only 14 km and it fell back to earth. Post flight investigation showed that spurious signals to the KORD engine monitoring system had caused deliberate shut down of two of the engines. Later, a pipe measuring fuel pressure broke and sprayed kerosene into the hot engine area and caught fire. The fire burned the insulation on wires carrying electrical power bundled with KORD signals. In response, KORD shut down the 28 operating engines.

In typical fashion, the Soviet Union did not publish the launch failure of the first N1 rocket for nearly 20 years. The crash site was about 50 km from the launch site so U.S. reconnaissance satellite observations did not pick up the catastrophe.

Second N1 flight

Several upgrades to the N1 systems were made in response to the problems encountered with the first launch, and a second N1 was readied for flight. The second flight was planned to send a modified L1 spacecraft on a circumlunar trajectory. The N1 was transported to the launch site in June 1969.

Launch of the second N1 was made on 3 July 1969. The spacecraft had ascended to about 200 meters when all but one engine shut down and the spacecraft tilted to the side and fell back to the launch area and exploded. The explosion was said to have been the largest nonnuclear explosion the world had ever seen.

Post flight investigation disclosed that a liquid oxygen pump for one engine had suffered an internal explosion and the explosion severed fuel and oxygen lines to other engines resulting in a massive fire. The KORD engine diagnostic system then shut down all engines but one. That one firing engine tilted the spacecraft to the side before the entire spacecraft fell to the launch pad.

The massive destruction at the launch pad was captured on film by a US reconnaissance satellite. It signaled the beginning of the end of the Soviet-U.S. space race.

Third N1 flight

It was nearly two years later that the principals of the Soviet N1 program thought that the reliability of the launcher had been improved to the point that a third flight could be attempted. The space race had been lost by then and the improvement program likely proceeded at a more deliberate pace.

Launch of the third N1 was made on 27 June 1971. The payload consisted of mass mockups of the LOK orbiter and LK lander. All engines seemed to be firing normally at launch, but the vehicle began rolling about its longitudinal axis a few seconds after liftoff. Even with full thrust of the six vernier engines to counter the roll, the roll continued. The uncontrolled roll caused the upper stage to break away from the third stage at 48 seconds after liftoff. At 50 seconds after liftoff, the roll angle had increased to 200 degrees and the KORD system received an emergency command from the gyros and it shut off the engines.

Investigation of the failure disclosed that the vehicle lost control because the designers had misjudged the influence of the pyrotechnic starter exhaust tubes. Those tubes were located asymmetrically on the 30 engines. The two rings of engines on the bottom of the N1 caused formation of two zones of air depression behind the booster. The asymmetric positioning of the starter exhaust tubes caused a high torque rotating force on the borders of the air depression zones. The reason that this problem had not been seen on the two previous launches was that not all of the engines in the outer ring had been firing and that left a gap to diminish the effect of the depression zones.

Fourth N1 flight

There were calls to terminate the N1 program after the third consecutive launch failure. However, engineers believed that they were close to success and the program continued. Changes were made to the booster configuration and the vernier thrusters were replaced by new liquid propellant vernier engines. In addition, flight control would be performed by a new onboard S-530 digital computer. Newly designed improved engines were available by that time, but the decision was to proceed with the present engines for one more flight.

Launch of the fourth N1 took place on 23 November 1972. The payload was an operational LOK lunar orbiter and a mockup of the LK lunar lander. The engines were firing normally and all systems looked good until 106 seconds after launch when an explosion in the tail section of the booster breached the oxygen tank and the booster exploded.

Investigation into the cause of the failure was contentious, but it was finally established that one of the engines had exploded causing damage to the aft compartment, and that resulted in the massive explosion that destroyed the vehicle.

The Aftermath

A fifth N1 launch vehicle was assembled using the newly designed engines and other improvements to the booster. The payload was to be operating versions of the LOK and LK. The flight plan was to orbit the moon, perform maneuvers, and return to earth without making a landing on the moon.

The United States closed down the Apollo program in December 1972 after successfully completing six landings on the moon that featured far-ranging exploration on later flights. As a result, Soviet Union leaders began questioning the purpose of continued effort to make flights to the moon.

Finally, along with a massive shakeup of the Soviet Space program in May 1974, Vasili Mishin was removed as head of the premier design bureau, OKB-1. He was blamed for failure of the manned circumlunar program and the manned lunar landing program under his leadership. He had become leader of OKB-1 upon the death of Sergei Korolev in 1966. Korolev had led OKB-1 brilliantly, and he had started the circumlunar and manned landing programs before his death. In a book published several years after his dismissal, Mishin penned a telling sentence: "We, the successors to Korolev, did everything we could, but it was not enough."

Mishin was succeeded as head of OKB-1 by Valentin Glushko who had clashed bitterly with Korolev. Glushko stopped all work on the N1 program in May 1974. He had detested the N1 since his quarrel with Korolev over engines. Glushko had been elevated to a member of the Central Committee of the Communist Party and that gave him a lot of clout. The manned lunar landing program effectively died when work on the N1 was suspended. The N1 was officially canceled in March 1976.

Thus ended the Soviet Union's dream of space supremacy with cosmonauts on the moon.

Bibliography

Grahn, Sven, *The Kontakt Rendezvous and Docking System*, http://www.svengrahn.pp

Harvey, Brian, *Soviet and Russian Lunar Exploration*, Praxis Publishing, Chichester, UK, 2007

Lindroos, Marcus, *The Soviet Manned Lunar Program,* https://fas.org/sp/eprint/lindroos_moon1.htm

LK (Russian manned lunar lander), http://astronautix.com

NASA-NSSDCA-Spacecraft-Details, *Cosmos 434*, NSSDCA/COSPAR ID: 1971-069A

Soyuz 7K-LOK (Russian manned lunar orbiter), http://astronautix.com

Zak, Anatoly, *LK Lunar Module for the L3 Project*, *LOK Spacecraft*, *L3 System* russianspace-web.com

Index

© Springer Nature Switzerland AG 2018
T. Lund, *Early Exploration of the Moon*, Springer Praxis Books,
https://doi.org/10.1007/978-3-030-02071-2